ETHICS AND PLANETARY BOUNDARIES

The planetary boundaries framework examines the profound risks human actions pose to Earth's stability and resilience. Since its introduction in 2009, and through subsequent updates, the framework has become one of the most influential ideas of our age, yet it has not been put to close ethical scrutiny. This book takes a multidisciplinary approach to the ethics of the planetary boundaries, ranging from international law to Indigenous knowledge and from science to art, and political ecology. The editors introduce each boundary before two chapters examine the reach, limits, and ethical stakes of each of the nine planetary boundaries. This volume comes at a critical moment, when unprecedented environmental challenges demand new approaches, tools, and perspectives to address questions of epistemology and justice. It is a valuable resource for students, citizens, and academics concerned with relationships of knowledges, ethics, and environments. This title is available Open Access through Cambridge Core.

SIMON P. MEISCH is Senior Lecturer at the International Centre for Ethics in the Sciences and Humanities at the University of Tübingen. Meisch's research focuses on co-production, ethics, and epistemic justice in transdisciplinary sustainability research, particularly in relation to climate and water-related challenges.

JEREMY J. SCHMIDT is Reader in Environmental Geography in the Department of Geography and Environmental Science at Queen Mary University of London. Schmidt's research examines the relationships of ethics and policy to a world increasingly shaped by human impacts on the Earth system.

ETHICS AND PLANETARY BOUNDARIES

Edited by

SIMON P. MEISCH
University of Tübingen

JEREMY J. SCHMIDT
Queen Mary University of London

CAMBRIDGE
UNIVERSITY PRESS

CAMBRIDGE
UNIVERSITY PRESS

Shaftesbury Road, Cambridge CB2 8EA, United Kingdom

One Liberty Plaza, 20th Floor, New York, NY 10006, USA

477 Williamstown Road, Port Melbourne, VIC 3207, Australia

314–321, 3rd Floor, Plot 3, Splendor Forum, Jasola District Centre, New Delhi – 110025, India

Cambridge University Press is part of Cambridge University Press & Assessment, a department of the University of Cambridge.

We share the University's mission to contribute to society through the pursuit of education, learning and research at the highest international levels of excellence.

www.cambridge.org
Information on this title: www.cambridge.org/9781009443579

DOI: 10.1017/9781009443586

First published 2026

Cover image: Cover illustration by Anna Voigtländer.

A catalogue record for this publication is available from the British Library

Library of Congress Cataloging-in-Publication Data
Names: Meisch, Simon editor | Schmidt, Jeremy J.
Title: Ethics and planetary boundaries / Edited by Simon P. Meisch, University of Tübingen, Jeremy J. Schmidt, Queen Mary University of London.
Description: New York, NY : Cambridge University Press, 2026. | Includes bibliographical references and index.
Identifiers: LCCN 2026013279 (print) | LCCN 2026013280 (ebook) | ISBN 9781009443579 hardback | ISBN 9781009443586 epub
Subjects: LCSH: Environmental geography | Environmental ethics | Climatic changes–Political aspects
Classification: LCC G143 .E84 2026 (print) | LCC G143 (ebook)
LC record available at https://lccn.loc.gov/2026013279
LC ebook record available at https://lccn.loc.gov/2026013280

ISBN 978-1-009-44357-9 Hardback

Contents

Contributors

Andreas Bachmann, Scientific Officer at the Federal Office for the Environment, Berne, Switzerland

Alice Benessia, Pianpicollo Selvatico Foundation

Neelke Doorn, Antoni van Leeuwenhoek Professor, Department of Values, Technology and Innovation, School of Technology, Policy and Management, Delft University of Technology

Leslie-Anne Duvic-Paoli, Reader in Environmental Law, Dickson Poon School of Law, King's College London

Erle Ellis, Professor, Department of Geography & Environmental Systems, University of Maryland, Baltimore County

James Elser, Research Professor, School of Sustainability and Sustainable Phosphorus Alliance, Arizona State University and Bierman Professor of Ecology, Flathead Lake Biological Station, University of Montana

Uta Eser, Biologist and Ethicist, Office for Environmental Ethics, Tübingen

Rachel Haug Fossbakk, Chief Scientist, SINTEF Ocean, Department of Fisheries and New Biomarine Resources, Trondheim

Miriam Gay-Antaki, Assistant Professor, Department of Geography and Environmental Studies, University of New Mexico

Jeanne-Marie Gherardi, Associate Professor, University of Versailles Saint Quentin / CEARC

Carlota Houart, PhD Candidate, Department of Environmental Sciences, Water Resources Management, Wageningen University, the Netherlands

Simon P. Meisch, Senior Lecturer at the International Centre for Ethics in the Science and Humanities, University of Tübingen

Konrad Ott, retired Professor for Philosophy and Ethics of the Environment, Department of Philosophy, Christian Albrechts University Kiel, Germany

Joana Castro Pereira, Assistant Professor of Politics and International Relations, Faculty of Arts and Humanities, University of Porto

Tanguy Sandré, Post-doctoral Fellow, CEARC, UVSQ, University Paris-Saclay and SVT, University of Bergen

Jeremy J. Schmidt, Reader in Environmental Geography, Department of Geography and Environmental Science, Queen Mary University of London

Makere Stewart-Harawira, Professor in Indigenous, Environmental, and Global Studies, Faculty of Education, College of Social Sciences and Humanities, University of Alberta

Katerina Teaiwa, Professor, School of Culture, History and Language, College of Asia and the Pacific, Australian National University

Louise du Toit, Lecturer, Southampton Law School, University of Southampton

Ariane Willemsen, Executive Secretary of the Federal Ethics Committee on Non-Human Biotechnology, Berne, Switzerland

Nicole J. Wilson, Assistant Professor and Canada Research Chair in Arctic Environmental Change and Governance, Environment and Geography and Centre for Earth Observation Science, University of Manitoba

Christine Winter, (Ngāti Kahungunu ki the Wairoa, Pākehā) Senior Lecturer in Environmental, Climate Change, Multispecies, Planetary, and Indigenous Politics, Politics Programme, Otago University Ōtakou Whakaihu Waka

Acknowledgements

A volume of this scope is invariably the fortunate outcome of collaboration among many individuals, to whom we would like to express our sincere gratitude. Foremost among them are the contributors to this volume, who joined us on the long and often challenging journey that such an ambitious editorial project entails. We are deeply thankful for their insightful contributions as well as for their enduring patience throughout the process.

We are also grateful for the generous financial support provided by the Tübingen–Durham Seedcorn Fund and the Universitätsbund Tübingen e.V. as well as by research funds from Durham University and the International Centre for Ethics in the Sciences and Humanities (University of Tübingen). Their support made it possible for this volume to be published Open Access.

Behind the scenes, we have been fortunate to receive help from several dedicated people. We warmly thank Matt Lloyd and Maya Zakrzewska-Pim at Cambridge University Press for their guidance and support, Anna Voigtländer for creating the cover image, and Natalie Halseband for producing the index. Our sincere thanks also go to our student assistants, Lukman Mezdour and – sine qua non – Valerie Weber, whose skill and diligence ensured a steady grasp of an ever-expanding manuscript.

Part I

Introduction

1

The Planetary Boundaries

Thinking across Science and Ethics

JEREMY J. SCHMIDT

1.1 Introduction

In 2009, scientists introduced the planetary boundaries, a set of nine variables designed to portray the degree to which human activities are pushing Earth into unsafe territory. Nearly 15 years later, in 2023, all 9 boundaries were quantitatively measured for the first time (see Richardson et al. 2023). Six exceeded what is required to retain the stability and resilience of the Earth system. Those eclipsed were the boundaries for climate change, biosphere integrity, land system change, freshwater change, biogeochemical flows, and novel entities. Three others were at levels safe for the planet, but not always trending in the right direction: ocean acidification, atmospheric aerosol loading, and stratospheric ozone depletion. Then, in 2025, a seventh boundary was crossed: ocean acidification (Findlay et al. 2025). Between its original announcement and its collective quantification, the planetary boundaries framework has been challenged, its measurements refined, and its content updated as it moved from a novel idea to a mainstay of sustainability science. It has also been used to justify new approaches to sustainable development and become the basis for alternative economic models. It has inspired policies at urban and regional scales and provided a paradigm for new ideas in ecological law and governance. In 2021, the planetary boundaries framework even had its own special on Netflix narrated, in part, by David Attenborough – *Breaking Boundaries: The Science of Our Planet.*

As the planetary boundaries framework has gained influence, however, its ethical aspects have received comparatively less attention. This is peculiar given that, at its centre, the framework has a normative target of maintaining planetary conditions that provide a 'safe operating space' for humanity (Figure 1.1). Recently, authors and advocates of the planetary boundaries framework have taken up ethical issues more directly. In that work, the planetary boundaries are used to scope matters of justice within what the planet might permit (Gupta et al. 2021,

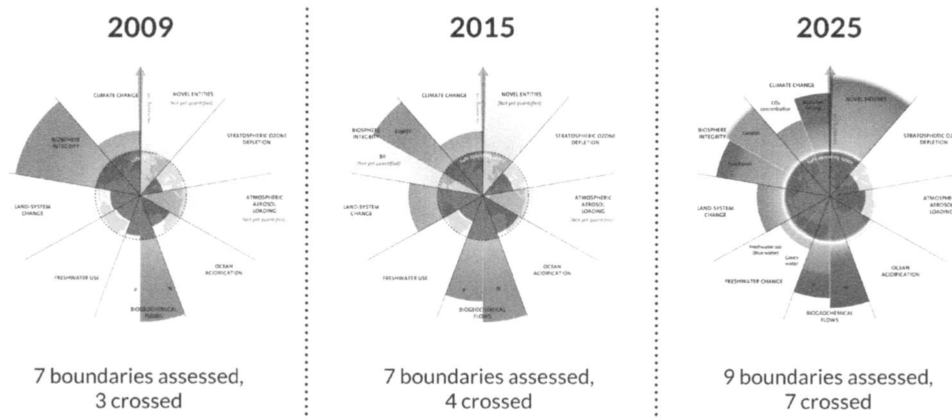

Figure 1.1 The planetary boundaries framework. This figure tracks the progression of the planetary boundaries framework, including some of the changes to how boundaries are defined, in its three assessments in 2009, 2015, and 2023. Licensed under CC BY-NC-ND 3.0. Credit: Azote for Stockholm Resilience Centre, based on analysis in Sakschewski et al. 2025.

2023a, 2023b; Rockström et al. 2023). Proponents argue that the framework guides 'humanity's future on Earth' because truly sustainable development is defined by staying within planetary boundaries (Rockström et al. 2024: 773). Those who use the planetary boundaries framework in this way share a common premise: that preserving the planetary conditions that have made complex societies possible marks a perimeter for justice. The question then becomes: what approach to justice best suits the ways that the planetary boundaries framework discloses issues as diverse as climate change, mass extinction, water scarcity, or 'forever' chemicals? This question sits at the heart of this book, but not because it is one this collection of essays sets out to answer. Instead of treating the planetary boundaries as a perimeter within which to pursue justice, this book examines how the framework not only describes interactions of people and planet but also influences how connections among humans, other species, and Earth's biophysical processes are interpreted and understood.

Ethics matter to the planetary boundaries framework beyond just applications of the framework that use it to stake out parameters for justice. This is because the planetary boundaries framework directs attention – scientific and ethical – to some concerns and not others. For instance, the authors of its first fully quantified assessment make clear that 'the planetary boundaries framework is only concerned with the stability and resilience of the Earth system, i.e. not about human or ecosystem health' (Richardson et al. 2023: 6). This is a clear restriction on the scale of scientific concern. It also entails ethical considerations. This is because, as the original authors of the planetary boundaries framework state, boundaries themselves do not represent facts about planetary thresholds for Earth's climate or biodiversity (or any other planetary boundary). Rather, when the planetary boundaries were introduced in 2009, they were described as 'human-determined values of the control variable set at a "safe" distance from a dangerous level'; as such, 'determining a safe distance involves normative judgments of how societies choose to deal with risk and uncertainty' (Rockström et al. 2009b: 3). This prominent role for values means that ethics are never far from the science of planetary boundaries. By 2024, however, this explicit role for values and judgements had been expunged. Instead, proponents stated that 'it is a fundamental property of the PB framework that boundaries are set purely on biophysical grounds' and without reference to human considerations (Rockström et al. 2024: 774). How did the intrinsic role of values and judgement ossify into inhuman boundaries? They were intrinsic to the framework initially, yet now, when that framework is used to establish the perimeter for justice, the boundaries are presented as value-free facts. It is not uncommon for scientific frameworks to evolve, but it is also important to situate them socially, especially when they make claims about humanity's future writ large. When and how, for instance, did the initial value judgements used to establish planetary boundaries evolve into a basis for defining the perimeter for justice?

This book offers a set of different views on ethics and planetary boundaries. Rather than taking the planetary boundaries as a perimeter within which to think about ethics or justice, it probes matters of ethics across a range of concerns and perspectives, including those critical of the planetary boundaries framework. To grapple with ethics and the planetary boundaries we invited experts to a dialogue with the editors, and where possible with each other, on each of the nine planetary boundaries. These conversations often entailed thinking about ethics in multiple ways, and there is no effort in the book to adjudicate the different ways that concerns of fairness, equity, moral standing, relations, or justice are approached. Instead, we embraced a diversity of approaches so that a richer discussion might be had about the ethics of planetary boundaries. Often, these discussions also entailed thinking about how one planetary boundary is entangled with others. For example, it is difficult (and arguably a bit contrived) to talk about anomalously extreme water events of flood or drought without discussing climate change. In what follows, scientists, artists, anthropologists, philosophers, as well as political scientists, geographers, lawyers, and Indigenous scholars discuss what the planetary boundaries mean to them, their work, and the ways concerns over what to do are entangled with ways of knowing.

The chapters that comprise this book do not all pull in the same direction, and they account for only some potential positions on ethics and the planetary boundaries. This reflects the aim of the volume, which is not comprehensive coverage but to collectively show why ethics are critical to situating the planetary boundaries framework with respect to scholarship, policy, and concerns with relations among people, other species, and Earth system processes. To this end, the remainder of this introduction does three things: First, it explains the planetary boundaries framework in more detail, both in terms of the sciences it relies on and how its authors and advocates have used it to shape spaces of ethics and justice. Second, it discusses some of the ways planetary boundaries have sparked new ideas among scholars and governance practitioners – some of these ideas have been subsequently incorporated into later iterations of the planetary boundaries itself. Third, it discusses criticisms of the planetary boundaries framework and why further ethical engagement with them is warranted. With these considerations in hand, the conclusion to this chapter situates the book's dialogues on each planetary boundary. It also identifies themes that emerge across different chapters – of both consensus and contention – discussed further in the book's conclusion.

1.2 The Planetary Boundaries Framework

When it was introduced, the planetary boundaries framework identified the 'environmental preconditions' for human development through nine 'control variables' of the Earth system that establish a safe operating space for humanity (Rockström et al. 2009a). These variables – the nine planetary boundaries – have subsequently been renamed, refocused, and refined in many cases. This includes changes to what is measured and how, and these choices matter ethically. This is because the planetary boundaries framework seeks an account of how anthropogenic (i.e., human) activities affect the planet. As a result, the choice of what to measure is also a choice about which activities, and responsibility for them, contribute to its planetary assessment. Further, those activities entail distinct histories, economies, politics, technologies, and inequalities. So, these measures matter morally even where the science stops short of assigning responsibility. This remains the case even when the scale of concern is restricted to the planet (i.e., not humans or ecosystems) because changes to planetary conditions have effects on the conditions of social life. So, in addition to the value judgements that, as discussed earlier, were initially acknowledged when determining boundaries themselves, matters of ethics are never far from the sciences the framework draws together either.

The planetary boundaries framework is based on bringing together 'three branches of scientific inquiry': (1) research on the scale of human action relative to Earth's capacity to sustain it; (2) research on 'Earth system processes', including changes humans have introduced to them; and (3) research on resilience and the ways complex systems regulate themselves in response to change and disturbance (Rockström et al. 2009b: 4). Students and scholars of environmental thought will notice straightaway that each of these areas has its own history, and the planetary boundaries framework is explicitly set in a history of Western environmental sciences. For instance, on the first issue of human impacts on the planet, the framework draws on earlier ideas of ecosystem capacity and debates over the 'limits to growth' that characterized twentieth-century environmental debates (see Sabin 2013). But the planetary boundaries framework also seeks to update earlier ideas. For instance, rather than taking planetary limits as 'fixed' or stable, it recognizes the ways that human activities are now of such scale that they affect how the planet regulates itself. In making these updates, the framework walks a careful line between planetary limits, on the one hand, and, on the other, boundaries that operate in proximity to those limits. This tension is discussed further next, but it is worth noting that the planetary boundaries framework is explicit about the fact that, even though 'boundaries' are not synonymous with planetary limits, the Earth system does have thresholds that exist 'irrespective of peoples' preferences, values, or compromises based on political and socio-economic feasibility' (Rockström et al. 2009b: 5).

With Earth's thresholds in mind, planetary boundaries began by defining 'the "planetary playing field" for the human enterprise' (Rockström et al. 2009b: 6). Those thresholds and the setting of boundaries that would keep human impacts a safe distance from them are anchored in Earth system sciences that treat the Earth as a single, integrated system – the Earth system. To treat the Earth as a single system is a relatively recent idea, one developed during the latter decades of the twentieth century (see Steffen et al. 2020). It departs from earlier notions of the planet as either a machine or an organic whole that is characterized by its stability – what is sometimes referred to as the 'balance of nature'. Instead, the Earth system is an ever-changing, self-regulating system in which complex feedback links the atmosphere, the biosphere, the hydrosphere (including the cryosphere), and the geosphere at different temporal and spatial scales. For instance, the cyanobacteria that evolved to produce oxygen radically remade the Earth's atmosphere in a complex set of biogeochemical relationships. Amid the deep history of planetary evolution, Earth system scientists now identify a new sphere dominated by humans – the anthroposphere – that cumulatively, albeit extremely unequally, has become another of the 'great forces' of nature (Steffen et al. 2007).

Crucial for the planetary boundaries framework is the idea that the complex processes at work across the Earth system can be understood in terms of its nine control variables. These control variables – the planetary boundaries – narrow how the Earth system's complex relationships are represented and give special attention to how different planetary configurations make for safe or unsafe conditions for humans. For instance, during approximately the past 12,000 years human civilizations have evolved under relatively stable geologic conditions known as the Holocene. Stability in this sense does not imply a static planet, but rather that variability in planetary conditions operated in an envelope of planetary thresholds conducive to the emergence of increasing social complexity. By contrast, when the planetary boundaries framework was introduced, its authors frequently referenced a new geological epoch, the Anthropocene, owing to how the collective impact of human activities now affects planetary evolution in ways that could destabilize planetary conditions (Rockström et al. 2009b, 2024). The capacity for humans to affect how the Earth system functions marks an important evolutionary shift, one in which the planet's future now turns on how human–Earth interactions proceed (Waters et al. 2016). As we consider below, however, both the idea and the practice of treating 'human impacts' collectively can mask inequalities regarding responsibility for these impacts.

The third and final scientific pillar of the planetary boundaries framework is the notion of resilience. In its original formulation by the ecologist C. S. 'Buzz' Holling (1973), resilience was part of a shift in understanding complex systems not as tending towards stability but as characterized by change and surprise. This was a major departure from dominant theories at the time, where the 'balance of nature' had been suffused to interpretations of ecological systems as seeking stability. Forests, for instance, were harvested according to their 'maximum sustainable yield' based on the assumption that their reproductive processes were stable over time and not naturally changing (Langston 1995). The idea of stability was well suited to policy approaches of 'command and control' but not to thinking about how human impacts on ecosystems were just one among many dynamics that affect them and through which they evolve (Holling & Meffe 1996). In this context, ecological resilience offered a paradigmatic shift: complex systems were not stable, but rather constantly changing and adapting through complex processes, some of which could be expected but others of which were unanticipated. As the idea developed, resilience came to be understood as the capacity of a system to respond to disturbances while still retaining its essential functions, feedback, and identity (Folke 2006).

Closely related to the concept of resilience was the idea of thresholds – another of Holling's key scientific ideas. Ecological thresholds are the points at which the relationships that make an ecological system resilient are radically, often irreversibly, reorganized (Holling 1986). These thresholds, for instance, are what would be surpassed when logging is so extensive that a forest no longer regrows and instead a new set of processes set in, such as when rainforest deforestation leads not to tropical regrowth but to savannah. Importantly, these thresholds are not entirely knowable in advance. This is because the complexity of ecological systems means that there are irreducible elements of uncertainty. Uncertainty with respect to thresholds makes it prudent to set boundaries at a 'safe' distance from potentially irreversible shifts, but also to appreciate the ways that rapid disturbances, like industrialized extraction, can have unforeseen, surprising effects when they accelerate at unprecedented rates (Gunderson & Holling 2002).

Taken together, the three scientific pillars of the planetary boundaries framework both update debates on limits and growth and incorporate new Earth system sciences that significantly alter thinking about the planet. The shifts in thought are immense. Donna Meadows (1999, 2008), one of the early authors of work on the limits to growth, argued that what was now required was to 'think in systems' and to identify 'leverage points' that are key to effecting change. Arguably, the idea that the complexity of the Earth system can be understood in terms of nine key variables – the planetary boundaries – takes the idea of leverage points out to the scale of Earth as a whole. That is, the planetary boundaries are proxies for complexity that capture important – not all – of the ways that human activities intersect with Earth's biophysical processes. As the planetary boundaries framework evolved, some boundaries like climate change became more closely aligned with biophysical determinations while others retained their association with areas of increasing risk (Rockström et al. 2024). The result is that not all 'boundaries' are boundaries of the same sort. Further, because they are not synonymous with 'limits' or thresholds, there remain important normative considerations regarding how to integrate Earth system science into assessments of the risks that arise from the impacts of 'the human enterprise' on the planet. In addition to shifting how human–Earth relationships are conceptualized, however, the planetary boundaries framework did not stop there. It has also been used to scope matters of ethics and justice.

1.2.1 Normative Space within Planetary Boundaries

The planetary boundaries framework has evolved over time from its initial contrast of value-laden boundaries and objectively independent thresholds. Today, it draws a sharper separation of facts and values. This creates a dynamic where matters of judgement once acknowledged within the framework are being reworked with respect to it. In this section we introduce some of the ways that this shift is explained by authors and advocates of the planetary boundaries framework. This is principally accomplished by using scientific assessments to mark a perimeter for a 'safe operating space' for humanity. Or, as Gupta (2016) put it, to define an 'eco-space' in which Earth's processes and resources can be shared fairly by humans and other species without risking the transgression of boundaries that would undermine planetary conditions. But this is not all that the planetary boundaries framework does. As the late ecologist James Kay (2000: 135) argued, recognition of large-scale scientific uncertainties has the effect of putting 'scientists in the role of narrators'. Narrators that, as they explain complex dynamics, also scope potential trade-offs and delimit possible futures. This section identifies how the use of the planetary boundaries framework to scope matters of justice similarly implies trade-offs that, as later chapters consider, do not always capture the full array of ethical options available.

First, there is what seems a conundrum: the planetary boundaries framework initially acknowledged that what constitutes a 'boundary' is determined through social values and judgements regarding risks, while Earth's biophysical processes remain indifferent to human ideas about them. One of the original authors of the planetary boundaries framework, Hans Schellnhuber (2000: 3), thought this tension could be resolved by subsuming ethics within science. As he argued in the same issue of the newsletter that first announced the Anthropocene: 'ultimately, Earth System analysis will even be able to address the challenge of sustainable development in a no-nonsense way by deducing macro-options for future ecosphere-anthroposphere co-evolution from first cognitive and ethical principles'. The reduction of ethics to science, however, is increasingly seen as unlikely given the ways that uncertainty and complexity mark the limits of what could be rationally deduced for ethics from Earth system science. However, there is a close conceptual cousin that is often relied upon to navigate the intersection where values meet biophysical indifference. This is the view that science does not settle matters of ethics but rather scopes the space of ethical possibilities. This latter view more fairly captures the position of the planetary boundaries framework, which does not offer a 'roadmap for sustainable development' but rather identifies the 'biophysical boundaries at the planetary scale within which humanity has the flexibility to choose a myriad of pathways for human well-being and development' (Rockström et al. 2009b: 7).

The premise that science marks a perimeter for ethics can be found in numerous articles that seek a more explicit place for justice in the planetary boundaries framework. In fact, these efforts have led to adjustments in terminology to reflect a more conscious incorporation of justice. In that adjusted terminology, planetary boundaries are replaced with ideas of 'safe and just' Earth system boundaries – a shift meant to demarcate what is 'safe' as referring to the resilience of the planet and what is 'just' referring to equitable distribution of benefits and risks among people (Rockström et al. 2021). For instance, Gupta et al. (2023a) build on the planetary boundaries framework to elaborate what they term safe and just Earth system boundaries. Here, the social values initially used to determine the planetary boundaries fade into the background and are instead presumed upon to shape how matters of justice consider both the limits on, and demands for, sufficient resources for a good life. Gupta et al. (2023a) find that lifting those worst off out of poverty cannot be done without exceeding the planetary boundaries unless there is also a radical redistribution of wealth from rich to poor. Still, Gupta et al. (2023b) find the idea of Earth system boundaries a promising way to rethink how sustainable development agendas and the global environmental assessments that shape them can be enhanced by attention to justice.

One challenge to explicitly incorporating justice into the planetary boundaries framework is that there are multiple ways of knowing and multiple approaches to justice. This is not a unique challenge. It affects global environmental assessments generally (Castree et al. 2021). A significant issue for the planetary boundaries framework, however, is that the background values used to determine boundaries do not stay in the background. Instead, explaining the relationship between different boundaries is a constant challenge. Initially, the planetary boundaries framework distinguished zones of ‘increasing’ versus ‘high’ risks. However, owing to the difficulty in quantifying where the line between the two could be drawn in a consistent way across all boundaries, the framework now adopts the ‘burning embers’ approach used by the Intergovernmental Panel on Climate Change (Richardson et al. 2023). In the ‘burning embers’ model, risks grade into one another instead of being sharply demarcated. However, this model is also highly contested. In 2007, climate negotiators dropped it altogether amid debates over what constitutes ‘dangerous’ climate change (see Mahony 2015). Likewise, when the planetary boundaries framework adopts the ‘burning embers’ model it raises questions over who should decide what is ‘safe’ and through what processes? In negotiations over dangerous degrees of climate warming by the IPCC, the difference between 1.5°C versus 2.0°C warming proved a crucial part of a larger socio-political context in which international negotiations took place (Cointe & Guillemot 2023; Randalls 2010). Not only does the planetary boundaries framework apply the ‘burning embers’ model to all nine boundaries, it does so in an international context that is not equal when it comes to deciding on how to define what is safe. For instance, IPCC processes perpetuate inequality when they ignore how even 1.5°C warming is dangerous for island nations and many peoples of colour, such as those in the Caribbean, who already bear the burden of environmental inequality (Sealey-Huggins 2018).

A concern closely related to how risks are navigated is how planetary boundaries are measured. This concern matters not only empirically, in terms of what may or may not be considered a ‘safe’ space, but also ethically, since what is measured in terms of anthropogenic impacts on the planet has implications for who is responsible for those impacts. Here, it is important to note that, over its assessments (in 2009, 2015, 2023, 2025), many of the planetary boundaries have evolved in terms of what is measured and how. This makes pinning down matters of responsibility difficult.

It also makes consistently calculating the risk trajectory of human impacts on different boundaries challenging. For instance, the first iteration of the framework identified 'chemical pollution' as a planetary boundary but wasn't yet able to quantify it (Rockström et al. 2009b). That boundary evolved into the idea of 'novel entities' (Steffen et al. 2015), which refer to 'truly novel anthropogenic introductions to the Earth system' like microplastics, nuclear waste, genetically modified organisms, and other 'direct human interventions into evolutionary processes' (Richardson et al. 2023: 6). These kinds of conceptual adjustments are not uncommon in the formation of scientific ideas. Yet in the planetary boundaries framework they affect questions of ethics, such as those regarding different social causes of environmental impacts. For instance, the boundary for land-system change was initially described as 'driven primarily by agricultural expansion and intensification' (Rockström et al. 2009b), but its most recent appraisal doesn't mention agriculture at all. Instead, land-system change is referenced to deforestation (Richardson et al. 2023).

Recent efforts to integrate considerations of justice with the planetary boundaries framework, or through terminological shifts to 'safe and just' Earth system boundaries, are efforts to craft ethical space. Initially, the planetary boundaries were designed to safeguard the 'freedom to pursue long-term social and economic development' (Rockström et al. 2009a: 475). More recent efforts continue to use the planetary boundaries to mark out a moral perimeter by employing judgements and metrics to calculate different parameters affecting Earth's resilience. In its old and new iterations, ethical challenges arise. For one, the power of the inference from independent planetary processes structuring safe conditions for humanity to the claim that planetary boundaries similarly structure ethics is intuitive to many. But it is entirely provisional. It is an inference premised on accepting the value judgements that determine the planetary boundaries themselves, from those initially acknowledged to those now used to negotiate different sorts of boundaries and their collective perimeter for justice. It is reasonable here to ask questions about how determinations of 'safe' were made, for whom, and by whom. These issues bump into those over categories of risk, whether strictly demarcated or which operate through a gradual, 'burning embers' approach. Replacing a system for risk calculation premised on measurement with one politically negotiated for climate change – not with other planetary boundaries necessarily so easily captured – is a significant ethical issue. It is one that can reproduce marginalization or suppress alternatives that may incorporate different values or approaches to risk. These are challenges returned to below and which later chapters address in different ways. First, however, it is worth exploring further how the planetary boundaries framework has inspired and influenced new ideas in economics, governance, and law. These, too, help show why ethics matter to the planetary boundaries.

1.3 The Promise of Planetary Boundaries

The planetary boundaries framework has shaped global discussions about science and sustainability and has also inspired new ideas that draw upon and extend its central ideas (Rockström et al. 2024). For instance, Jeffrey Sachs (2015), the President of the UN Sustainable Development Solutions Network, incorporated the planetary boundaries framework directly into his influential book, *The Age of Sustainable Development*. Ecological economists have similarly made reports to the United Nations that use the planetary boundaries to argue that the economy must be seen as just one aspect of social relationships that are all embedded in Earth's life-support system (Costanza et al. 2012). Here we discuss just two examples of how the planetary boundaries framework has inspired new ways of thinking in other fields concerned with human–Earth relationships. The first, popularized by Kate Raworth, complements the 'upper ceiling' of planetary boundaries with a 'social floor' that would establish minimum distributions of planetary goods for human security. The second employs the planetary boundaries framework in international law so that a system of agreements can be rationally developed based on a shared empirical framework. Both ideas draw on what might be termed the normative promise of the planetary boundaries framework – the idea that the framework offers terms upon which to link an empirical account of how humans affect the planet with scales of decision-making in which values can operate in ways consistent with Earth system science.

1.3.1 Donut Economics

Kate Raworth (2017a) has argued that while the planetary boundaries framework marks an upper ceiling for human impacts on the planet, what is also needed is a social floor to ensure minimum needs for all are met. Unlike climate change, or ocean acidification, or other boundaries that matter to the planet, the 'social floor' is about what matters inside the 'safe operating space' for humanity. Specifically, it is about both material needs for energy, clean water, adequate food, and health that exist alongside a suite of other socially relevant requirements for a good life, such as education, literacy, gender equality, income security, and work. The economy, Raworth argues, should be reconfigured to allocate resources between these two constraints in a way that is fair and just. Raworth's idea has caught on, with the city of Amsterdam beginning to experiment with it at a municipal level. It has also been influential in the planetary boundaries framework itself. For instance, Rockström et al. (2023) cite the idea explicitly in their effort to incorporate social goals alongside those that the planetary boundaries framework sets for the Earth system.

The idea that there is a 'social floor' for what might be considered a fair or just distribution of resources is not one that operates exclusively at a planetary scale. This is because many of the policies affecting societies and the environment are made by institutions operating at national or sub-national scales (i.e. regional or local). So, if the planetary boundaries are to gain purchase on sites of actual decision-making, finding ways to utilize them at other scales is necessary (Dearing et al. 2014). For this kind of 'down-scaling' it is critical to note that Raworth's model uses a host of indicators developed to capture social concerns, such as metrics of gender equality in earnings or gender representation in parliaments from the International Labour Organisation and the World Bank, respectively. Likewise, material requirements are determined through measures of access to drinking water from the World Health Organisation and UNICEF, and access to energy is measured in terms of electricity and cooking facilities data from the OECD (see Raworth 2017b). Here, the justification for what is measured (and how) within the donut is not itself referenced to a theory of justice or ethics, but to metrics developed by a variety of globally influential organizations and institutions.

The reach of the Raworth's idea in discussions of planetary boundaries stands equal to the challenge that it also points out. Namely, and as Gupta et al. (2023a) noted earlier, efforts to quantify a 'social floor' have found that ensuring equity in access to social and material resources would likely exceed the planetary boundaries. More than this, however, efforts to translate the planetary boundaries into a framework for justice can also create a moral hierarchy with science at the top. For instance, Häyhä et al. (2016) attempt to craft an integrated use of the planetary boundaries across scales but achieve this by prioritizing biophysical considerations before socio-economic concerns, both of which are addressed before ethical issues. Given the intrinsic normative concerns within the planetary boundaries framework and judgements used to benchmark risks, however, it is unclear why ethics is left to the last. Or, as in Raworth's model, why explanations for what ought to be aimed for are outsourced to metrics developed by institutions that are not free from their own normative positions. So, in a way similar to how the normative values of the planetary boundaries framework don't stay in the background, neither should the normative values of institutions be uncritically accepted in accounts of a 'social floor' designed to complement the ceiling provided by planetary boundaries. As many chapters in this book point out, there is more to consider.

1.3.2 International Environmental Law

Planetary boundaries have also inspired new approaches to environmental law and governance. Many scholars find significant promise in how the planetary boundaries can provide a basis for agreements that are not biased to any country or international actor. For instance, Geoffrey Garver's (2021: 4) arguments for ecological law are based on the idea that planetary boundaries 'must have primacy over economic, political, and other considerations'. Importantly, efforts to incorporate the planetary boundaries into legal or governance frameworks are not only theoretical. For instance, reference to planetary boundaries can be found in statements by the United Nations and the European Union (see Fernández & Malwé 2019). So, the promise of the planetary boundaries on matters of environmental governance is not confined to academia but is reaching practical, applied avenues of governance as well.

One prominent proposal for incorporating the planetary boundaries into law and governance is to treat the boundaries as a *grundnorm* – a concept with a longer legal history that identifies the ways that some norms come logically and empirically prior to others. Kim and Bosselmann (2015) argued that the environmental preconditions identified through the planetary boundaries framework satisfy the logical and empirical demands to be a *grundnorm* because they provide a framework for preserving the conditions of ecological and social life. Since those conditions are an empirical requirement for any system of law, they should also be appraised for whether they offer a rational basis for law. On this score they also find strong promise owing to the scientific basis of the planetary boundaries not in any society's views but in the science of how the Earth system functions. Since it is irrational for international agreements to find common ground in ways that would empirically undermine the conditions that universally provide for social well-being, the planetary boundaries framework is presented as a rationally sound and empirically defensible candidate for a *grundnorm*.

The promise of a *grundnorm* has been further extended by those who argue that treating the planetary boundaries in this way can provide a way to find convergence between Earth system science and global governance initiatives, like the Sustainable Development Goals (Kim 2016). This aim fits well with the initial promise of the planetary boundaries to provide a framework for human development without prescribing what form that might take (Rockström et al. 2009b). But it also repeats, albeit in a different way, the idea that there is a moral hierarchy for decision-making that can be mapped onto a hierarchy for maintaining planetary conditions. For instance, it could potentially delay ethical deliberation until after biophysical matters are scientifically 'settled', which is a significant issue given that every iteration of the planetary boundaries framework has changed something about how to measure different variables. It could also make some logical inferences intractable. This is because justifying a *grundnorm* through the planetary boundaries framework would also need a way to deduce, from the many non-linear and cross scalar dynamics that impact the Earth system, a rational account of cause and effect that could be used in decision-making (Schmidt 2019). As philosophers of science sometimes put it, however, there is a logical gap between hunting causes and using them in public policy (Cartwright 2007). So, despite the promise of a planetary *grundnorm*, issues of judgement, values, and ethics remain salient.

1.4 Criticisms of the Planetary Boundaries

In different ways, the aforementioned examples are inspired by the planetary boundaries and encounter the kinds of challenges that are not uncommon in cases of significant social and ecological complexity. There are also, however, critiques of the planetary boundaries framework. These come in forms that can be anticipated from some of what has already been implied – several authors of the chapters in this book develop their own critiques as well. Some of these critiques are partial in the sense that they target just one element (or some subset) of the planetary boundaries framework. Others are more holistic. Here, we survey some prominent criticisms of the planetary boundaries framework alongside some of the responses to them.

In an in-depth, holistic critique of the planetary boundaries framework, Biermann and Kim (2020: 3) note that, although the normative aspects of determining boundaries are highlighted, 'this normativity had no consequences for the assessment process itself'. Instead, they argue, the definitions of boundaries proceeded instead from a 'purely scientific, expert-driven perspective'. This critique gets at an important consideration: the role of judgements in determining boundaries is not explained well in the planetary boundaries framework, and neither was determining those boundaries an open social process. Instead, judgements relied on appraisals by a particular scientific community that formed around the planetary boundaries concept. These formations are sometimes referred to as 'epistemic communities' because they tend to share deeply held commitments about how to appraise sources of knowledge and forms of reasoning (Haas 1992). That is, they share a way of knowing. This way of knowing in the planetary boundaries framework, however, is further critiqued by Biermann and Kim (2020) for prioritizing epistemic practices of the Global North and claiming a kind of planetary authority for them.

Epistemic communities can share ethical commitments too, such as those Naomi Oreskes (2020) describes as a 'duty to warn' that arises from the specialized knowledge scientists hold. But what the critique of Biermann and Kim (2020) points out is something different. Namely, if commitments to a shared way of knowing do not translate into an open explanation of how normative judgements inform scientific assessments, then the framework is not operating with a clear ethical justification. It is important here to note that measures of risk and questions of viability regarding different future scenarios do not operate in a social vacuum, and themselves can also be affected by institutional, cultural, or other constraints (Paprocki 2022). As with the insight from James Kay (2000) earlier regarding the role of the scientists as narrators of different risks and trade-offs, what this means is that ethical implications should receive more up-front considerations rather than being found on the last branches of decision trees.

A second criticism is that the idea of a 'safe operating space for humanity' is anthropocentric. That is, it situates humans as the central, potentially exclusive subjects of moral consideration. Rejections of, and responses to, anthropocentrism have a long history in environmental ethics: from claims of intrinsic value for non-human individuals or species through to 'weak' anthropocentrism that views the natural world as worthy of attention owing to the importance of healthy environments for people (see Norton 1984). In the case of how the planetary boundaries place humans at the framework's centre, two critiques are often in play. The first is that humans are relational in the sense that they exist in interdependent relations with multiple species both within and beyond their persons (for example, Heikkurinen et al. 2021). From gut bacteria to food supply chains to the biosphere's role in biogeochemical cycles, people are entangled in multispecies worlds. This critique targets the singling out of one species from those it is interdependent with. The second idea is that non-humans are important to consider in their own right but also that creating a safe space for humanity must entail respecting the ways that other species and processes act in ways that are relevant to more than only the human enterprise (Eckersley 2017). This critique recommends respect for everything happening on the planet that isn't about people, which is quite a lot. Taken together, critiques of anthropocentrism are not merely about ensuring attention is paid to more than humans alone but about whether the framework has at its core an adequate account of why planetary conditions matter morally given the many ways that other species – the biosphere – relate to one another, people, and the planet.

A third criticism is that the planetary boundaries framework isn't capacious enough when it comes to other ways of knowing, or the social worlds in which those ways of knowing matter. For instance, it is often the case that achieving justice for the poor is not among the key priorities of large-scale environmental assessments (Kashwan et al. 2020). Farhana Sultana (2023) argues that planetary crises demand social as well as scientific explanations so that matters of inequities at a planetary scale can be seen as outcomes of social forces and political choices, such as colonialism. Another aspect of this critique is that the kinds of risk-forecasting that anchors the planetary boundaries framework do not respect how many peoples have already lived through environmental apocalypse and have practices – ethical and epistemic – for maintaining their lifeways (Davis & Todd 2017; Ghosh 2016). The worry here is amplified when, as Kyle Powys Whyte (2020) argues in the case of Indigenous injustices, the opportunities to address long-standing inequalities are overridden by the urgency of acting quickly on planetary concerns; to take one example, avoiding a climate tipping point may come at the expense of transgressing a relational tipping point that could address planetary crises in terms ethically acceptable to Indigenous peoples.

Not everybody is convinced by these (or other) critiques, and in this book, we do not seek to adjudicate them. For example, Clive Hamilton (2017) argues that human impacts on the Earth system burdens humanity with a unique, geologically exceptional form of moral responsibility. The exceptional planetary condition identified through Earth system science, on Hamilton's view, presents a powerful case for reconceiving of what 'ethics' even means when decisions have far-reaching and difficult to predict consequences. In this volume, the aim is to prompt further discussions given the role of values in different aspects of the framework itself and the implications of using (or ignoring) the ways that it discloses relationships between people and planet on each of the variables that it treats discretely but which collectively are meant to capture core, collective concerns of human relations to the Earth system.

References

Biermann, F., & Kim, R. E. (2020). The Boundaries of the Planetary Boundary Framework: A Critical Appraisal of Approaches to Define a 'Safe Operating Space' for Humanity. *Annual Review of Environment and Resources*. 45(1), 497–521. DOI: 10.1146/annurev-environ-012320-080337.

Cartwright, N. (2007). *Hunting Causes and Using Them: Approaches in Philosophy and Economics*. Cambridge: Cambridge University Press.

Castree, N., Bellamy, R., & Osaka, S. (2021). The Future of Global Environmental Assessments: Making a Case for Fundamental Change. *The Anthropocene Review*. 8(1), 56–82.

Cointe, B., & Guillemot, H. (2023). A History of the 1.5°C Target. *WIREs Climate Change*. 14(3), e824.

Costanza, R., Alperovitz, G., Daly, H., et al. (2012). *Building a Sustainable and Desirable Economy-in-Society-in-Nature*. New York: United Nations Division for Sustainable Development.

Davis, H., & Todd, Z. (2017). On the Importance of a Date, or Decolonizing the Anthropocene. *ACME: An International Journal for Critical Geographies*. 16(4), 761–780.

Dearing, J. A., Wang, R., Zhang, K., et al. (2014). Safe and Just Operating Spaces for Regional Social-Ecological Systems. *Global Environmental Change*. 28, 227–238. DOI: 10.1016/j.gloenvcha.2014.06.012.

Eckersley, R. (2017). Geopolitan Democracy in the Anthropocene. *Political Studies*. 65(4), 983–999. DOI: 10.1177/0032321717695293.

Fernández, E. F., & Malwé, C. (2019). The Emergence of the 'Planetary Boundaries' Concept in International Environmental Law: A Proposal for a Framework Convention. *Review of European, Comparative & International Environmental Law*. 28(1), 48–56. DOI: 10.1111/reel.12256.

Findlay, H., Feely, R. A., Jiang, L.-Q., Pelletier, G., & Bednaršek, N. (2025) Ocean Acidification: Another Planetary Boundary Crossed. *Global Change Biology*. 31(6), e70238. https://doi.org/10.1111/gcb.70238

Folke, C. (2006). Resilience: The Emergence of a Perspective for Social–Ecological Systems Analyses. *Global Environmental Change*. 16(3), 253–267. DOI: 10.1016/j.gloenvcha.2006.04.002.

Garver, G. (2021). *Ecological Law and the Planetary Crisis: A Legal Guide for Harmony on Earth*. London: Routledge.

Ghosh, A. (2016). *The Great Derangement: Climate Change and the Unthinkable*. Gurgaon: Penguin Books.

Gunderson, L., & Holling, C. S., eds. (2002). *Panarchy: Understanding Transformations in Human and Natural Systems*. Washington, DC: Island Press.

Gupta, J. (2016). Toward Sharing Our Ecospace. In S. Nicholson & S. Jinnah, eds., *New Earth Politics: Essays From the Anthropocene*. Cambridge, MA: MIT Press, 271–292.

Gupta, J., Liverman, D., Bai, X., et al. (2021). Reconciling Safe Planetary Targets and Planetary Justice: Why Should Social Scientists Engage with Planetary Targets? *Earth System Governance*. 10, 100122. DOI: 10.1016/j.esg.2021.100122.

Gupta, J., Liverman, D. M., Prodani, K., et al. (2023a). Earth System Justice Needed to Identify and Live within Earth System Boundaries. *Nature Sustainability*. 6(6), 630–638. DOI: 10.1038/s41893-023-01064-1.

Gupta, J., Prodani, K., Bai, X., et al. (2023b). Earth System Boundaries and Earth System Justice: Sharing the Ecospace. *Environmental Politics*. 33(7), 1286–1305. DOI: 10.1080/09644016.2023.223479.

Haas, P. M. (1992). Introduction: Epistemic Communities and International Policy Coordination. *International Organization*. 46(1), 1–35.

Hamilton, C. (2017). *Defiant Earth: The Fate of Humans in the Anthropocene*. Cambridge: Polity Press.

Häyhä, T., Lucas, P. L., Van Vuuren, D. P., et al. (2016). From Planetary Boundaries to National Fair Shares of the Global Safe Operating Space: How Can the Scales Be Bridged? *Global Environmental Change*. 40, 60–72. DOI: 10.1016/j.gloenvcha.2016.06.008.

Heikkurinen, P., Clegg, S., Pinnington, A. H., et al. (2021). Managing the Anthropocene: Relational Agency and Power to Respect Planetary Boundaries. *Organization & Environment*. 34(2), 267–286. DOI: 10.1177/1086026619881145.

Holling, C. S. (1973). Resilience and Stability of Ecological Systems. *Annual Review of Ecology and Systematics*. 4, 1–23.

Holling, C. S. (1986). The Resilience of Terrestrial Ecosystems: Local Surprise and Global Change. In W. C. Clark & R. E. Munn, eds., *Sustainable Development of the Biosphere*. Cambridge: Cambridge University Press, 292–316.

Holling, C. S., & Meffe, G. K. (1996). Command and Control and the Pathology of Natural Resource Management. *Conservation Biology: The Journal of the Society for Conservation Biology*. 10(2), 328–337.

Kashwan, P., Biermann, F., Gupta, A., et al. (2020). Planetary Justice: Prioritizing the Poor in Earth System Governance. *Earth System Governance*. 6, 100075. DOI: 10.1016/j.esg.2020.100075.

Kay, J. (2000). Ecosystems as Self-Organizing Holarchic Open Systems: Narratives and the Second Law of Thermodynamics. In S. Jørgensen & F. Müller, eds., *Handbook of Ecosystem Theories and Management*. Boca Raton, FL: Lewis Publishers, 135–160.

Kim, R. E. (2016). The Nexus between International Law and the Sustainable Development Goals. *Review of European, Comparative & International Environmental Law*. 25(1), 15–26. DOI: 10.1111/reel.12148.

Kim, R. E., & Bosselmann, K. (2015). Operationalizing Sustainable Development: Ecological Integrity as a Grundnorm of International Law. *Review of European, Comparative & International Environmental Law*. 24(2), 194–208. DOI: 10.1111/reel.12109.

Langston, N. (1995). *Forest Dreams, Forest Nightmares: The Paradox of Old Growth in the Inland West*. Seattle: University of Washington Press.

Mahony, M. (2015). Climate Change and the Geographies of Objectivity: The Case of the Ipcc's Burning Embers Diagram. *Transactions of the Institute of British Geographers*. 40(2), 153–167.

Meadows, D. (1999). *Leverage Points: Places to Intervene in a System*. Hartland: Sustainability Institute.

Meadows, D. (2008). *Thinking in Systems: A Primer*. White River Junction, VT: Chelsea Green Publications.

Norton, B. (1984). Environmental Ethics and Weak Anthropocentrism. *Environmental Ethics*. 6(2), 131–148.

Oreskes, N. (2020). What Is the Social Responsibility of Climate Scientists? *Daedalus*. 149(4), 33–45.

Paprocki, K. (2022). On Viability: Climate Change and the Science of Possible Futures. *Global Environmental Change*. 73, 102487. DOI: 10.1016/j.gloenvcha.2022.102487.

Randalls, S. (2010). History of the 2°C Climate Target. *WIREs Climate Change*. 1(4), 598–605.

Raworth, K. (2017a). A Doughnut for the Anthropocene: Humanity's Compass in the 21st Century. *The Lancet: Planetary Health*. 1(2), e48–e49. DOI: 10.1016/S2542-5196(17)30028-1.

Raworth, K. (2017b). *Doughnut Economics: Seven Ways to Think Like a 21st-Century Economist*. New York: Random House.

Richardson, K., Steffen, W., Lucht, W., et al. (2023). Earth beyond Six of Nine Planetary Boundaries. *Science Advances*. 9(37), 2458. DOI: 10.1126/sciadv.adh2458.

Rockström, J., Donges, J., Fetzer, I., et al. (2024). Planetary Boundaries Guide Humanity's Future on Earth. *Nature Reviews Earth & Environment*. 5(11), 773–788. DOI: 10.1038/s43017-024-00597-z.

Rockström, J., Gupta, J., Lenton, T. M., et al. (2021). Identifying a Safe and Just Corridor for People and the Planet. *Earth's Future*. 9(4), e2020EF001866. DOI: 10.1029/2020EF001866.

Rockström, J., Gupta, J., Qin, D., et al. (2023). Safe and Just Earth System Boundaries. *Nature*. 619(7968), 102–111. DOI: 10.1038/s41586-023-06083-8.

Rockström, J., Steffen, W., Noone, K., et al. (2009a). A Safe Operating Space for Humanity. *Nature*. 461(7263), 472–475. DOI: 10.1038/461472a.

Rockström, J., Steffen, W., Noone, K., et al. (2009b). Planetary Boundaries: Exploring the Safe Operating Space for Humanity. *Ecology & Society*. 14(2): Article 32. DOI: 10.5751/ES-03180-140232.

Sabin, P. (2013). *The Bet: Paul Ehrlich, Julian Simon, and Our Gamble over Earth's Future*. New Haven, CT: Yale University Press.

Sachs, J. (2015). *The Age of Sustainable Development*. New York: Columbia University Press.

Sakschewski, B., Caesar, L., Andersen, L. S., et al. (2025). *Planetary Health Check*. Potsdam Institute for Climate Impact Research (PIK), Potsdam, Germany. DOI: 10.4s8485/pik.2025.017

Schellnhuber, J. (2000). The Waikiki Principles: Rules for a New GAIM. *Global Change Newsletter*. 41, 3–4.

Schmidt, J. J. (2019). The Moral Geography of the Earth System. *Transactions of the Institute of British Geographers*. 44(4), 721–734.

Sealey-Huggins, L. (2018). The Climate Crisis Is a Racist Crisis: Structural Racism, Inequality and Climate Change. In A. Johnson, R. Joseph-Salisbury, & B. Kamunge, eds., *The Fire Now: Anti-Racist Scholarship in Times of Explicit Racial Violence*. London: Zed Books, 99–113.

Steffen, W., Crutzen, P. J., & McNeill, J. R. (2007). The Anthropocene: Are Humans Now Overwhelming the Great Forces of Nature. *Ambio*. 36(8), 614–621. DOI: 10.1579/0044-7447(2007)36[614:TAAHNO]2.0.CO;2.

Steffen, W., Richardson, K., Rockström, J., et al. (2015). Planetary Boundaries: Guiding Human Development on a Changing Planet. *Science*. 347(6223), 1259855. DOI: 10.1126/science.1259855.

Steffen, W., Richardson, K., Rockström, J., et al. (2020). The Emergence and Evolution of Earth System Science. *Nature Reviews Earth & Environment*. 1(1), 54–63. DOI: 10.1038/s43017-019-0005-6.

Sultana, F. (2023). Whose Growth in Whose Planetary Boundaries? Decolonising Planetary Justice in the Anthropocene. *Geography and Environment*. 19(2), e00128. DOI: doi.org/10.1002/geo2.128

Waters, C. N., Zalasiewicz, J., Summerhayes, C., et al. (2016). The Anthropocene Is Functionally and Stratigraphically Distinct from the Holocene. *Science*. 351(6269), aad2622. DOI: 10.1126/science.aad2622

Whyte, K. (2020). Too Late for Indigenous Climate Justice: Ecological and Relational Tipping Points. *WIREs Climate Change*. 11(1), e603.

Part II

Stratospheric Ozone Depletion

2

Introduction to the Stratospheric Ozone Depletion Boundary

SIMON P. MEISCH

The Earth is covered by five atmospheric layers, composed mainly of nitrogen and oxygen, but also of other gases. In the stratosphere, the second lowest of these layers, a high concentration of the gas ozone (O_3) absorbs most of the sun's ultraviolet radiation, which, unfiltered, would be most harmful to biological organisms on the planet's surface. The planetary boundary on stratospheric ozone is, therefore, concerned with the ability of this ozone layer to filter UV radiation. This ability is measured in Dobson Units (DU), which is the concentration of a gas in a column of air. To determine the ozone concentration, imagine taking out all the ozone in a column of air over a given area, creating standardized conditions (of temperature and air pressure) and checking how thick this layer of pure ozone would be (European Environment Agency 2023). Under these conditions, 100 DU are 1 millimetre, which is the equivalent of a stack of 10 sheets of writing paper. The planetary boundaries framework takes the pre-industrial Holocene baseline of 290 DU as a reference and, assuming a range of uncertainty of less than 5 per cent of this baseline, sets the limit at 276 DU. In 2023, the actual value was 284.6 DU, making the ozone boundary one of three that have not been crossed (Richardson et al. 2023: 5). 220 DU or less is considered an ozone hole, which, contrary to its name, is not the complete absence of ozone in an area, but a concentration so depleted that it does not absorb enough UV radiation (European Environment Agency 2023).

Of all the planetary boundaries, stratospheric ozone depletion stands out because it is associated with an oft-repeated success story in international environmental politics. In the mid-1980s, scientists discovered a hole in the ozone layer over Antarctica and identified groups of halogenated hydrocarbons as the cause of stratospheric ozone depletion. Back then, many of these gases were found as chlorofluorocarbons, or CFCs, in everyday items such as refrigerators and air conditioners. Alarmed, the global political community moved quickly to negotiate and ratify the 'Vienna Convention for the Protection of the Ozone Layer' (signed in 1985, entered into force in 1987) and the 'Montreal Protocol' (signed in 1987, entered into force in 1989). In essence, the treaties set out a process by which substances defined as harmful to stratospheric ozone are phased out and the resulting burdens are shared equitably. The Vienna Convention and the Montreal Protocol were the first international environmental treaties to be signed by all member states of the United Nations. Furthermore, this ozone regime has proven to be effective, as stratospheric ozone concentrations have increased. Against this background, Rockström et al. (2009: 12) concluded that '[o]n balance, the case of stratospheric ozone is a good example where concerted human effort and wise decision making seem to have enabled us to stay within a planetary boundary'.

This narrative has been challenged on epistemological and ethical grounds. Grundmann (2006), for example, stresses that the scientific consensus was not as great as the subsequent success story makes it appear. Despite initial scientific uncertainty, decision-makers recognized the moral importance of the ozone hole for human survival and acted anyway. Gareau (2013) points out that the ozone regime follows a neoliberal logic. As a result, it struggles to respond adequately to questions of global justice.

In their respective contributions, Louise du Toit and Christine Winter add to and transcend these critiques. While recognizing the pioneering nature of the ozone regime, they raise ethical questions about the nature of its success. Louise du Toit stresses the importance of creating political institutions with the mandate and capacity to protect morally relevant goods such as the ozone layer. However, du Toit also raises the issue of 'problem-shifting' in the context of the ozone regime. Such problem-shifting occurs when the successful resolution of some moral or empirical problems within the regime's legal framework creates other problems outside the scope of that framework. She goes on to explain the tensions that arise when contested ethical principles, such as the precautionary principle or the principle of sustainability, need to be specified by legal systems. She critically discusses the narrow utilitarian and anthropocentric moral bias of the ozone regime, which fails to assess the full extent of the problems associated with the ozone hole.

Writing from a relational and anti-colonial perspective, Christine Winter advances an approach to justice that seeks to truly grasp planetary relations. Winter argues against thinking of ozone merely as a bounded and delimitable molecule, as the planetary boundaries framework does. Instead, she argues for understanding and relating to ozone in all its natural relationships with living and non-living matter. She then asks what moral rights and obligations arise in and through these relationships. Winter also points to the colonial implications of planetary boundaries in general and ozone in particular. In her view, the concept of planetary boundaries provides an argument for enclosing and 'colonizing' a common good. In doing so, it creates the right for some people to use and pollute the stratosphere at the expense of others. Instead, Winter argues for 'an anticolonial planetary justice [that] proposes a planet on which a life sustaining planetary net of overlapping (local) relationships are the subject of justice'.

References

European Environment Agency. (2023). What Is the Current State of the Ozone Layer? www.eea.europa.eu/en/topics/in-depth/climate-change-mitigation-reducing-emissions/current-state-of-the-ozone-layer

Gareau, B. J. (2013). *From Precaution to Profit: Contemporary Challenges to Environmental Protection in the Montreal Protocol*. New Haven, CT: Yale University Press.

Grundmann, R. (2006). Ozone and Climate: Scientific Consensus and Leadership. *Science, Technology, & Human Values*. 31(1), 73–101. DOI: 10.1177/0162243905280024.

Richardson, K., Steffen, W., Lucht, W., et al. (2023). Earth beyond Six of Nine Planetary Boundaries. *Science Advances*. 9(37), 2458. DOI: 10.1126/sciadv.adh2458.

Rockström, J., Steffen, W., Noone, K., et al. (2009). Planetary Boundaries: Exploring the Safe Operating Space for Humanity. *Ecology & Society*. 14(2): article 32. DOI: 10.5751/ES-03180-140232.

3

The Planetary Boundary on Stratospheric Ozone Depletion and the Evolution of Values under the Ozone Regime

LOUISE DU TOIT

3.1 Introduction

Stratospheric ozone, or ozone in the stratospheric layer of the atmosphere, plays a crucial role in absorbing the sun's ultraviolet (UV) radiation. Destruction of stratospheric ozone allows UV radiation, primarily UV-B, to reach the Earth's surface. Excessive UV-B radiation can cause skin cancer and cataracts in humans, stunting in plants, and harm to animals integral to the marine food chain (Nanda 1989: 489–90). In recognition of the severe and irreversible effects of UV-B radiation on human health and ecosystems, Rockström and colleagues identified stratospheric ozone depletion as one of the nine planetary boundaries (Rockström et al. 2009a, 2009b; see also: Steffen et al. 2015).

Stratospheric ozone depletion is one of the few planetary boundaries for which a relevant international law regime – consisting of the Vienna Convention for the Protection of the Ozone Layer (1985) and the Montreal Protocol on Substances that Deplete the Ozone Layer (1987)[1] – is readily identifiable; and it is widely considered to be one of the most effective international environmental law (IEL) regimes to date. However, the implementation of the ozone regime has had the effect of shifting problems (Kim & van Asselt 2016), thereby creating new, unregulated problems. Furthermore, based as it is on principles such as the principle of sustainable development and the principle of permanent sovereignty over natural resources, the ozone regime is human-centred, and concern for the protection of ozone is arguably linked to its value to humans. While several concerns discussed in this chapter could potentially be addressed by 'tinkering at the margins' (Kotzé 2020: 77), this would not address the underlying values and principles of the ozone regime (which underlie international environmental law more generally: see, e.g., Garver 2019; Kotzé & Kim 2019; Kotzé et al. 2020); and it is argued in this chapter that broader shifts are required in order to be able to respond the types of challenges being faced in the

[1] In this chapter, the Vienna Convention and the Montreal Protocol are together referred to as 'the ozone regime'.

Anthropocene (Crutzen & Stoermer 2000; Steffen et al. 2007), a new geological epoch, in which humans have become 'geological agents' (Oreskes 2007: 93; see also Raworth 2014; Moore 2016).

The chapter first briefly outlines the international law regime for ozone depletion. The section following highlights various socio-ecological impacts and injustices that arise from the implementation of the ozone regime. Thereafter, the chapter begins to explore how values underlying the ozone regime could potentially evolve in order to address these types of concerns and highlights the importance of shifting away from the human-centred and utilitarian values that underlie the ozone regime and IEL more generally.

3.2 The International Regulation of Stratospheric Ozone Depletion

Chlorofluorocarbons (CFCs) were invented in 1928 as a safer (non-toxic) and more stable alternative to other substances, such as sulphur dioxide and ammonia, for use as home refrigerants (Sinclair 2002: 4).[2] Their use on a large scale began in the 1930s (Goyal et al. 2019: 1), and, by 1985, total CFC production exceeded one million tonnes. Of note, per capita consumption of CFCs was 10 times greater in developed countries than in developing countries (Gillespie 2006: 21). While there was growing concern regarding the increasing presence of CFCs in the atmosphere and their potential role in depleting stratospheric ozone (Molina & Rowland 1974), there was also considerable uncertainty throughout the 1970s and early 1980s regarding the causes of ozone depletion (with many trying to attribute it to natural causes) as well as the risk that ozone-depleting substances (ODSs) actually posed[3] (DeSombre 2000: 49). There was also disagreement over whether a precautionary approach should be followed, or whether conclusive evidence of ozone depletion should first be required (Gillespie 2006: 152–57).

It was thus significant that the Vienna Convention was agreed to in March 1985. The negotiation process – with regard to both the Vienna Convention and the Montreal Protocol (agreed to two years later) – has been described as 'particularly impressive' since negotiations were carried out and agreement reached 'under conditions of uncertainty, both over the existence and extent of environmental harm and the costliness of taking action to mitigate it' (DeSombre 2000: 49). Parties to the Vienna Convention, in the Preamble, noted the potentially harmful impact of the modification of the ozone layer on human health and the environment and indicated their determination to protect human health and the environment from adverse impacts resulting from such modification.

[2] The discussion in this section draws on Du Toit (2021).

[3] Scepticism regarding the 'CFC-ozone depletion theory' persisted into the 1990s: see, e.g., Lieberman (1994).

In 1987, the same year in which the link between CFCs and ozone depletion was conclusively established (Sinclair 2002: 22), the Montreal Protocol was concluded. Parties more explicitly recognized that emissions of certain substances can deplete the ozone layer and they resolved to control the emission of such ODSs 'with the ultimate objective of their elimination' (Montreal Protocol: Preamble). Significantly, the Montreal Protocol introduced control measures for the phase-down of ODSs. Initially, only CFCs and halons were regulated by the Montreal Protocol as these were the only substances that had been identified as being ozone-depleting. However, over time, more substances were identified as being responsible for ozone depletion (Gillespie 2006: 164; Roberts 2017: 221), and they were accordingly also brought under the control of the Montreal Protocol through several amendments (articles 2A–2J). The Montreal Protocol now covers 114 chemicals in 10 chemical groups, including hydrochlorofluorocarbons (HCFCs), carbon tetrachloride (CTC), and hydrofluorocarbons (HFCs); and it provides for the near-complete phase-out of the substances under its control.

These substances are used in various applications, including air conditioners, refrigerators, and fire extinguishers, and as aerosol propellants and fumigants (DCCEEW 2018). It is notable that many of the substances controlled under the Montreal Protocol are also powerful greenhouse gases and have global warming potentials[4] (GWPs) hundreds to thousands of times greater than that of carbon dioxide (WMO 2018: Table A-1).

The ozone regime takes specific account of the special needs and circumstances of developing country parties (Vienna Convention, Preamble; Montreal Protocol, Preamble and article 5) and includes several mechanisms in this regard, including delayed compliance with the Montreal Protocol's control measures (article 5(1) and article 5(8 *bis*)–article 5(8 *qua*)) and the allowance of a certain amount of production of controlled substances to 'satisfy the basic domestic needs' of developing country parties (articles 2D–2J). Furthermore, in 1990, the Montreal Protocol was amended to provide for the establishment of a financial transfer mechanism (article 10); and the Multilateral Fund – which provides funding to developing country parties to assist them in complying with their obligations under the Montreal Protocol – was established in 1991. This Fund is managed by an Executive Committee, which has equal numbers of Article 5 (developing) and non-Article 5 (developed) country parties. There is also a double majority voting system, in terms of which decisions require a two-thirds majority vote, which represents a majority amongst developed and developing country parties. On this basis, it has been argued that the Montreal

[4] The GWP is 'a metric for determining the relative contribution of a substance to climate warming' (WMO 2018: ES.13).

Protocol has given effect to the principle of common but differentiated responsibilities (Gonzalez et al. 2015: 124–25).

More generally, the Montreal Protocol is the first international treaty to have been universally ratified (Godin-Beekmann et al. 2018: 331); and it has been highly effective in addressing its central concern. Indeed, it was described by former United Nations Secretary-General Kofi Annan (Environmental Investigation Agency 2017) as '[p]erhaps the single most successful international agreement to date'. Rockström et al. argue that '[o]n balance, the case of stratospheric ozone is a good example where concerted human effort and wise decision making seem to have enabled us to stay within a planetary boundary' (2009a: 12). Nevertheless, the implementation of the ozone regime has given rise to several concerns, which are now turned to.

3.3 Concerns Arising from the Implementation of the Ozone Regime

First, and more narrowly, several concerns arise due to 'gaps' in the ozone regime (discussed in detail in Du Toit, 2023). For example, 'production' is defined so as to exclude controlled substances that are used entirely as feedstock to produce other chemicals (Montreal Protocol, article 1(5)). This means that harmful substances, such as CFCs and HCFCs, may continue to be produced in order to manufacture other substances, including plastics (Andersen et al. 2021), which are associated with their own adverse socio-ecological impacts (see, among many others, Galloway & Lewis 2016; Azoulay et al. 2019; Hamilton et al. 2019; Whitchurch 2019; Revell et al. 2021).

Furthermore, the Montreal Protocol is focused on phasing out substances that are highly ozone-depleting, which are then replaced by substances that are less ozone-depleting – without regard being had to other adverse impacts. Indeed, 'alternative substances' are narrowly defined in the Vienna Convention as 'substances which reduce, eliminate or avoid adverse effects on the ozone layer' (article 1(4)). Following the phase-down of CFCs and HCFCs, the production and consumption of HFCs increased significantly (Roberts 2017: 224). While HFCs are largely not ozone-depleting, they are powerful greenhouse gases and some HFCs are thousands of times more powerful than carbon dioxide (WMO 2018: Table A-1). Increasing HFC emissions were therefore contributing to climate change. This was a classic example of problem shifting, which 'occurs when a solution for one problem backfires and generates one or more new problems at different times or locations' (Kim & van Asselt 2016).

Significantly, in 2016, parties agreed to the addition of HFCs to the list of substances controlled by the Montreal Protocol (UNEP 2016). Consequently, HFCs are being replaced by other substances, including hydrofluoroolefins

(HFOs), which while having low ozone-depleting potentials[5] (ODPs) *and* low GWPs, degrade to produce trifluoroacetic acid (TFA) and ozone. TFA is a 'persistent toxic chemical' (Montzka and Velders 2018: 2.4), and its presence has been detected in humans; plants, including crops consumed by humans; aquatic environments; and in the Arctic region (Pickard et al. 2020: 2). For its part, ozone is a greenhouse gas and pollutant at ground level and contributes to more than one million pollution-related (human) deaths per year. Ozone is also toxic to many plants and contributes to significant crop losses (Climate and Clean Air Coalition n.d.), which has implications for, at the very least, human health and well-being. While their long-term impacts are not yet clear (Montzka and Velders 2018: 2.34), it has been argued that HFOs 'cannot be regarded as overall sustainable alternatives' (De Graaf et al. 2021: 9). These examples demonstrate that there is the potential for problem shifting to arise again in the future. However, these new problems – created through the implementation of the ozone regime – would fall beyond the regulation of the ozone regime.

At a deeper level, various values and principles underlying the ozone regime specifically, and international law more broadly, are problematic, and possibly underpin the types of concerns identified earlier. It has been argued generally that international environmental law is incapable of responding to the types of socio-ecological crises that are being experienced in the Anthropocene, including because it is anthropocentric, and it does not embrace an earth system perspective (Kotzé & Kim 2019, 2021).

The ozone regime was established in the 1980s, and, while it has developed, it remains human-centred. Its concern, reflected throughout, is with humans first and then the environment. Furthermore, the Vienna Convention in its Preamble specifically recalls the 'pertinent provisions' of the Declaration of the United Nations Convention on the Human Environment (the Stockholm Declaration), particularly 'principle 21, which provides that "States have, in accordance with the Charter of the United Nations and the principles of international law, the sovereign right to exploit their own resources pursuant to their own environmental policies"'. Although the principle of permanent sovereignty over natural resources is linked to decolonization and addressing the North–South divide (Atapattu 2015: 80), it clearly promotes the exploitation of natural resources (Kotzé et al. 2020). More generally, the primary concern of the Stockholm Declaration is with preserving the 'human environment' due to its utility to humans and it does not recognize the intrinsic value of nature. The Declaration emphasizes that '[e]conomic and social development is essential for ensuring a favourable living and working environment for man and for creating conditions on earth that are necessary for the

[5] The ODP is 'a metric for determining the relative strength of a chemical to destroy ozone' (WMO 2018: ES.13).

improvement of the quality of life' (United Nations 1972: principle 8) and, furthermore, that '[o]f all things in the world, people are the most precious' (United Nations 1972: Preamble). The Stockholm Declaration is clearly focused on enhancing the quality of human life, even at the expense of non-human life. The explicit recalling of the provisions of the Stockholm Declaration suggests that the ozone regime, too, takes a utilitarian approach to the environment.

Furthermore, while the ozone regime is concerned with phasing out substances that deplete the ozone layer, it is not concerned with limiting the activities that rely on these substances. Taken together with its recalling of the provisions of the Stockholm Declaration, the ozone regime arguably supports 'sustainable' development, which has been described as an 'epistemology of mastery' (Adelman 2015: 22) and as 'so conveniently malleable that it enables its proponents to mask the negative socio-ecological impacts of relentless growth-driven development promoted by states and corporations' (Kotzé & Adelman 2023: 3). Problematically, sustainable development is central to environmental law and governance (Kotzé & Adelman 2023; Schmidt 2022).

As the world warms, demand for air conditioning and refrigeration (the primary applications of HFCs) will increase and coolants will become increasingly important to health, well-being, and life itself, particularly in developing countries. However, coolants such as HFCs contribute to global warming which will, in turn, further increase the demand for coolants (the 'cooling paradox' (IISD 2021)). Such uses should be contrasted against the use of refrigeration for the transport of food to wealthy countries and the use of air conditioning in private motor vehicles. Recent research shows that almost 20 per cent of greenhouse gas emissions from the food sector are generated by the transport of food across and between countries, with wealthy countries making a disproportionately large contribution to the problem (Li et al. 2022). In addition, temperature-controlled transport is required for vegetables and fruit which further increases their food miles emissions. Low-income countries were responsible for generating just 20 per cent of food transport emissions (Malik & Li 2022). Such 'luxury' uses are not questioned by the ozone regime.

To the extent that the ozone regime implicitly supports continued 'sustainable' development (and the continued use of substances that are potentially harmful to earth system components), socio-ecological harms and injustices will likely continue. Adelman points to the challenges of reconciling the welfare of humans and the planet under mainstream forms of development (Adelman 2021: 68); and others argue that achieving equity (between humans) within planetary boundaries will require reduced consumption in the North (Steffen & Stafford Smith 2013; Hickel 2019). It is thus increasingly recognized that alternatives to mainstream forms of development (such as sustainable development), which are centred around *growth*, are required (see, e.g., Hickel et al. 2022; Kotzé & Adelman 2021).

A further concern is that the development of newer technologies and the tightening of standards in developed countries have resulted in inefficient equipment, containing ODSs due to be phased down or phased out, being offloaded in developing countries, which will bear the associated environmental, economic, and social costs. Developing countries have accordingly called for policies and measures to prevent such dumping (Agyarko et al. 2021; UNEP 2022).

Furthermore, the ozone regime is grounded in science, and the Montreal Protocol has been referred to as an example of 'the power of Earth systems science in facilitating international environmental lawmaking' (Boyd 2017: 486). The ozone regime provides that measures to protect the ozone layer should be based on scientific knowledge, and there are references throughout the Vienna Convention and Montreal Protocol to scientific and technical information. Interestingly, around the time that the Vienna Convention and Montreal Protocol were concluded, economic growth came to be linked to scientific and technological progress; and Allan argues that the 'tight link between growth and scientific and technological progress continues to be reproduced in the present international order' (Allan 2018: 261). In practice, knowledge is developed by several scientific and technical assessment panels, which are largely dominated by men from northern countries (see, e.g., UNEP n.d.).

Although the problem of stratospheric ozone depletion was caused by human-created substances, ozone depletion and other problems arising due to the implementation of the ozone regime have implications for all life on earth. The absence of references in the ozone regime to natural ecosystems (apart from references to 'human health and the environment') and non-state actors – in contrast to other IEL agreements – is striking.

The following section begins to explore how some of these concerns might be addressed.

3.4 The Evolution of Values under the Ozone Regime

While some of the aforementioned concerns might be addressed through the tightening of definitions and narrowing of exclusions (see Du Toit 2023), the focus here is on deeper shifts that might be required. At a broader level, there have been calls for the transformation of 'modern society's anthropocentric and colonial worldview through which nature exists to serve humankind' (Piccolo et al. 2022: 1). In the context of international law, it has been argued that reforms of international laws and institutions are required in order to 'make them less anthropocentric and more responsive to the imperatives of the global ecological crisis' (Lawrence 2022: 577). However, there is no consensus on how this could be achieved.

A growing trend, prominent at the national level, is the granting of rights to nature generally or to specific natural entities, such as rivers (see, amongst others,

Boyd 2017; Kotzé & Villavicencio Calzadilla 2017; Villavicencio Calzadilla & Kotzé 2018; Tănăsescu 2020). While the recognition of rights for nature might present a response to concerns around anthropocentrism in environmental law, several problems have been highlighted. For example, granting rights to nature may further entrench divisions between human and non-human earth system components (Biber 2022; O'Donnell 2022). Furthermore, Tănăsescu challenges the assumption that rights of nature are ecocentric or of Indigenous origin and argues that extending personhood to natural entities 'reproduces the centrality of the human' (Tănăsescu 2020: 448), which clearly does not address concerns regarding the anthropocentrism of IEL. Spitz and Penalver (2021: 81) also question whether granting personhood to natural entities actually makes a difference in terms of legal outcomes; and Kurki (2022) goes further, arguing that granting personhood to natural entities is simply a legal fiction (even though it may have symbolic value). While not possible to explore fully here, the value of granting procedural rights to natural entities has also been highlighted (Schapper et al. 2022; see also Tănăsescu 2020; Winter 2021).

It has also been proposed that ecological integrity or 'the integrity of Earth's life-support systems' should be established as a *grundnorm* of international law, 'similar to the *grundnorm* character that human rights or the rule of law have in domestic and international law' (Kim & Bosselmann 2015: 194). Adelman argues further that 'ecological integrity should be both a precondition and the ultimate objective of [international environmental law] in order to stay within the safe operating space of the planetary boundaries' (Adelman 2021: 77). Some IEL agreements have recently started to recognize the importance of ecological integrity. For example, the Paris Agreement (UNFCCC 2015: Preamble) emphasizes 'the importance of ensuring the integrity of all ecosystems, including oceans, and the protection of biodiversity, recognized by some cultures as Mother Earth'. The Glasgow Climate Pact contains similar language, and it notes 'the importance of ensuring the integrity of all ecosystems, including in forests, the ocean and the cryosphere, and the protection of biodiversity, recognized by some cultures as Mother Earth'. While not binding, the 1982 World Charter for Nature (UNGA 1982) arguably goes even further through its acknowledgement of the intrinsic value of nature in the following terms: '[hu]mankind is a part of nature and life depends on the uninterrupted functioning of natural systems', and '[e]very form of life is unique, warranting respect regardless of its worth to [humans]'. Attention could be paid to including similar language within the preambular text of the ozone regime.

In regard to concerns around the predominant reliance on (Western) scientific knowledge, Garver (2019: 167) argues that achieving a 'mutually enhancing human-Earth relationship' requires that human choices are informed by contemporary science as well as 'traditional ecological knowledge of people with long-standing connections to place, including an understanding of the behavior of complex adaptive systems'. More recently, some IEL treaties have started to refer to non-Western knowledge. For example, under the climate change regime, the Paris Agreement (UNFCCC 2015: article 7(5)) acknowledges that 'adaptation action should follow a country-driven, gender-responsive, participatory and fully transparent approach, taking into consideration vulnerable groups, communities and ecosystems, and should be based on and guided by the best available science and, as appropriate, traditional knowledge, knowledge of indigenous peoples and local knowledge systems'. Similarly, the Glasgow Climate Pact (UNFCCC 2021: article 93) emphasizes 'the important role of indigenous peoples' and local communities' culture and knowledge in effective action on climate change and *urges* Parties to actively involve indigenous peoples and local communities in designing and implementing climate action'. It would also be possible to include this type of language within the preambular provisions of the ozone agreements.

More broadly, it will be necessary to move beyond 'mainstream forms of development' (Adelman 2021: 68) and the current 'growth-insistent paradigm' (Garver 2019). Proposals include embracing non-Western perspectives such as sumac kawsay or buen vivir (Acosta 2017) and implementing degrowth strategies (Hickel et al. 2022). Buen vivir translates directly as 'living well'; although it is said to translate more accurately as 'harmonious coexistence' (Calisto Friant & Langmore 2015: 1). For its part, the degrowth concept originated in ecological economics, and it is described as 'a planned reduction of energy and resource use designed to bring the economy back into balance with the living world in a way that reduces inequality and improves human well-being' (Hickel 2021: 1105). Embracing such perspectives is viewed as important to efforts to shift away from the anthropocentric basis of law and governance and to tackle the socio-ecological challenges of the Anthropocene (see, e.g., Calisto Friant and Langmore 2015; Kotzé & Adelman 2023). However, their realization would go beyond the scope of the ozone regime, and indeed IEL, and would require wide-ranging societal and political transformations.

In the interim, measures like those described earlier, such as the recognition of ecological integrity and non-Western perspectives within the ozone regime, could represent starting steps in shifting away from anthropocentric and utilitarian values, and they could also bring the ozone regime in line with more recently concluded IEL agreements.

3.5 Concluding Remarks

The ozone regime is widely considered to be one of the most effective international law regimes to date. While it has been very successful in regard to phasing out the substances under its control and in ensuring that we remain within the planetary boundary on stratospheric ozone depletion, the earlier discussion has highlighted several concerns. Apart from several technical gaps, it has been argued that the ozone regime (similarly to IEL in general) is based on a utilitarian and human-centred approach and implicitly supports ongoing 'sustainable' development, which has given rise to various socio-ecological harms. Furthermore, although it has been argued that the ozone regime gives effect to the principle of common but differentiated responsibilities, as discussed earlier, power imbalances between developed and developing countries remain, which enable the exacerbation of injustices. While not presenting definitive answers, this chapter has attempted to explore the evolution of values underlying the ozone regime so that it may continue to effectively protect stratospheric ozone, while also contributing to efforts to equitably protect all earth system components.

References

Acosta, A. (2017). Rethinking the World from the Perspective of Buen Vivir. P2P Foundation. blog.p2pfoundation.net/degrowth-movements-buen-vivir/2017/03/20.

Adelman, S. (2015). Epistemologies of Mastery. In A. Grear & L. J. Kotzé, eds., *Research Handbook on Human Rights and the Environment*. Research Handbooks in Human Rights. Cheltenham: Edward Elgar, 9–27.

Adelman, S. (2021). Planetary Boundaries, Planetary Ethics and Climate Justice in the Anthropocene. In D. French & L. J. Kotzé, eds., *Research Handbook on Law, Governance and Planetary Boundaries*. Research Handbooks in Human Rights. Cheltenham: Edward Elgar, 65–83.

Agyarko, K. A., Andersen, S. O., Ferris, R., et al. (2021). The Importance of Stopping Environmental Dumping in Ghana: The Case of Inefficient New and Used Cooling Appliances with Obsolete Refrigerants. *Duke Environmental Law & Policy Forum*. 32(1), 51–106.

Allan, B. B. (2018). *Scientific Cosmology and International Orders*. Cambridge: Cambridge University Press.

Andersen, S. O., Gao, S., Carvalho, S., et al. (2021). Narrowing Feedstock Exemptions under the Montreal Protocol Has Multiple Environmental Benefits. *Proceedings of the National Academy of Sciences*. 118(49), e2022668118. DOI: 10.1073/pnas.2022668118.

Atapattu, S. (2015). The Significance of International Environmental Law Principles in Reinforcing or Dismantling the North–South Divide. In S. Alam, S. Atapattu, C. G. Gonzalez, & J. Razzaque, eds., *International Environmental Law and the Global South*. New York: Cambridge University Press, 74–108.

Azoulay, D., Villa, P., Arellano, Y., et al. (2019). Plastic and Health: The Hidden Costs of a Plastic Planet. Centre for International Environmental Law. www.ciel.org/wp-content/uploads/2019/02/Plastic-and-Health-The-Hidden-Costs-of-a-Plastic-Planet-February-2019.pdf.

Biber, E. (2022). Will Rights of Nature Save The World? LegalPlanet. legal-planet.org/2022/06/23/will-rights-of-nature-save-the-world/

Boyd, D. R., ed. (2017). *The Rights of Nature: A Legal Revolution that Could Save the World*. Toronto: ECW Press.

Calisto Friant, M., & Langmore, J. (2015). The Buen Vivir: A Policy to Survive the Anthropocene? *Global Policy*. 6(1), 64–71. DOI: 10.1111/1758-5899.12187.

Climate and Clean Air Coalition. (n. d.). Hydrofluorocarbons (HFCs). www.ccacoalition.org/short-lived-climate-pollutants/hydrofluorocarbons-hfcs.

Crutzen, P. J., & Stoermer, E. F. (2000). The 'Anthropocene'. *Global Change Newsletter*. 41, 17–18.

De Graaf, D., Elsner, C., Hoffmann, G., et al. (2021). Hydrofluorocarbon Emission Reduction: A Crucial Contribution to Climate Protection. Umweltbundesamt. Scientific Opinion Paper. www.umweltbundesamt.de/sites/default/files/medien/2546/publikationen/2021-05-04_scientific_opinion_paper_hfcs_climate_protection_contribution_final.pdf.

Department of Climate Change, Energy, the Environment and Water (DCCEEW). (2018). Ozone Depleting Substances. www.dcceew.gov.au/environment/protection/ozone/publications/ozone-depleting-substances-factsheet.

DeSombre, E. R. (2000). The Experience of the Montreal Protocol: Particularly Remarkable, and Remarkably Particular. *UCLA Journal of Environmental Law and Policy*. 19(1), 49–81. DOI: 10.5070/L5191019217.

Du Toit, L. (2021). Stratospheric Ozone Depletion. In D. French & L. J. Kotzé, eds., *Research Handbook on Law, Governance and Planetary Boundaries*. Research Handbooks in Environmental Law Series. Cheltenham: Edward Elgar Publishing, 261–76.

Du Toit, L. (2023). Strengthening the Global Regulation of Hydrofluorocarbons under the Montreal Protocol. In K. Kulovesi, Y. Yamineva, & E. Recio, eds., *Reducing Emissions of Short-Lived Climate Pollutants: Perspectives on Law, Policy and Science*. Brill, 90–124.

Environmental Investigation Agency. (2017). Happy UN Ozone Day: Celebrating 30 Years of Ozone and Climate Protection! eia-international.org/news/happy-un-ozone-day-celebrating-30-years-ozone-climate-protection/.

Galloway, T. S., & Lewis, C. N. (2016). Marine Microplastics Spell Big Problems for Future Generations. *Proceedings of the National Academy of Sciences*. 113(9), 2331–33. DOI: 10.1073/pnas.1600715113.

Garver, G. (2019). A Systems-Based Tool for Transitioning to Law for a Mutually Enhancing Human-Earth Relationship. *Ecological Economics*. 157, 165–74. DOI: 10.1016/j.ecolecon.2018.09.022.

Gillespie, A. (2006). *Climate Change, Ozone Depletion and Air Pollution: Legal Commentaries with Policy and Science Considerations*. Leiden: Martinus Nijhoff Publishers.

Godin-Beekmann, S., Newman, P. A., & Petropavlovskikh, I. (2018). 30th Aniversary of the Montreal Protocol: From the Safeguard of the Ozone Layer to the Protection of the Earth's Climate. *Comptes Rendus Geoscience*. 350(7), 331–33. DOI: 10.1016/j.crte.2018.11.001.

Gonzalez, M., Taddonio, K. N., & Sherman, N. J. (2015). The Montreal Protocol: How Today's Successes Offer a Pathway to the Future. *Journal of Environmental Studies and Sciences*. 5(2), 122–29. DOI: 10.1007/s13412-014-0208-6.

Goyal, R., England, M. H., Sen Gupta, A., & Jucker, M. (2019). Reduction in Surface Climate Change Achieved by the 1987 Montreal Protocol. *Environmental Research Letters*. 14(12), 124041. DOI: 10.1088/1748-9326/ab4874.

Hamilton, L. A., Feit, S., Muffett, C., et al. (2019). Plastic and Climate: The Hidden Costs of a Plastic Planet. Centre for International Environmental Law. www.ciel.org/wp-content/uploads/2019/05/Plastic-and-Climate-FINAL-2019.pdf.

Hickel, J. (2019). Is It Possible to Achieve a Good Life for All within Planetary Boundaries? *Third World Quarterly*. 40(1), 18–35. DOI: 10.1080/01436597.2018.1535895.

Hickel, J. (2021). What Does Degrowth Mean? A Few Points of Clarification. *Globalizations*. 18(7), 1105–11. DOI: 10.1080/14747731.2020.1812222.

Hickel, J., Kallis, G., Jackson, T., et al. (2022). Degrowth Can Work: Here's How Science Can Help. *Nature*. 612(7940), 400–3. DOI: 10.1038/d41586-022-04412-x.

International Institute for Sustainable Development (IISD). (2021). Cooling Paradox Must Be Addressed: Montreal Protocol Task Force. sdg.iisd.org/news/cooling-paradox-must-be-addressed-montreal-protocol-task-force/.

Kim, R. E., & Bosselmann, K. (2015). Operationalizing Sustainable Development: Ecological Integrity as a Grundnorm of International Law. *Review of European, Comparative & International Environmental Law*. 24(2), 194–208. DOI: 10.1111/reel.12109.

Kim, R. E., & van Asselt, H. (2016). Global Governance: Problem Shifting in the Anthropocene and the Limits of International Law. In E. Morgera & K. Kulovesi, eds., *Research Handbook on International Law and Natural Resources*. Cheltenham: Edward Elgar, 473–95.

Kotzé, L., Du Toit, L., & French, D. (2020). Friend or Foe? International Environmental Law and Its Structural Complicity in the Anthropocene's Climate Injustice. *Oñati Socio-Legal Series*. 11(1), 180–206. DOI: 10.35295/osls.iisl/0000-0000-0000-1140.

Kotzé, L. J. (2020). Earth System Law for the Anthropocene: Rethinking Environmental Law alongside the Earth System Metaphor. *Transnational Legal Theory*. 11(1–2), 75–104. DOI: 10.1080/20414005.2020.1776556.

Kotzé, L. J., & Adelman, S. (2023). Environmental Law and the Unsustainability of Sustainable Development: A Tale of Disenchantment and of Hope. *Law and Critique*. 34(2), 227–48. DOI: 10.1007/s10978-022-09323-4.

Kotzé, L. J., & Kim, R. E. (2019). Earth System Law: The Juridical Dimensions of Earth System Governance. *Earth System Governance*. 1, 100003. DOI: 10.1016/j.esg.2019.100003.

Kotzé, L. J., & Kim, R. E. (2021). Exploring the Analytical, Normative and Transformative Dimensions of Earth System Law. *Environmental Policy and Law*. 50(6), 457–70. DOI: 10.3233/EPL-201055.

Kotzé, L. J., & Villavicencio Calzadilla, P. (2017). Somewhere between Rhetoric and Reality: Environmental Constitutionalism and the Rights of Nature in Ecuador. *Transnational Environmental Law*. 6(3), 401–33. DOI: 10.1017/S2047102517000061.

Kurki, V. A. (2022). Can Nature Hold Rights? It's Not as Easy as You Think. *Transnational Environmental Law*. 11(3), 525–52. DOI: 10.1017/S2047102522000358.

Lawrence, P. (2022). Justifying Representation of Future Generations and Nature: Contradictory or Mutually Supporting Values? *Transnational Environmental Law*. 11(3), 553–79. DOI: 10.1017/S2047102522000176.

Li, M., Jia, N., Lenzen, M., et al. (2022). Global Food-Miles Account for Nearly 20% of Total Food-Systems Emissions. *Nature Food*. 3(6), 445–53. DOI: 10.1038/s43016-022-00531-w.

Lieberman, B. C. (1994). Stratospheric Ozone Depletion and the Montreal Protocol: A Critical Analysis. *Buffalo Environmental Law Journal*. 2(1), 1–32.

Malik, A., & Li, M. (2022). The World's Affluent Must Start Eating Local Food to Tackle the Climate Crisis, New Research Shows. The Conversation. theconversation.com/

the-worlds-affluent-must-start-eating-local-food-to-tackle-the-climate-crisis-new-research-shows-185410.

Molina, M. J., & Rowland, F. S. (1974). Stratospheric Sink for Chlorofluoromethanes: Chlorine Atom-Catalysed Destruction of Ozone. *Nature*. 249(5460), 810–12. DOI: 10.1038/249810a0.

Montreal Protocol on Substances that Deplete the Ozone Layer 1522 UNTS 3.

Montzka, S. A., & Velders, G. J. M. (Lead Authors), Krummel, P. B., et al. (2018). Hydrofluorocarbons (HFCs). In World Meteorological Organization (WMO), ed., *Scientific Assessment of Ozone Depletion: 2018*. Global Ozone Research and Monitoring Project – Report No. 58. Geneva: WMO, 2.1–2.56.

Moore, J. W. (2016). *Anthropocene or Capitalocene? Nature, History, and the Crisis of Capitalism*. Oakland: PM Press.

Nanda, V. (1989). Stratospheric Ozone Depletion: A Challenge for International Environmental Law and Policy. *Michigan Journal of International Law*. 10(2), 482–525.

O'Donnell, E. (2022). Constitutionality and Nature's Rights. Dulluh Omar Institute Next Generation Constitutionalism Webinar Series, June 10. https://dullahomarinstitute.org.za/events/ngc-webinar-series-natures-constitutional-right-to-exist-as-nature-10-june-2022.

Oreskes, N. (2007). The Scientific Consensus on Climate Change: How Do We Know We're Not Wrong? In J. F. C. Dimento & P. Doughman, eds., *Climate Change: What It Means for Us, Our Children, and Our Grandchildren*. Cambridge, MA: MIT Press, 105–48.

Piccolo, J. J., Taylor, B., Washington, H., et al. (2022). Nature's Contributions to People' and Peoples' Moral Obligations to Nature. *Biological Conservation*. 270, 109572. DOI: 10.1016/j.biocon.2022.109572.

Pickard, H. M., Criscitiello, A. S., Persaud, D., et al. (2020). Ice Core Record of Persistent Short-Chain Fluorinated Alkyl Acids: Evidence of the Impact From Global Environmental Regulations. *Geophysical Research Letters*. 47(10), e2020GL087535. DOI: 10.1029/2020GL087535.

Raworth, K. (2014). Must the Anthropocene Be a Manthropocene? *The Guardian*, October 20. www.theguardian.com/commentisfree/2014/oct/20/anthropocene-working-group-science-gender-bias.

Revell, L. E., Kuma, P., Le Ru, E. C., Somerville, W. R. C., & Gaw, S. (2021). Direct Radiative Effects of Airborne Microplastics. *Nature*. 598(7881), 462–67. DOI: 10.1038/s41586-021-03864-x.

Roberts, M. W. (2017). Finishing the Job: The Montreal Protocol Moves to Phase Down Hydrofluorocarbons. *Review of European, Comparative & International Environmental Law*. 26(3), 220–30. DOI: 10.1111/reel.12225.

Rockström, J., Steffen, W., Noone, K., et al. (2009a). Planetary Boundaries: Exploring the Safe Operating Space for Humanity. *Ecology & Society*. 14(2), 32.

Rockström, J., Steffen, W., Noone, K., et al. (2009b). A Safe Operating Space for Humanity. *Nature*. 461(7263), 472–75. DOI: 10.1038/461472a.

Schapper, A., Hoffmann, C., & Lee, P. (2022). Procedural Rights for Nature: A Pathway to Sustainable Decarbonisation? *Third World Quarterly*. 43(5), 1197–216. DOI: 10.1080/01436597.2022.2057293.

Schmidt, J. J. (2022). Of Kin and System: Rights of Nature and the UN Search for Earth Jurisprudence. *Transactions of the Institute of British Geographers*. 47(3), 820–34. DOI: 10.1111/tran.12538.

Sinclair, L. (2002). The Science of Ozone Depletion: From Theory to Certainty. In S. O. Andersen & K. M. Sarma, eds., *Protecting the Ozone Layer: The United Nations History*. London: Earthscan, 1–41.
Spitz, L., & Penalver, E. M. (2021). Nature's Personhood and Property's Virtues. *Harvard Environmental Law Review*. 45(1), 67–97.
Steffen, W., Crutzen, P. J., & McNeill, J. R. (2007). The Anthropocene: Are Humans Now Overwhelming the Great Forces of Nature. *Ambio*. 36(8), 614–21. DOI: 10.1579/0044-7447(2007)36[614:TAAHNO]2.0.CO;2.
Steffen, W., Richardson, K., Rockström, J., et al. (2015). Planetary Boundaries: Guiding Human Development on a Changing Planet. *Science*. 347(6223), 1259855. DOI: 10.1126/science.1259855.
Steffen, W., & Stafford Smith, M. (2013). Planetary Boundaries, Equity and Global Sustainability: Why Wealthy Countries Could Benefit from More Equity. *Current Opinion in Environmental Sustainability*. 5(3), 403–8. DOI: 10.1016/j.cosust.2013.04.007.
Tănăsescu, M. (2020). Rights of Nature, Legal Personality, and Indigenous Philosophies. *Transnational Environmental Law*. 9(3), 429–53. DOI: 10.1017/S2047102520000217.
United Nations. (1972). Declaration of the United Nations Conference on the Human Environment. UN Doc. A/Conf.48/14/Rev.1; 11 ILM 1416.
UNEP (United Nations Environment Programme). (n. d.). EEAP Members. ozone.unep.org/science/assessment/eeap/eeap-members.
UNEP (United Nations Environment Programme). (2016). Report of the Twenty-Eighth Meeting of the Parties to the Montreal Protocol on Substances that Deplete the Ozone Layer. ozone.unep.org/sites/default/files/2019-08/MOP-28-12E.pdf UNEP/OzL.Pro.28/12.
UNEP (United Nations Environment Programme). (2022). Draft Decision on Stopping the Harmful Dumping of New and Used Inefficient Refrigeration and Air Conditioning Appliances Using Obsolete Ozone-Depleting Substances and Hydrofluorocarbon Refrigerants. ozone.unep.org/meetings/thirty-fourth-meeting-parties/session-documents UNEP/OzL.Pro.34/CRP.2.
UNFCCC (United Nations Framework Convention on ClimateChange). (2015). Adoption of the Paris Agreement. FCCC/CP/2015/L.9/Rev/1. https://unfccc.int/resource/docs/2015/cop21/eng/l09r01.pdf.
UNFCCC (United Nations Framework Convention on ClimateChange). (2021). Glasgow Climate Pact: Advance Version. FCCC/PA/CMA/2021/L.16. unfccc.int/sites/default/files/resource/cma2021_L16_adv.pdf.
UNGA (United Nations General Assembly). (1982). World Charter for Nature. A/RES/37/7. www.un.org/documents/ga/res/37/a37r007.htm.
Vienna Convention for the Protection of the Ozone Layer 1513 UNTS 293.
Villavicencio Calzadilla, P., & Kotzé, L. J. (2018). Living in Harmony with Nature? A Critical Appraisal of the Rights of Mother Earth in Bolivia. *Transnational Environmental Law*. 7(3), 397–424. DOI: 10.1017/S2047102518000201.
Whitchurch, A. (2019). Fossil Plastics. *Geoscientist*. 29(9), 5. DOI: 10.1144/geosci2019-048.
Winter, C. J. (2021). A Seat at the Table. *Borderlands Journal*. 20(1), 116–39. DOI: 10.21307/borderlands-2021-005.
World Meteorological Organization (WMO), ed. (2018). *Scientific Assessment of Ozone Depletion: 2018*. Global Ozone Research and Monitoring Project – Report No. 58. Geneva.

4

Are Ozone's Relationships of Normative Importance?

CHRISTINE WINTER

4.1 Introduction: Why Should You Care?

A blanket of ozone, an unstable bond of three promiscuous oxygen atoms, cossets the globe from harmful UV radiation. Without it 'life on Earth would not be possible' (Mohr 2021: 31). At ground level, however, this same bond of oxygen atoms is hazardous – interfering with the flourishing of human and plant life caught in its presence. Industrial-age human activity has done two things: destroyed beneficial stratospheric ozone and created destructive atmospheric ozone. The activities of some humans, therefore, interfere with the well-being of themselves, other human beings, animal and plant life. Furthermore, they interfere with the oxygen bonds that make ozone itself and ozone's natural relationships. The focus of this chapter is on stratospheric ozone, and it is, I think, important to have noted the contradictory impacts of ozone at stratospheric and atmospheric levels on human and other planetary life forms. It is also important to note that the natural balance of ozone at both levels is disrupted by the activities of those industrial/Western/modern human beings who pay scant heed to any form of responsibility to maintain harmonious relationships with the non-human realm.

It is this lack of a sense of responsibility to the natural realm that the idea of the planetary boundaries seeks to address – not in an ethical frame but a scientific one. Put simply, while the ozone layer high in the stratosphere supports 'life on earth' in its multiple meanings, a particular anthropocentric 'life on earth' has reciprocated not in a relationship of care and nurture, but one that seeks to define acceptable limits to ozone's destruction. That is, as with all the planetary boundaries articulated in Rockström et al.'s seminal works (2009; Steffen et al. 2015), the idea of an ozone boundary rests on (at least) two assumptions. First, the natural systems that support human well-being can be 'extracted' from for the accumulation of human property and/or wealth. By 'extracted' I mean may be interfered with by human activity in a way that reduces their natural extent and/or functions. Second, natural systems can be viewed as bounded, that is as non-relational.[1] In what follows I will argue that, while undoubtedly an agent of good, the idea of planetary boundaries rests nested in an anthropocentric, colonial ontology that silently undergirds the epistemic community (Haas 1991)[2] of Western science, a community that deliberately expunges (or claims to expunge) values from its reckonings. Furthermore, the scientific endeavour of planetary boundaries is wrought from that same ontological furnace. The question then is, can you resolve a problem from within the same ontological foundations whence it arose?

My goal here is to try to (un)think planetary boundaries and (re)think planetary relationships. My intuition is that to describe a set of boundaries up to which 'we'[3] may pollute, assumes the same basket of rights that is leading the planet and all that draws sustenance from it to a precipice. The assumptions underpinning the concept of planetary boundaries are another example of a colonial relationship, or to paraphrase Max Liboiron (2021), *planetary boundaries are colonialism*. That is, they have imbedded an ethics that legitimizes the enclosure of a commons (in this case stratospheric ozone) for the benefit of a privileged few, who are often, but not exclusively, the heirs of historical colonial appropriation of others' Land-based[4] territory – the wealthy Western states and their citizens (although not all citizens equally). My claim is ozone destruction and the ozone boundary might be framed as a property right founded in an ethics that legitimates the way that some human behaviour messes with the natural relations within the air, within the atmosphere and stratosphere. This chapter is less about how much we can pollute to be safe (the Montreal Protocol has successfully eliminated the main culprits (Weber et al. 2022)), but, rather, it is a challenge to that framing – a framing that suggests (some) humans have a legitimate right to interfere in atmospheric and stratospheric relationships to the point/level that is just below one that is harmful *to humans*. My claim is that rather than a simple duty to protect humans, there needs to be a reconfiguration of the conception of duties and responsibility to support the well-being of ozone and ozone's (and all other planetary boundaries') basket of

[1] On 29 March 2022, I attended an Australian Earth Laws Alliance presentation by the late Will Steffen in which he talked of Planetary Boundaries 3.0 (to see a YouTube of the presentation go to www.youtube.com/watch?v=B1vSWPsjknI). This new iteration recognizes the inter-relationships between each of the previously separated planetary systems and their related 'boundaries'. At the time of writing this chapter, however, the article on Planetary Boundaries 3.0 has not been published.

[2] Thanks to Dr Lena Tan for directing me to Peter Haas's work. Haas says, 'An epistemic community is a knowledge-based transnational network of specialists whose members share common views about the causes of a problem and the policies which should be adopted to manage it' (1991: 226).

[3] 'We' is placed in scare quotes to identify that it is not all humans who bear responsibility for ozone's destruction and production, and that to project responsibility onto those who bear none, which is what a universalizing 'we' can do, is irresponsible and unethical.

[4] Following Libroiron, Land with a capital 'L' denotes the land, soil, geologic formations, waters, plants, animals, insects, fish, spirits, human beings, and so on as a complex of relationships and a subject, in contrast to land (with a lower case 'l'), which denotes an object over which humans claim property rights.

relationships – with humans and other planetary beings, the air, elements, and stratosphere.

An anticolonial approach challenges the root assumption that humans can legitimately 'extract' oxygen atoms from ozone up to a certain level. I suggest that chemical pollution does harm to the natural relationships that exist within the blanket of ozone that cossets the earth – it is an extractive relationship that generates negative externalities: burnt skin or scorched crops say. Secondly, I use the idea of extraction to indicate that the governance frameworks and the planetary boundaries literature represent ozone as independent from its relationships (and from any human obligations and duties to ozone itself) and that this representation is a function of governance regimes and epistemic and ontological formations – ones that isolate elements one from the other and create an illusion of boundaries. These are the same epistemic and ontological foundations from which the need for a scholarship on planetary boundaries arose. Equally I challenge the legitimacy of creating ozone at terrestrial levels and thus burdening human, plant, and animal life with reduced welfare and intervening in natural atmospheric chemical relationships and elemental processes and relationships. However, and to reiterate, here the focus is on stratospheric ozone the subject/object of the planetary boundary.

This chapter takes the following form. I assume you have read the introduction to this section and are familiar with the basics of the 'ozone hole' and the Montreal Protocol. However, the chapter stands alone without that background. I briefly outline the ozone success story, that is that the ozone hole is the only planetary boundary that has been rescued from a tipping point.[5] I will then introduce the idea of ozone as relationship, followed by a more comprehensive discussion of planetary boundaries as colonialism. I conclude by proposing an anti colonial alternative in which I posit that humans have a responsibility to enhance ozone's natural relationships and demonstrate that in so doing the need for (imaginary) planetary boundaries collapses.

4.2 Ozone as Success Story

Limiting depletion of stratospheric ozone is rightfully lauded as a global success story: one of exemplary scientific and political cooperation. There were tailwinds, of course, that facilitated this success: scientists willing to implement the precautionary principle quickly;[6] politicians prepared to respect scientific advice;

[5] Although it may be too early to make this triumphalist claim as I will discuss later.

[6] The Wingspread Statement of 1998 created a widely accepted definition of the Precautionary Principle: 'When an activity raises threats of harm to the environment or human health, precautionary measures should be taken even if some cause and effect relationships are not fully established scientifically.' Retrieved on 22 July 2022 from www.healthandenvironment.org/environmental-health/social-context/history/precautionary-principle-the-wingspread-statement.

the euphoria of 'the end of history'[7] as the Cold War ceased and international cooperation heralded a new future; a small number of producers of the key chemicals (chlorofluorocarbons or CFCs) responsible for stratospheric ozone's depletion; and suitable, cost-effective replacement chemicals readily at hand. Scientists identified a problem that industrial science could resolve under the right regulatory conditions, and politicians acted quickly to enact that regulation.

First, the Vienna Convention of 1985 and then the Montreal Protocol, 1987, provided the right regulatory conditions. From the time at which it was realised that man-made/manipulated chemicals, those produced in industrial and agricultural processes and used in consumer-ware, were destabilizing the fragile bonds of ozone's oxygen atoms to the implementation of the Montreal Protocol was rapid. The chemistry and risk were illuminated, for instance, by Crutzen (1971, 1972, 1974), and repeatedly thereafter. Pressure for action was mounting on and in political circles when in 1985 Farman et al. (1985) provided irrefutable evidence of a hole in the ozone layer over Antarctica. The Montreal Protocol was adopted swiftly thereafter.[8] So successful has been the regulatory regime that the UNEP Ozone Secretariate wrote in 2019:

> Thanks to the ozone treaties, which now control production and consumption of a range of ozone-depleting and global-warming substances, the ozone layer is healing, and is expected to recover by around the middle of this century. (UNEP Ozone Secretariat 2019)

And yet in April 2020 *Nature* reported a hole in the ozone layer above the Arctic (Witze 2020), a big hole the headline said, possibly the biggest ever recorded. The cause? Unusually cold temperatures. However, temperature was not the direct cause: the blame must still be placed on anthropogenically produced and emitted pollutants. The cold allowed high-altitude clouds to form and the pollutants reacted with the clouds to destabilize ozone's fragile bonds or, as the article claims, to 'chew away at ozone' (Witze 2020: 19). In 2021 NOAA announced the ozone hole above the Antarctic was enlarged – in the top 25 per cent recorded (National Oceanic and Atmospheric Administration 2021). The culprit they were quick to note was not chemical alone but rather again a matter of colder than usual Antarctic temperatures. The suggestion is that the Protocol may be said to remain a success story. However, simultaneously the ozone layer remains vulnerable. These stories highlight two features relevant to my argument: (1) the Montreal Protocol limited emissions of ozone 'chewing' chemicals but did not *eliminate* them from either production or release; and (2) ozone exists as a set of spatial, temporal, elemental, and chemical relationships, not as an individuated chemical bond.

[7] As Francis Fukuyama triumphantly labelled the moment later (Fukuyama 2002).
[8] I won't detail this process further. See Chapter 3 by Louise du Toit for details on the regulatory regime.

Let's circle back to the previous quote from the Ozone Secretariat (is it quaint to think a chemical blanket 15 kilometres above the earth's surface should have its own secretariat? Perhaps not). They say the agreements 'control production and consumption of . . . ozone depleting . . . substances'. Despite the science of harm, the ethics underscoring the agreements rests in an assumption that some production and consumption of these chemicals may and will continue. That is, there is agreement that some human beings are entitled to consume a portion of the global commons that is stratospheric ozone. I suggest it is possible to draw a direct line from three historical moments which continue to foist great weight on some – some people, some environments, some elements – and the ethics of the Montreal Protocol and the logic of planetary boundaries: the enclosures of common lands in Europe; the appropriation of lands from the indigenous First Peoples in the colonies and their transfer to private title; and Locke's rationale for legitimating such enclosures (Laslett & Locke 1997: § 5). That enclosure was/is legalized does not make it just. At each juncture what was previously held in common for the good of all was re-labelled 'property' and assigned to the benefit of the owner.

I approach the ozone boundary as an object of colonization because it seems to embed a property right to pollute the global commons, an assumed right to 'eat into' or 'extract from' the atmosphere and stratosphere. Here I attempt to weave an anti colonial approach to human responsibilities *to ozone*.

4.3 Ozone's Relationships

The planetary boundaries literature describes a 'safe operating space for humanity' (Steffen et al. 2015), and the Montreal Protocol is held as *the* case study for successfully navigating one boundary's limits. I trouble this notion by focusing on the relational entanglements of the planetary system – human, animal, vegetable, mineral, and elemental. I am suggesting that ozone cannot, and should not, be seen as separate from its own relationships – its own kinships as Anja Kanngieser and Zoe Todd might have it (Kanngieser & Todd 2020). I am steering you towards understanding ozone not as individual molecules, nor as a cossetting blanket, but rather as a set of relationships. Furthermore, I am also suggesting ozone's protection is ethically something more than the safety of humanity. All life on the planet depends to some extent on stratospheric ozone. Ozone exists as a set of spatial, temporal, elemental, and chemical relationships, not simply as an individuated chemical bond, a molecule. The very existence of the 'ozone layer' suggests a relational coming-togetherness predilection in ozone. The layer is not 'just a layer' but a(n) (o)zone of relationship. It exists despite humanity and humanity exists *because* of ozone. It has liveliness of its own that is worthy of respect: it is not simply a passive victim of human misadventure.

Ozone is not a 'case-study' (Kanngieser & Todd 2020). Oxygen is a promiscuous element – in and out of relationships constantly. This, admittedly, is an observation from someone who is neither a chemist nor a physicist. Nevertheless, this seems an appropriate description for an element that readily detaches from one chemical bond to form another. 'Stratospheric ozone is formed and consumed naturally by photochemical reactions'; its oxygen atoms move in and out of relationships such that '[a]t any time, the formation and consumption of ozone proceed simultaneously' (Freedman 2021: 3272). It is this promiscuity that might be said to have led to the ozone hole. The love triangle of three oxygen atoms that form ozone quickly collapses once chlorine, fluorine, or nitrogen, and other pollutants, join the party. It is the relational instability between the three oxygen atoms of an ozone molecule that the Montreal Protocol acknowledges and that the concept of planetary boundaries wishes to (re)stabilize by limiting the number of disruptive chemicals in the stratospheric commons.

But what of ozone's origin stories – these too are relational. The earth's oxygen, the precious life-giving oxygen of the atmosphere, originated in a relationship between cyanobacteria – blue-green algae – sunshine, water, and carbon dioxide. A planet without these tiny organisms, energy from the sun, and the carbon, hydrogen, and oxygen caught in bonds of water and carbon dioxide is a planet without oxygen. Atmospheric oxygen originates from a critical relationship between plant life and the sun, and yet the concept of the ozone planetary boundary does not directly protect the critical 'feed stock'[9] life-form required for ozone – oxygen. The boundary focuses on the direct destruction of ozone in stratosphere.

Burning plant life, or more accurately the death and destruction of large ecosystems, is attributed to the destruction of stratospheric ozone in 2020–21. Directly linked to the wildfires that razed one-third of Australia's natural forests, the ozone layer shrank in those years. Particulate matter, dead trees, dead animals, and dead insects not only insinuated their way into the lungs of Australia's human and surviving non-human animals but rode an atmospheric elevator into the stratosphere. The figures of death and destruction from those Black Summer fires are now legend. What is less heralded is their effect on the ozone layer. '1 million tons of smoke particles [were slung] into the atmosphere, reaching up to 35 kilometres above Earth's surface' (Chu 2022: n.p.). These catastrophic fires and others, incinerating forests around the globe in what seems like an endless seasonal global procession, insert their presence into the stratosphere and play cuckold to the ozone bonds. Such fires enter a *relationship* with ozone, albeit a destructive relationship. Researchers 'estimate that [a] smoke-induced chemistry depleted the column of ozone by 1 percent' which

9 To borrow a term from the fossil fuel industry deployed to validate ongoing extraction and production of oil, gas, plastics, and petrochemicals.

exactly cancels out ozone's gains over the past 10 years (Chu 2022: n.p.). The Ozone Secretariat's expectation that the ozone layer will be fully repaired by the 'middle of this century' now seems hopeful at best, naïve at worst because it addresses only *one* of the colonizing relationships of modernity; that of industrial chemicals. Land use, land abuse, the tumults of climate change triggered by colonial extractivism[10] are also members of the ozone attack-gang.

Forest (wild) fires, though, are related to other 'modern' practices and are intimately entangled with the other great atmospheric threat, climate change. In Australia (and the west coast of North America) these forest fires sear a straight line of culpability to colonial violence, land theft, land-use practices, and Indigenous repression. The philosophies of modernity severed practices of relating and stewardship and embedded those of dualism[11] and extraction. The idea of property rights enclosed Land and created land from which Indigenous people (and their philosophies of entanglement with and care of Land (Celermajer et al. 2021; Kimmerer 2013; Stewart-Harawira 2005; Whyte 2017; Winter 2022)), native animals, and plant life are excluded and replaced by extractivist farming,[12] forestry, and mining. The Lockean philosophy that underscored land-based enclosure now underpins the idea of planetary boundaries – the right to 'enclose' the atmosphere (and land, waterways, etc.) up to a 'safe operating limit' to facilitate progress, productivity, and GNP. Chemicals production continues, climate is changing, forests are ablaze, and the ozone layer at least doubly damned. Damned first by CFCs and now, on the cusp of restoration, damned by the entangled relationships between intensive agricultural extractivism, colonialism, climate change (and thus also fossil fuel mining/extraction), and wildfires.

Furthermore, the relationship between oxygen atoms forming an ozone molecule depends on prior *elemental* relationships. Ozone's origin story begins with cyanobacteria – blue-green algae which converts sunshine, water, and carbon dioxide, and produces oxygen. And beyond energy from the sun, ozone has another elemental relationship – with lightening and light. Lightening is known to create ozone – at both stratospheric and atmospheric levels – and ozone is produced also 'from the photoexcitation of the $(O_2)_2$ species' (Braslavsky & Rubin 2011: 1517). Its creation, in other words, relies on a complex web of relationships between living and non-living matter, elements and natural phenomena, sun and earth. Conversely, its destruction 'is the collateral damage of industrial society' (Haas 1991: 225).

[10] Extractivism here means the ongoing removal of soil, nutrients, forests, fish stocks, and mineral deposits at rates that exceed their natural reproduction or regeneration cycles.

[11] Dualism here refers to the idea that human/culture is something outside of and separate to non-human/nature.

[12] Artificial nitrogenous fertilizers that support intensive agriculture have their own planetary boundary and are complicit in – indeed are the primary current source for – ongoing depletion of the ozone layer.

4.4 Ozone's Planetary Boundary Is Colonialism

The description 'planetary boundaries' camouflages ongoing colonial practices of privatizing the commons: hidden within the fabric of the descriptor is an assumption that humans – some humans – have a right to pollute (up to some predetermined limit), a right to 'colonize' the stratosphere, to 'enclose' it. This does two things: it means only some humans get to access the limited availability to enclose (and I will expand on this concept fully next, please bear with me for now); and it makes ozone's relationship to thriving – for human, animal, and plant life – tenuous, precarious even, for others. The underlying assumption is some degree of environmental harm is tolerable if it enhances GDP and some human lives, even when it involves appropriation of a global commons.

Human-induced ozone depletion, which grows from a system that assumes nature is the 'sink' for human waste and waste products (Liboiron 2021), is a form of extractivism. Like the profligate use of fossil fuels, so the convenience of CFC propellants to eject hairsprays and whipped cream 'extract' oxygen atoms from ozone to created new bonds with chlorine and fluorine and carbon from the CFCs reaching the stratosphere. Likewise, the release of CFCs when refrigeration and air-conditioning units are dumped – as a product of planned obsolescence – ozone becomes the victim of corporate profits. The fragile bonds of three promiscuous oxygen atoms are easily shattered to soak up the unnatural by-product of human convenience, luxury, corporate bottom lines, and waste (Winter & Schlosberg 2024). Ozone's very being is then extracted from the stratosphere, its existence terminated. It is apparent, then, that in these processes we find there are no boundaries – everything, everyone, every inter- and intra-action (Barad 2012) is connected.

The stratosphere, that which might once have been imagined an infinite sink, a convenient 'away' for industrial externalities,[13] a repository so infinite that it need not be accounted for (in the multiple meanings of account), is now recognized (by some) to be otherwise, to be fragile and perilously finite. Scientists' realization that the ozone hole existed as a result of human activity is a vivid reminder there is no 'away'. Making this realization palpable has been one success of the planetary boundaries project. It has been a success too of the Montreal Protocol. However, presenting each domain of planetary overreach as bounded – as discrete and contained – means the relational is subsumed by the bordered. Subsumed too are the world's extant knowledge traditions that base their epistemologies within relationality.

[13] See Barry Commoner's (1971) account of the absurdity of liberal capital's notion that 'the environment' could infinitely absorb pollutants.

4.5 An Anticolonial Planetary Boundary for Ozone

> Doing theory requires being open to the world's aliveness, allowing oneself to be lured by curiosity, surprise, and wonder. Theories are not mere metaphysical pronouncements on the world from some presumed position of exteriority. Theories are living and breathing reconfigurings of the world.
>
> *(Barad 2012: 207)*

The message of this chapter is: boundaries are an illusion. The 'boundary' between you and me is broken the moment we share air mingled with our breath or as microbes jumped ship from me to you as we shook hands (Yong 2016). Even if there is not physical contact between us, our brains can be forever changed as we share ideas, knowledge, emotions, and theory, and our synapsis lay them down as memory. That is, as Barad suggests earlier, being on earth means being alive to the world in its wholeness. The idea of boundaries creates an illusion of separation where there is interaction, blending, sharing, blurring. The river that flows between states or countries contains and shares shards from each bank, blends them, redeposits them regardless of the designated boundary. The atmosphere is intimately entangled with plant life, humanity's offal and pollutants, with lightening, wildfires, volcanoes, winds, water, minute particulate matter, and gobbling chemicals. There are no boundaries to some of humanity's arrogance: it touches everything, and, in Karen Barad's words, '[t]ouch is never pure or innocent' (Barad 2012: 215).

One great good the Montreal Protocol and the literature on the planetary boundaries have wrought is redirecting attention from the divisiveness of the 'global', the discourse of borders, nationstates, competition, and multi-nationalism. As Dipesh Chakrabarty so convincingly argues, a global mindset is incapable of resolving the planetary crisis of climate change – that is, a global focus, born from the same the epistemological and ontological foundations as colonialism, bakes in anthropocentrism and competition (Chakrabarty 2021). Without detracting from the good of the planetary boundaries project, my argument here is similar to Chakrabarty's, that the threats to planetary thriving cannot be resolved solely from within the philosophy from which they arose. Importantly a decolonial planetary vision faces towards multi-species multi-elemental flourishing *in their/its own right.*

I am suggesting an anticolonial planetary (non)boundary for ozone requires a relational ethic in which duties and responsibilities are to the relationships that inhere to ozone. That, as Max Liboiron says, any level of pollution breaks the bonds of relationality and thus the responsibility between human and the planetary system: it is colonial. The subject of planetary justice is not simply humanity, not simply the sensate or living being, it extends to all matter. Anticolonial planetary boundary scholarship and policymaking require development of and engagement with relational epistemologies and ontologies. Anticolonial planetary studies decentre the human and Western thinking patterns, epistemologies, and ontologies, and include all living beings, the elemental, and geophysical. In an anticolonial boundary-less planetary vision there are no polluting property rights. An anticolonial planetary justice proposes a planet on which a life-sustaining planetary net of overlapping (local) relationships are the subjects of justice.

References

Barad, K. (2012). On Touching: The Inhuman that Therefore I Am. *differences*. 23(3), 206–23. DOI: 10.1215/10407391-1892943.

Braslavsky, S. E., & Rubin, M. B. (2011). The History of Ozone: Part VIII. Photochemical Formation of Ozone. *Photochemical & Photobiological Sciences*. 10(10), 1515–20. DOI: 10.1039/C1PP05121K.

Celermajer, D., Schlosberg, D., Rickards, L., et al. (2021). Multispecies Justice: Theories, Challenges, and a Research Agenda for Environmental Politics. *Environmental Politics*. 30(1–2), 119–40. DOI: 10.1080/09644016.2020.1827608.

Chakrabarty, D. (2021). *The Climate of History in a Planetary Age*. Chicago: University of Chicago Press.

Chu, J. (2022). Study Reveals Chemical Link between Wildfire Smoke and Ozone Depletion. *MIT News*, 28 February. news.mit.edu/2022/wildfire-smoke-ozone-depletion-0228.

Commoner, B. (1971). *The Closing Circle: Nature, Man, and Technology*. New York: Alfred A. Knopf.

Crutzen, P. J. (1971). Ozone Production Rates in an Oxygen-Hydrogen-Nitrogen Oxide Atmosphere. *Journal of Geophysical Research*. 76(30), 7311–27. DOI: 10.1029/JC076i030p07311.

Crutzen, P. J. (1972). SST's: A Threat to the Earth's Ozone Shield. *Ambio*. 1(2), 41–51.

Crutzen, P. J. (1974). Future Effects of Continued Use of Chloro-Fluoro-Methanes on Stratospheric Ozone. *Transactions American Geophysical Union*. 55(12), 1153.

Farman, J. C., Gardiner, B. G., & Shanklin, J. D. (1985). Large Losses of Total Ozone in Antarctica Reveal Seasonal ClOx/NOx Interaction. *Nature*. 315(6016), 207–10. DOI: 10.1038/315207a0.

Freedman, B. (2021). Ozone Layer Depletion. In K. H. Nemeh & J. L. Longe, eds., *Vol. V. of The Gale Encyclopedia of Science*, 6th ed., 3272–74.

Fukuyama, F. (2002). *The End of History and the Last Man*. New York: Perennial.

Haas, P. M. (1991). Policy Responses to Stratospheric Ozone Depletion. *Global Environmental Change*. 1(3), 224–34. DOI: 10.1016/0959-3780(91)90044-T.

Kanngieser, A., & Todd, Z. (2020). From Environmental Case Study to Environmental Kin Study. *History and Theory*. 59(3), 385–93. DOI: 10.1111/hith.12166.

Kimmerer, R. W. (2013). *Braiding Sweetgrass: Indigenous Wisdom, Scientific Knowledge and the Teachings of Plants*. New York: Milkweed Editions.

Laslett, Peter, & Locke, John, eds. (1997). *Two Treatises of Government*. Cambridge Texts in the History of Political Thought, 9th ed. Cambridge: Cambridge University Press.

Liboiron, M. (2021). *Pollution Is Colonialism*. Durham, NC: Duke University Press.

Mohr, C. (2021). When Science and Politics Come Together: From Depletion to Recovery of the Stratospheric Ozone Hole. *Ambio*. 50(1), 31–34. DOI: 10.1007/s13280-020-01427-4.

National Oceanic and Atmospheric Administration. (2021). Antarctic Ozone Hole Is 13th Largest on Record and Expected to Persist into November. www.noaa.gov/news/antarctic-ozone-hole-is-13th-largest-on-record-and-expected-to-persist-into-november.

Rockström, J., Steffen, W., Noone, K., et al. (2009). Planetary Boundaries: Exploring the Safe Operating Space for Humanity. *Ecology & Society*. 14(2), Article 32. www.jstor.org/stable/26268316.

Steffen, W., Richardson, K., Rockström, J., et al. (2015). Planetary Boundaries: Guiding Human Development on a Changing Planet. *Science*. 347(6223), e1259855. DOI: 10.1126/science.1259855.

Stewart-Harawira, M. (2005). *The New Imperial Order: Indigenous Responses to Globalization*. London: Zed Books.

UNEP Ozone Secretariat. (2019). *Handbook for the Vienna Convention for the Protection of the Ozone Layer*, 12th ed. Nairobi: United Nations Environmental Programme.

Weber, M., Arosio, C., Coldewey-Egbers, M., et al. (2022). Global Total Ozone Recovery Trends Attributed to Ozone-Depleting Substance (ODS) Changes Derived from Five Merged Ozone Datasets. *Atmospheric Chemistry and Physics*. 22(10), 6843–59. DOI: 10.5194/acp-22-6843-2022.

Whyte, K. P. (2017). Our Ancestors' Dystopia Now: Indigenous Conservation and the Anthropocene. In U. Heise, J. Christensen, & M. Niemann, eds., *The Routledge Companion to the Environmental Humanities*. London: Routledge, 206–15.

Winter, C. J. (2022). *Subjects of Intergenerational Justice: Indigenous Philosophy, the Environment and Relationships*. London: Routledge.

Winter, C. J., & Schlosberg, D. (2024). What Matter Matters as a Matter of Justice. *Environmental Politics*. 31(2) 1205–24. DOI: 10.1177/00905917221128833.

Witze, A. (2020). Rare Ozone Hole Opens over Arctic – and It's Big. *Nature*. 580(7801), 18–19. DOI: 10.1038/d41586-020-00904-w.

Yong, E. (2016). *I Contain Multitudes: The Microbes within Us and a Grander View of Life*. London: Bodley Head.

Part III

Climate Change

5

Introduction to the Climate Change Boundary

SIMON P. MEISCH

As a phenomenon, climate change is hard to capture. While its science is mostly uncontested, its politics is not. What we do know is that climate change is warming the atmosphere, with direct and indirect impacts on people, societies, and the natural environment. Like humans, flora and fauna physically suffer from rising temperatures, which also dry out soils and increase the risk of forest and bush fires. Meanwhile, warmer air absorbs more water vapour, leading to extreme weather events with heavy rainfall. Droughts and heavy rains both contribute to human suffering by negatively affecting agriculture (and thus food security) or by challenging and destroying existing infrastructure. The negative impacts of climate change are thus many and varied.

The main cause of global warming is the massive burning of fossil energy sources such as coal, oil, and gas since the start of the Industrial Revolution in the 1750s. Such burning produces gases (mainly CO_2) that accumulate in the atmosphere (and also in the oceans). There, they create the same effect as in a greenhouse: short-wave solar radiation can pass through these gases and reach the Earth's surface, where it is converted into long-wave thermal radiation. These longer waves, however, cannot radiate through greenhouse gases back out to space and so remain in the atmosphere as heat. In addition to burning fossil fuels, also other activities cause greenhouse gases (GHGs), for instance, ploughing up green land, which releases soil-bound GHGs, such as CO_2 or nitrous oxide (N_2O), or animal husbandry as cattle emit methane (CH_4).

The planetary boundaries framework has two control variables related to climate change. The first is the amount of CO_2 in the atmosphere, measured in parts per million (ppm). With the 2015 update of the planetary boundaries concept, the second is the amount of energy remaining in the Earth's greenhouse (Steffen et al. 2015). This effect is called radiative forcing and is measured in watt per square metre (W/m^2). Until the industrial revolution, the thermal balance between the energy received from the sun and the energy radiated by the earth was roughly equal. The greenhouse effect means that much more energy is being retained in the atmosphere. The values of the two control variables are well above their pre-industrial reference values. The boundary value for the CO_2 concentration is 350 ppm, while the latest update of the concept puts the current value at 417 ppm. The radiative forcing boundary value is 1 W/m^2, with an actual value of 2.91 (Richardson et al. 2023).

It is clear from what has been said that climate change is a moral wrong, the causes of which must be mitigated and to which both individuals and societies must adapt to avoid unacceptable harms. Now, GHG emissions are largely the result of the ways humans produce and use energy and food, and as such are deeply embedded in the daily lives of all people. What is needed are 'forms of life' (Jaeggi 2018) that emit as little GHGs as possible. This is already a huge ethical challenge, as it involves intense and controversial social deliberations about what constitutes a good human life in such radically different circumstances. Then there are the costs and burdens of climate change, a challenge exacerbated by the fact those causing climate change and those most severely affected often belong to different groups. This leads to debates about restorative and distributive justice (Kalfagianni & Meisch 2020). Current climate change is caused by GHGs emitted in previous decades by people in the Global North and now harming people, often the most vulnerable social groups, in the Global South. So, people today need to correct past social choices while ensuring that future people can live good lives. Moreover, climate change interacts with most of the other challenges addressed in the planetary boundaries framework (Richardson et al. 2023). For this reason, the philosopher Stephen Gardiner (2011) calls it the 'perfect moral storm'. The STS scholar Reiner Grundmann (2016: 562) describes climate change as a 'wicked social problem', that is, 'one that cannot be solved, but must instead be re-solved and renegotiated, over and over again'.

Both chapters on climate change in this volume address the ethical challenge of such a constant renegotiation of climate change, and more specifically, how we know about climate change and whose knowledge contributions count in policy responses to (the causes as well as the effects of) climate change. Both raise ethical questions of epistemic justice, that is, dealing with the unjust aspects of knowledge production processes (Kidd et al. 2017). Gay-Antaki's and Sandré and Gheradi's texts challenge the specific understanding of Western science as objective and universal, when, in fact, it is culturally embedded and as such carries specific value judgements, for example, about the social role of science, forms of acceptable harm, or human–nature relations. While cultural embeddedness is not in itself a problem, it can become one if the underlying normativity remains unquestioned. Such a problem is the question of which knowledge contributions are considered relevant for describing the challenges of climate change or how to respond to them.

As both chapters argue, Western climate science unduly values some forms of knowledge over others, thereby silencing relevant knowledge holders and producing morally problematic knowledge for action.

Consequently, in their own ways, both chapters address the role of scholars as co-producers of knowledge. In her conceptual chapter, Gay-Antaki argues from a feminist decolonial perspective for ‘world travelling’ as a different mode of knowledge production that engages deeply with other ways of knowing about the world. For their part, Sandré and Gheradi draw on their experience of working on a project in the Arctic and ask how, as researchers, they have developed an epistemic virtuousness that allows them to amplify plural voices within science.

References

Gardiner, S. M. (2011). *A Perfect Moral Storm: The Ethical Tragedy of Climate Change*. Environmental Ethics and Science Policy Series. Oxford: Oxford University Press.

Grundmann, R. (2016). Climate Change as a Wicked Social Problem. *Nature Geoscience*. 9(8), 562–563. DOI: 10.1038/ngeo2780.

Jaeggi, R. (2018). *Critique of Forms of Life*. Cambridge, MA: The Belknap Press of Harvard University Press.

Kalfagianni, A., & Meisch, S. P. (2020). Epistemological and Ethical Understandings of Access and Allocation in Earth System Governance: A 10-Year Review of the Literature. *International Environmental Agreements: Politics, Law and Economics*. 20(2), 203–221. DOI: 10.1007/s10784-020-09469-5.

Kidd, I. J., Medina, J., & Pohlhaus, G., eds. (2017). *The Routledge Handbook of Epistemic Injustice*. New York: Routledge.

Richardson, K., Steffen, W., Lucht, W., et al. (2023). Earth beyond Six of Nine Planetary Boundaries. *Science Advances*. 9(37), eadh2458. DOI: 10.1126/sciadv.adh2458.

Steffen, W., Richardson, K., Rockström, J., et al. (2015). Planetary Boundaries: Guiding Human Development on a Changing Planet. *Science*. 347(6223), 1259855. DOI: 10.1126/science.1259855.

6

Climate Boundaries and Border Crossers

MIRIAM GAY-ANTAKI

6.1 Introduction: Against the Age of 'Man'

Planetary boundaries were devised to call into action concerted global efforts to preserve 'Earth system stability for the survival of humanity' (Rockström et al. 2009). As discussed in Chapter 1 of this volume, planetary boundaries required identifying and quantifying biophysical targets that should not be surpassed to maintain the Holocene state, a time where 'environmental change occurred naturally, and Earth's regulatory capacity maintained the conditions that enabled human development' (Rockström et al. 2009: 472). The Anthropocene, or 'the age of man', has been described as the most current epoch where man's activities have unsettled the balance of the Holocene and is traced to the beginning of the industrial revolution. Given this clear marker of the Anthropocene that can be traced to a specific space and time, that was inclusive of specific exploitative activities of both humans and the environment (the industrial revolution and consequent global capitalism occurred in tandem with and necessitated colonization and slavery), the question of 'who is the human in the Anthropocene' (Collard & Dempsey 2018) should be central to this formulation, even more so as we seek to move away from global environmental harm. Yet, the most popular solutions to this problem are devised under the same operating frameworks causing the Anthropocene. The threat of looming thresholds shifts important conversations away from ideology and values toward quick technoscientific fixes that fail to question whether these might be the best, most encompassing long-term approaches to resolve global environmental change.

Specifically, in this short chapter, I consider the social, epistemological, and ethical dimensions of 'climate change', one of the nine planetary boundaries, and the challenges and opportunities that the framing of planetary boundary represents. The objective of creating these boundaries was that we do not surpass them and do all we can to help prevent human activities from causing unacceptable environmental change. Climate change was one of three boundaries portrayed as already transgressing the 'safe operating space for humanity' when the concept of planetary boundaries was operationalized (the other two transgressors were rate of biodiversity loss, and human interference with the nitrogen cycle). While three was the 'baseline' planetary boundary assessment, others are frequently being reported as crossed (Persson et al. 2022). Given that such clear-cut boundaries and specific thresholds were identified and quantified, my feminist decolonial training interrogates how vague terms such as 'human activities' or 'unacceptable' could be used to advance this concept. Surely, not all 'human activities' are responsible for climate change, and many would disagree what changes may count as acceptable or unacceptable. Yet the conversation of respecting boundaries, even though centred on measurable and quantifiable markers, inevitably brings up moral and ethical questions regarding what constitutes appropriate or desirable human behaviour on this Earth (Schmidt 2022: 1088).

Mary O'Brien, a feminist biologist, stresses the importance of feminist perspectives around issues that concern humanity. For instance, mainstream attempts to address climate change employ a seemingly objective scientific method to figure out how much more we can contaminate, exterminate, abuse, or warm up, to remain on Earth (O'Brien 1993). The way that climate change has been addressed, mostly by scientists from the Global North, has failed to reduce global warming or differential impacts. Often proposed mechanisms to address the crisis, such as carbon markets, only exacerbate social and environmental inequalities and fail to halt the use of fossil fuels (Westholm & Arora-Jonsson 2018). For these reasons, the need to develop and integrate approaches that challenge the structures and mechanisms that created the crisis in the first place is becoming increasingly apparent. Incorporating different visions to combat problems as complex as climate change may seem insurmountable to those who live comfortably in the normative strata of society and who believe in the universal quality of their knowledge. But for many others, knowing and navigating different worlds and ways of knowing has always been a matter of survival. Many lessons can be learned from those who 'cross borders' (Anzaldúa 2003), and who are hybrid and bilingual (Cusicanqui 2012). As Collins (1986) explains, marginalized groups have had to learn to navigate a world full of oppressive systems (patriarchy, capitalism, colonialism) as a way of survival by creating a vision that allows them to see the world more clearly, enabling them to navigate it safely.

In this chapter I offer border crossing as an approach to challenge power relations that produce and legitimize western scientific knowledge and, along with its distribution, the way in which it systematically disadvantages people because of their gender, geography, race, religion, nationality, sexuality, colonial status, and so on. By delineating power relations that transform people into hierarchical categories, we can begin to challenge socio-environmental relations that contribute to the climate crisis to cultivate more just and resilient pathways.

6.2 Western (Climate) Science as Objective and Universalizing

While the historical dichotomy of science being objective, impersonal, and masculine, and 'feminine' being all the opposite seems outdated, the majority of women and men still believe that men are better suited for science (Shen 2013). This construction of science naturalizes gender and gender differences within the sciences and explains that many women simply decide to not pursue science. This formulation hides the historical exclusionary practices of science and elucidates that there are power relations within science where gender relations have been central in defining the normative knowledge producer as male, white, and western (Harding 1986; Maynard 1997; Merchant 1990). The ahistorical and apolitical understanding of western science is one of the main mechanisms that allow this way of thinking not only to dominate how we understand the world, but not to question why western scientific practice is still so exclusionary of women and people of colour. To borrow from Grosfoguel (2007), the domination of western science as the only legitimate way to know things means that even if we do see an increase of different faces participating in science, it does not mean that these scientists will automatically challenge oppressive western scientific structures (Gay-Antaki, 2022). On the contrary, members that are welcome in western scientific spaces, such as the Intergovernmental Panel on Climate Change (IPCC) or policy spaces such as the Conference of the Parties (COP), will most likely be trained in and promote western epistemology and worldviews even if they inhabit non-western spaces. This is precisely one of the biggest victories of western epistemology, by removing its locus of enunciation – it's western roots – by equating western, masculine, and white to neutral and universal, by claiming objectivity, western science has placed itself as the only legitimate way to know the world (Mignolo 1999).

As western science has claimed to be the universal way to know things, so has western climate change science become the universal way to understand and address global environmental change. Even though a very small stratum of society has the privilege of ahistorical and apolitical understandings of global climate change, this stratum has the most power and voice guiding global efforts to adapt and mitigate climate change. For instance, at the beginning of the global climate talks, during the United Nations Framework Convention on Climate Change (UNFCCC) COPs, after the Rio Summit in 1992, industrialized nations clearly recognized their role in producing most of the historical global emissions leading to global warming. Actions forward were to represent this differential responsibility where Annex 1 countries (industrialized countries in UN language) would lead the reduction of greenhouse gas (GHG) emissions followed later by non-Annex countries (or Global South countries) (Vanhala & Hestbaek 2016). However, this acknowledgement of common but differentiated responsibility around climate change became very contentious and was quickly stifled, even though it remains a central issue at every COP meeting. Instead, mechanisms such as the Talanoa Dialogue, which seek to escape conflict by avoiding holding certain parties accountable, are fostered during the climate talks so that certain parties can be protected from being singled out during the negotiations (Lesniewska & Siegele 2018).

Under an Anthropocene frame, moral and ethical accountability, and thus actions to cut emissions, remain a distant goal as responsibility for emissions becomes blurred and narratives of a sinking ship (meaning we are all in this together) take precedence. Few would argue against urgent global action to address climate change threats; equally important is the everyday reality of the disproportionate impact that certain geographies and populations are already experiencing due to global warming. Yet, global calls to action usually fail to address climate change threats against the most vulnerable and at risk. The prevalent concept of planetary boundaries by Rockström et al. (2009) is a case in point; global environmental problems are described in measurable terms that set thresholds and boundaries trying to delimit how much more we can contaminate, or warm up, and still remain in an inhabitable Earth. The global and urgent framing allows apolitical and ahistorical understandings of the problem and thus solutions that fail to challenge epistemic and power hierarchies that are responsible for ongoing environmental change. For instance, as argued in Gay-Antaki (2022), the agreement to limit global warming by the end of the century to 2° Celsius, one of the most successful treaties to come out of the COP negotiations during Paris in 2015, was proposed by powerful countries consisting of the G8 (the USA, the UK, Germany, etc.) who believed this was a good threshold. Island Nation States have had to fight to make their voices heard since while a 2° Celsius threshold might seem reasonable to the G8, this threshold would put many Island States under the rising sea. As a matter of survival, most actions to adapt and mitigate climate change are already occurring and must take on a very localized response, which most global initiatives fail to do so.

6.3 Western (Climate) Science as Socially and Culturally Embedded

Even though climate change science carries hegemonic power, it is also highly contested as it is a science highly driven by national and international politics of climate change (Demeritt 2001). This positioning of climate science stands in sharp contrast with the idea of science as neutral, objective, and universal. This public understanding of science gives way for discrediting both knowledge claims and public policy decisions based upon these claims when the social relations through which climate science is constructed are disclosed (Grasswick 2014). Climate sceptics use the narrative of western hegemonic science to discount the science of climate change; as it is politicized such that science is seen as no longer objective and thus no different from folk, or traditional knowledge. This idealized understanding of science also impedes the practice of science becoming more democratic, gatekeeping what science looks like and who can practice it.

Climate change poses a seemingly insurmountable conundrum, while it will require a concerted global response, its impacts are grounded and embodied; actions are necessarily local, but policies and big technologies are devised and formulated far away from where they are intended to be implemented. Concerted, gradual, global action is necessary, but so is immediate, grounded action. While these debates lead many climate scientists to attempt to convince climate sceptics by throwing more science at them (that will still be embedded in social relations), we would be better off by trying to change public understandings of science and scientific practice. As '[s]cience on its own offers no moral vision, no ethical stance, and no political architecture for delivering the sort of world people desire' (Hulme 2020: 310), it seems key to participate with critical perspectives and historical understandings of the power that science wields in our society. As Hillary Rose, a feminist sociologist, has long argued, to exclude ourselves from spaces where such power is yielded allows the status quo to continue without any criticism or effort to incorporate equity and justice dimensions (Rose 2007 [1994]). Tuana underscores that 'the point, then, is not to eliminate politics or interests from science, but to understand which interests advance knowledge and for whom, and which obstruct knowledge, again for whom' (2013: 18).

An example of social scientists engaging with planetary boundaries is offered by Gupta et al. (2021), who ask that social scientists engage with planetary boundaries to 'ensure transformative, planetary justice' (2021: 3). They propose that adding justice targets to each biophysical target might 'avoid significant (irreversible and existential) harm to humans' (Gupta et al. 2021: 3). Gupta et al. explain that '[e]arth system stability for the survival or even wellbeing of humanity is not necessarily just for all humans and that safe targets may even make things worse for some' (2021: 3). They explain that some of the 'safe' targets may be at odds with social environmental justice and may disproportionately impact the most vulnerable. For instance, 'setting aside large areas ... from human use for biodiversity protection, without addressing systemic issues, such as inequality in land tenure and the food consumption habits of the rich, could have potentially devastating impacts on the world's poor and food security' (2021: 3). Gupta et al.

propose to add a justice perspective within the planetary boundary perspective. While this might speak to those in power, the addition of justice in such a narrow framework limits the work and impact that social scientists might have if they were to truly push the boundaries of the planetary boundary perspective.

6.4 Border Crossing: Feminist Decolonial Approaches and Climate Change

Climate change presents an opportunity for border crossing like no other problem: as a localized and embedded problem, it underscores the embodiment of climate change knowledge and fights against the idea of universal knowledge – a view from nowhere – as it enunciates the location of the knower. This localization stresses the social practice of science and its inseparability from sociopolitical positionings encouraging questioning preconceptions about expertise, legitimacy, and knowledge.

Insights from feminist, queer, Black, and decolonial scholars elucidate the relationship between the domination of nature, women, and people of colour. The bifurcation of nature from culture and the consequent association of white, masculine, and western to culture, and Brown, feminine, and non-western to nature, have sanctioned an ongoing domination of western epistemological practices over all others, claiming to be universal. In other words, western scientific thought pretends to be the only legitimate way to know things about this world. It might be more fruitful to disrupt, challenge, and underscore the narrow operating framework of this vision and the epistemic violence against other ways of knowing. Border crossing as a feminist geographical decolonial approach can offer alternatives to build a truly inclusive and transformative climate change science.

Feminism is a useful approach to diversify our knowledge base as it has long challenged the logic and rational procedures that seek objectivity. Feminists have long worked to highlight gendered power dynamics that have often been invisible in our society. Rather than believing we can be 'objective observers' or as Donna Haraway expressed having 'an all-seeing eye', the introspective approach offered by feminism doubts that there is an external reality that represents the truth waiting to be discovered. Feminism stresses that our sociopolitical position determines what we know and are capable of understanding (Harding 1986). 'Border crossing develops a view from somewhere, that requires constant dialogue to work through tensions, difference, and contradictions to get closer to understanding climate change' (Gay-Antaki 2022: 118). Importantly, underscoring the location of the knower has the potential to flatten knowledge hierarchies and operationalize the knowledge that is needed for the specific problem to be addressed.

In a recent article, I stress the importance of centring feminist analyses at the intersection of race and location to find sustainable pathways to climate change.

Centring race and location when attempting to better understand and address climate change asks that the solutions we propose do not perpetuate structures that continue to disadvantage and marginalize those who have long been silenced by western science and are particularly vulnerable to climate change. Alongside women of colour and Global South feminists I underscore that one of the most powerful ways to challenge oppressive climate structures is to transform who participates in the climate debate. To centre women of colour and Global South voices as agents of change in the climate debate will demand a complete restructuring of how we prioritize climate knowledge and action. This centring would elucidate just how similar or different well-meaning mainstream concepts aimed to mitigate and adapt to climate change are from those that perpetuate epistemic hierarchies, paternalistic approaches, modernization, and marketization ideologies that continue to exploit those from the periphery even in the struggle to address climate change. Concepts such as planetary boundaries and the Anthropocene would take on different meanings and understandings that would be relatable for most of the world's population. Shouldn't this be one of the driving forces of a universal approach to addressing one of the world's most pressing concerns?

Border crossing will not automatically discard useful knowledge just because of a label. On the contrary, border crossing as a method can expand our epistemological canon; contrary to western science's requirement of an objective observer, border crossing values knowledge that is produced by people who are situated and informed by their sociopolitical position while challenging colonial categories (all people are situated; thus, all knowledge is situated). For many who navigate comfortably in the normative strata of society, shifting the centre might seem like an impossible challenge, or might even perceive such a shift as a threat. For many others, who experience the normative strata as 'minorities' or 'outsiders', knowing and navigating the normative space has been a matter of survival. In the words of Maria Lugones, a feminist decolonial scholar, most of us 'who are outside the mainstream of, for example, the U.S. dominant construction or organization of life are "world travelers" as a matter of necessity and of survival . . . inhabiting more than one "world" at the same time and "traveling" between "worlds" is part and parcel of our experience and our situation' (Lugones 1987). For this reason, 'I believe that there is much to gain from learning from those who are "world travelers" (Lugones 1987) "border crossers" (Anzaldúa 2002), hybrid and bilingual (Cusicanqui 2012) . . . . This hybridity grants a vantage point that can better understand the world around us' (Gay-Antaki 2022). Significantly, it is not that there are knowledges that have been ignored and could simply be added into the mix. As Collins (1990) underscores, these knowledges are subjugated and violently repressed. This repression is what has given the 'west' its power over non-western epistemologies. Importantly, this silencing or indifference toward other knowledges has limited the western perspective. As Hill-Collins explains, 'it is not that marginalized groups have a privileged perspective, but rather, they have had to learn to navigate many different oppressive systems (here in particular Hill-Collins was referring to black American women) providing a vantage point by which to see the world more clearly and navigate it safely' (Gay-Antaki 2022).

The western assumption that we can have an 'all seeing eye' limits our ability to resolve global environmental problems such as climate change. The complexity and scale of climate change make it impossible to understand it fully from a single standpoint, and, thus, catch-all solutions proposed from this single standpoint, such as planetary boundaries, will be severely limited. It is becoming increasingly clear that 'global' concerted efforts that fail to challenge western epistemological structures are incapable of addressing climate threats against those who are most vulnerable (including more than human species) and might even be making certain populations and geographies more vulnerable. The shift away from addressing 'common but differentiated responsibility' at the beginning of the COPs toward 'loss and damage' currently underscores a major problem of narratives that claim universal qualities but that clearly represent western and Global North priorities.

Feminist objectivity underscores that the spaces that we inhabit influence what we know and what we can know as subjects of knowledge. Under this framework, no solution can be universal or global. Border crossing can foster a constant dialogue among equal but different subjects of knowledge inspiring us to ask different questions, conduct different research, and build different partnerships, producing more complete knowledge about the world around us that will be better equipped to address both the urgent and long-term impacts of climate change. Border crossing challenges the dominant positivist disciplines in understanding anthropogenic climate change that offer 'little or no sense of humans as diverse, interpretive creatures who frequently disagree about values, means and ends' (Castree et al. 2014: 765). Border crossing underscores how climate change science is socially and culturally embedded resulting in many possible meanings and political trajectories.

By collecting diverse and partial views of global climate change into a collective subject position, border crossing challenges the superior vantage point claimed by western science. By underscoring the situated positions of diverse knowers, border crossing seeks to create a vision from somewhere to better work within the limits and contradictions of climate science to elucidate the complex ways in which gendered and racialized assumptions and power relations that discriminate against women and other marginalized groups are produced, sustained, negotiated, and challenged in the IPCC, the climate debate, and science more broadly (Gay-Antaki

2021). Through border crossing we can better identify and address barriers and opportunities to access and produce quality climate research while fighting against overly narrow conceptions of what is possible and desirable in the fight against climate change.

References

Anzaldúa, G. (2002). Now Let Us Shift. . .the Path of conocimiento. . .Inner Work, Public Acts. In G. Anzaldúa & A. Keating, eds., *This Bridge We Call Home: Radical Visions for Transformation*. New York: Routledge, 540–78.

Anzaldúa, G. (2003). La Conciencia de la Mestiza: Towards a New Consciousness. In C. R. McCann, ed., *Feminist Theory Reader: Local and Global Perspective of Women's Studies*. New York: Routledge, 179–88.

Castree, N., Adams, W. M., Barry, J., et al. (2014). Changing the Intellectual Climate. *Nature Climate Change*. 4(9), 763–68. DOI: 10.1038/nclimate2339.

Collard, R.-C., & Dempsey, J. (2018). Accumulation by Difference-Making: An Anthropocene Story, Starring Witches. *Gender, Place & Culture*. 25(9), 1349–64. DOI: 10.1080/0966369X.2018.1521385.

Collins, P. H. (1986). Learning from the Outsider Within: The Sociological Significance of Black Feminist Thought. *Social Problems*. 33(6), S14–32. DOI: 10.2307/800672.

Collins, P. H. (1990). Black Feminist Thought in the Matrix of Domination. In P. H. Collins, ed., *Black Feminist Thought: Knowledge, Consciousness, and the Politics of Empowerment*. Boston: Unwin Hyman, 221–38.

Cusicanqui, S. R. (2012). Ch'ixinakax utxiwa: A Reflection on the Practices and Discourses of Decolonization. *South Atlantic Quarterly*. 111(1), 95–109. DOI: 10.1215/00382876-1472612.

Demeritt, D. (2001). The Construction of Global Warming and the Politics of Science. *Annals of the Association of American Geographers*. 91(2), 307–37. DOI: 10.1111/0004-5608.00245.

Gay-Antaki, M. (2021). Stories from the IPCC: An Essay on Climate Science in Fourteen Questions. *Global Environmental Change*. 71, 102384. DOI: 10.1016/J.GLOENVCHA.2021.102384.

Gay-Antaki, M. (2022). Border Crossers: Feminist Decolonial Geography and Climate Change. *Progress in Environmental Geography*. 1(1–4), 115–36. DOI: 10.1177/27539687221114887.

Grasswick, H. (2014). Climate Change Science and Responsible Trust: A Situated Approach. *Hypatia*. 29(3), 541–57. DOI: 10.1111/hypa.12090.

Grosfoguel, R. (2007). The Epistemic Decolonial Turn. *Cultural Studies*. 21(2–3), 211–23. DOI: 10.1080/09502380601162514.

Gupta, J., Liverman, D. M., Bai, X., et al. (2021). Reconciling Safe Planetary Targets and Planetary Justice: Why Should Social Scientists Engage with Planetary Targets? *Earth System Governance*. 10, 100122. DOI: 10.1016/j.esg.2021.100122.

Harding, S. (1986). *The Science Question in Feminism*. Ithaca, NY: Cornell University Press.

Hulme, M. (2020). One Earth, Many Futures, No Destination. *One Earth*. 2(4), 309–11. DOI: 10.1016/j.oneear.2020.03.005.

Lesniewska, F., & Siegele, L. (2018). The Talanoa Dialogue: A Crucible to Spur Ambitious Global Climate Action to Stay within the 1.5°C Limit. *Carbon & Climate Law Review*. 12(1), 41–49. DOI: 10.21552/cclr/2018/1/8.

Lugones, M. (1987). Playfulness, 'World'-Travelling, and Loving Perception. *Hypatia*. 2(2), 3–19. DOI: 10.1111/j.1527-2001.1987.tb01062.x.
Maynard, Mary, ed. (1997). *Science and the Construction of Women*. New York: Routledge.
Merchant, C. (1990). *The Death of Nature: Women and Ecology in the Scientific Revolution*. New York: HarperOne.
Mignolo, W. D. (1999). I Am Where I Think: Epistemology and the Colonial Difference. *Journal of Latin American Cultural Studies*. 8(2), 235–45. DOI: 10.1080/13569329909361962.
O'Brien, M. H. (1993). Being a Scientist Means Taking Sides. *BioScience*. 43(10), 706–8.
Persson, L. M., Carney Almroth, B. M., Collins, C. D., et al. (2022). Outside the Safe Operating Space of the Planetary Boundary for Novel Entities. *Environmental Science & Technology*. 56(3), 1510–21. DOI: 10.1021/acs.est.1c04158.
Rockström, J., Steffen, W., Noone, K., et al. (2009). A Safe Operating Space for Humanity. *Nature*. 461(7263), 472–75. DOI: 10.1038/461472a.
Rose, H. (2007 [1994]). *Love, Power and Knowledge: Towards a Feminist Transformation of the Sciences*. Cambridge: Polity Press.
Schmidt, J. J. (2022). Geography and Ethics I: Placing Injustice in the Anthropocene. *Progress in Human Geography*. 46(4), 1086–94. DOI: 10.1177/03091325221097104.
Shen, H. (2013). Inequality Quantified: Mind the Gender Gap. *Nature*. 495(7439), 22–24. DOI: 10.1038/495022a.
Tuana, N. (2013). Gendering Climate Knowledge for Justice: Catalyzing a New Research Agenda. In M. Alston & K. Whittenbury, eds., *Research, Action and Policy: Addressing the Gendered Impacts of Climate Change*. Dordrecht: Springer Netherlands, 17–31.
Vanhala, L., & Hestbaek, C. (2016). Framing Climate Change Loss and Damage in UNFCCC Negotiations. *Global Environmental Politics*. 16(4), 111–29. DOI: 10.1162/GLEP_a_00379.
Westholm, L., & Arora-Jonsson, S. (2018). What Room for Politics and Change in Global Climate Governance? Addressing Gender in co-Benefits and Safeguards. *Environmental Politics*. 27(5), 917–38. DOI: 10.1080/09644016.2018.1479115.

7

An Arctic Bounded by Whiteness

*Ethics of Amplification**

TANGUY SANDRÉ AND JEANNE-MARIE GHERARDI

7.1 Introduction: Converging Gazes

In this chapter, we mobilize our transdisciplinary research within Arctic communities to critique the concept of planetary boundaries (PB) and associated 'anthropocenic narratives'. We were invited to cross our distinct, but interdisciplinary, gazes to scrutinize the ethical and epistemological tensions that loomed from the encounter of anthropocenic narratives and the ethnographic practices we deploy. As a PhD Research Fellow in social sciences, with a background in climate change adaptation, Tanguy investigated the expression of the plurality of knowledges, experiences, and values within the climate crisis' narrative, rooting his research in place-based stories. The critical considerations for the science being done and the narrative being supported emerged early and widely in this research pathway. For the second co-author, Jeanne-Marie, it was very little discussed alongside her early training as a geologist, whether it was a blind spot or by virtue of science assumed as unquestionable. For her, it is only recently, in the context of climate emergency and her involvement in interdisciplinary climate research that the question of ethics and justice started being central. Throughout our participation, as project leader and PhD fellow, within the SeMPER-Arctic project (Sense Making, Place attachment and Extended networks as sources of Resilience in the Arctic), we paid attention to place-based stories of changes and upheavals in the Arctic. By transmitting concepts from one discipline to another and bringing in our perspectives, we put ourselves in position to develop a critical, interdisciplinary stance in response to the univocal narrative of the climate crisis which critics argue is often deaf to non-Eurocentric voices. Our exchange is fed by our repetitive fieldwork (seven stays for three to nine weeks between 2021 and

* The authors wish to acknowledge the participation and invaluable insights and meaning offered by the inhabitants of Ittoqqortoormiit. The authors were funded through Belmont Forum project SeMPER-Arctic (French National Agency Grant).

2023) in Ittoqqortoormiit (Kalaallit Nunaat/Greenland), where we have identified salient expressions of lived experience with community members.

The reflexive conversation we engage in this chapter aims to highlight power imbalances and the position of researchers in order to discuss the possibilities of an ethic of amplification that addresses equity in the shaping of narratives within the 'anthropocenic moment'. Our conversation begins with a discussion of the concept of whiteness in the Arctic that we use both as a physical parameter intrinsically linked to Arctic amplification and global climate dynamics for climate scientists and metaphorically as a description of the climate science carried out there amplifying Anthropocene discourse. Then, we suggest that the plurality of voices, perspectives, or concerns needed to address climate change in the Arctic requires not having only the singular story amplified by climate science, that is, the Anthropocene, because that would unethically keep amplifying whiteness at the expense of continued 'othering' and disempowering of those in Ittoqqortoormiit and in the rest of the Arctic region. Finally, acknowledging the contestation of 'anthropocenic narratives' as expressed among non-Western communities, we consider tensions with the concept of PB, and suggest using it only cautiously, for instance with regard to historical dominations including over other-than-human beings.

7.2 The Whiteness of the Arctic: Amplification and Dampening

Amplification and dampening in the Arctic are fertile and cross-cutting images to frame the conversation. These are both connected to whiteness, whose polysemy is used here either to describe a physical object connected to the larger climate system or to unfold the system that reproduces the dominant way of knowing. Therefore, the polysemic conversation here is guided by climate change–induced shifts in albedo and other positive feedback to ultimately achieve climate-coloniality.

7.2.1 Whiteness as a Key Physical Parameter of the Arctic Climate System

The Arctic is often studied by climate scientists for its key role within the larger climate system. This is notably explained by the phenomenon of Polar or Arctic Amplification, by which change in the net radiation balance tends to produce larger changes in temperature in polar regions than, for instance, in mid-latitudes. This process of amplification is observable in the past climate and glacial period, and is expected to become stronger in coming decades (Serreze & Barry 2011; Previdi et al. 2021). The phenomenon of amplification is strongly associated with positive feedback such as albedo effects, which refer to the fraction of incoming solar energy or light that a surface reflects back to space (Fox-Kemper et al. 2021; Intergovernmental Panel on Climate Change 2022). This reflected energy is a fundamental component of the earth's energy balance. The albedo has values between 0 for a black surface that absorbs all energy and reflects none, and 1 for a white surface that behaves like a mirror and reflects all light without absorbing any. The albedo of snow-covered sea ice or of a pristine white is therefore higher than that of the darker open sea. The whiteness of the Arctic, the snow coverage and sea ice extent, plays a major role into the planetary albedo and is therefore of the utmost interest for climate scientists to observe climate change. For instance, the progressive decline of sea ice extent and ice thinning is predicted to reach a seasonal ice-free Arctic Ocean by 2050. The region amplification also produces impacts beyond the Arctic, such as in storm tracks, the jet stream, and planetary waves and their associated energy propagation, as well as accentuated impacts of Pacific variability (Cohen et al. 2014; Overland et al. 2015; Preece et al. 2023). It must be noted here that the physical whiteness of the Arctic is itself a narrow representation of the regions which are predominantly and for most of the year uncovered by white surfaces.

7.2.2 Whiteness of the Gaze

> Naming whiteness is necessary; it is the 'invisibility' of white people, who are presented just as 'people', the default norm from which everyone else deviates, that is part of its normative power making. Yet the more you state and claim your 'whiteness', without doing any further work to unpack what that means, the more you become fixed to that articulation of self, the more you become wedded to whiteness . . . . We have to set whiteness up, to name it, to frame it, in order to dissemble it.
>
> (Dabiri 2021: 9–10)

Independent of the processes associated with the physical polar whiteness, social notions of whiteness are pervasive in the making of the science itself. Firstly, Arctic climate science remains mostly produced in mid-latitudes, and the field is largely characterized by 'fly in, fly out' practices, devoid of any form of local engagement (Skjervedal 2021). But it also submerges the framing of the climate change narrative on Arctic regions. Multifaceted changes and upheavals are mainly analysed in a global and science-based approach of climate and environmental sciences that tends to make invisible local contexts and specificities (Bremer et al. 2019; Krauß 2020). This is especially significant in the Arctic (Bravo 2009; Holm 2010; Baztan et al. 2017; Stephen 2018), where local and Indigenous communities are also often portrayed as passive 'victim-heroes' (Roosvall & Tegelberg 2013).

Even when these approaches acknowledge injustice and recognize agency, they remain views from the outside that preserve a certain kind of narrative. In other words, the conduct of the research itself imposes a restriction on the kind of knowledge involved – for instance, restrained to the realm of the quantifiable (for example, albedo) – in making and sustaining a narrow understanding of the Anthropocene.

With the concept of whiteness at the centre of their analysis, Baldwin and Erickson (2020) underline the inherent coloniality of the Anthropocene. They argue that the whiteness of the Anthropocene 'determines not only who can be heard in that space but also the very terms on which the Anthropocene debate unfolds as an object of public concern' (Baldwin & Erikson 2020: 6). In the Arctic, the regional – but globalizing – science-based narrative tends to annihilate expressions of place-attachment, cultural values, or local knowledge, and hence neglects local agencies through epistemic injustice and violence while strengthening pre-existing power imbalances. By contrast, community responses to a changing climate and environment 'is deeply implicated in their sense of who they were and are, also in relation to a larger world' (Hastrup 2018: 1) and the knowledge they hold (Krupnik & Jolly 2002). For instance, in Ittoqqortoormiit, sea ice changes can only be understood in nonlinear manifestations whose narrations articulate epistemic humility, place-attachment, belonging, and relation with more-than-human beings. Furthermore, the presence or the absence of sea ice shapes seasonality, whose occurrence directly connects with the ability to move and travel, hunting practices and expressions of freedom (Sandré et al. 2024).

How do we navigate with this critical view when we apply it to ourselves? We need to acknowledge that we embody the white gaze, which is deployed through whiteness, and that our position is unsatisfactory to address the aforementioned critiques. The SeMPER-Arctic project, for instance, aims at decolonizing concepts such as resilience but was itself developed with limited community engagement. Once the project was funded, we, as critical ethnographers, were committed to 'explicitly consider how (our) own acts of studying and representing people and situations are acts of domination' (Noblit et al. 2004: 3). This contradiction pushes us to step back and propose an ethic of amplification, where we aim to decentre our practices and to pluralize the telling of research stories.

7.3 Ethics of Amplification: How to Perform Plural Voices within Science?

We argue that the dominance of white- and science-informed shaping of the climate narrative demands an enhanced practice of softness, care, and listening. While different narratives are at play, with some being historically dominant and hegemonic, we potentially see a space within science to unmute local discourses which so far have systematically been placed at the margins. The uniformity of Anthropocenic narratives, such as the PB framework, intrinsically denies plurality in terms of ways of knowing, living, and responding to change – changes in which climate is not always the dominant framing. The volition to reach objective description and prescription opens an abyss of possibility when it comes to the processes of domination and unfairness to be corrected. Conversely, to the hard ball of natural sciences, following Ingold's (2018: 101) metaphor, we propose a softer science, a 'squishy ball, (that) by contrast, bends and deforms when it encounters other things, taking into itself some of their characteristics while they, in turn, bend to its pressure in accordance with their own inclinations and dispositions'. The metaphor is translated within our practices through an intention to care for place-based communities' narratives that potentially address processes of empowerment and fairness, while promoting plurality and alterity as crucial ethical lenses for engagement within transdisciplinary environmental research.

7.3.1 Amplifying Voices: Research as a Relation to Support the Expression of Plurality

Science is a powerful tool for good as well as for evil, for emancipation as well as for exploitation. How scientists use their time and talent, their training at public expense, their public research funds, and the public trust are not matters to be brushed aside lightly.

(Bleier 1986 cited in Denzin & Lincoln 2018: 1558)

In order to address, but not easily resolve, the criticisms outlined earlier, a new radical way of engaging with research is required, especially when working with groups represented as if on the margins, from yonder, that is, from the non-Western world. We suggest that by employing a critical ethnography framework, we are in a position to endorse 'an ethical responsibility to address processes of unfairness or injustice' (Madison 2020: 30). This view aims to take care of the sufferings of living beings and commits us to supporting changes towards an improved situation or to minimize sufferings. The critical approach to climate scholarship is all the more relevant as it allows us to emphasize how vulnerability to climate change in the Arctic is constructed through discourses that ignore the effects of colonialism and its ongoing impacts. Its formulation itself extends colonial forms of knowledge and practice (Cameron 2012). For instance, the PB framework relies on measurable and quantitative indicators (e.g., atmospheric CO_2 concentration levels or rates of biodiversity loss) and proxies (e.g., satellite-derived vegetation indices such as NDVI or nitrogen fertilizer application rates). Such global and unanchored metrics are constitutive of a particular, that is, Western, *episteme,* which, intentionally or not, excludes unquantifiable criteria salient to groups operating within different epistemic frameworks (e.g., spiritual, cultural, or relational values to ecosystems that resist quantification).

The tools used to measure and objectify PB research are abstractions, and these metrics are often very alien to the human population in general (Karlsdóttir et al. 2023). As critical ethnographers, we subvert concepts such as neutrality and taken-for-granted assumptions. This is not a pleasant journey, a fortiori, but not only for researchers like us from the settler's world, knowing that 'we have all internalised dominant ideas about what constitutes "good" research and "acceptable" research practice' (Brown & Strega 2005: 2). A common postulate among research methodologies is that the researcher is an outsider 'able to observe without being implicated in the scene' (Smith 2021: 157). Neutrality, objectivity, and distance are valued as part of the 'good science' framework. Nevertheless, being an outsider experiencing the inside clearly means being part of the landscape, entangled in multiple relationships and taking part in 'the care enveloping every social interaction' (field notes, 2021, p. 2). 'In Indigenous frameworks, relationships matter' (Smith 2021: 137). To deploy an ethical practice oriented towards plurality, it requires us to direct our intention around showing respect to the community we are working with and to walk softly: '(Y)ou need to prove that you can actually function as a positive member of that community; there's no way to do that without becoming a part of that community. That takes time' (NCAI & MSU 2012: 18).

7.3.2 Ethics towards Plurality: Why Does It Matter?

What makes plurality an ethically important consideration? We distance our views from those that support a plurality of views (on climate change, for example) only when they seem appropriate for dominant policy decisions but still fail to recognize local contexts. Our aims are not an instrumental plurality directed towards mastering, but to address a process of fairness. Within environmental justice and transdisciplinary research, narrative-centred approaches have shown their value to support the capacity of marginalized communities and groups to speak for themselves (Ottinger 2017) and to make sense of changes in the light of their own knowledge, beliefs, and experiences (Vanderlinden et al. 2020). Sultana (2022) argues for the crucial place of lived experiences to counterbalance hegemonic framings and structural forces that perform climate coloniality. A narrative-centred methodology allows for reframing categories and for recognizing that the science-based or policy-oriented way of categorizing the world is not necessarily salient and often insufficient, especially but not only in postcolonial contexts. In that way, narratives elicit the plurality of worlds and alternative worldviews existing outside and alongside the Western world.

As Jonsson (2016: 1) argues, 'It is only when the Indigenous attains the place of the subject telling the story that we can seriously claim to have a postcolonial narrative of the Arctic.' In Ittoqqortoormiit, alongside more formal interviews, the gesture of decentring our gaze, adopting a relational approach, pushed us to consider stories that were not initially in the scope of our research. While investigating the impacts of climate change through narratives of lived experience, we opened up space for other salient stories to emerge. One of those was associated with the central importance of *mattak* (narwhal skin) and threats to the future of narwhals hunting. That led us to the co-production of a short film in the native language and with community members. With this short film, we also tried to correct the inherent lack of engagement in the design and conduct of our research project. Feedback and evaluations by community members following the screening and the diffusion within the community – for example, repeated comments such as 'we've been telling that for many years' and 'you/we need to spread it out to the world' – suggest that an important step towards inclusive and fair research practices has been reached (Karlsdóttir et al. 2023). We believe that place-based stories that capture lived experiences of the acute environmental crisis have to be heard and contribute to pluralize Anthropocene discourse.

7.4 Discussion: Embodying an Ethic in a More-Than-Human Earth

The Anthropocene and PB are both familiar concepts for environmental scholars. In contrast, we open up discussions towards what they depict and how those depictions act discursively on the world. The ethical tensions we put into dialogue were towards the idea of quantifying to describe or to master, and the active ignorance regarding the non-quantifiable. On one hand, we would rapidly agree (and hence not further develop here) that Anthropocenic concepts are valuable to produce a certain panorama of the environmental crisis. PB, for instance, provide a glimpse on nine quantifiable parameters and their evolution over time (Richardson et al. 2023) – six of them have surpassed the safe operating system. Therefore, the latter have the capacity to be visually transposed which might help to foster awareness and provoke transformative actions. On the other hand, we further debate the limitations of such concepts which are not inherently neutral, while largely pretending to be so, and may tend to reproduce multifaceted domination and violence, including on other-than-human beings (Gupta et al. 2023; Rockström et al. 2023).

7.4.1 Unbounding the Anthropocene Discourses

The Anthropocene can be captured as an expression of a dualist framing in the form of a gendered and racialized Man pursuing the liberation from a passive and changeless Nature (Grove & Chandler 2017: 79). Since its introduction to depict or label a new geological epoch (Crutzen & Stoermer 2000; Crutzen 2006), the concept of the Anthropocene has been extended beyond geology and gained legitimacy in multiple fields of academia as well as in the public arena. Beyond the debate around the date for the Anthropocene (see for instance critical reflections from Lewis & Maslin 2015; Davis & Todd 2017), the diffusion of the concept has been accompanied with tense critiques that both cover a problem-framing – for instance tending to homogenize humanity (see Baldwin & Erickson 2020; Luke 2020) – and its embeddedness in the ideology of control and mastering. The Anthropocene was also extensively criticized as hiding root causes of global change found both in imperialism and in racial capitalism (Sealey-Huggins 2018). Alternative terms were suggested to re-root the climate and ecological crisis in the historical development of capitalism (Capitalocene) or enslavement (Plantationocene) (see for instance Haraway 2015; Bonneuil & Fressoz 2016). Whatever term one uses, the Anthropocene and more widely the anthropocenic framing are to be understood both as a moment that directly threatens (certain) ways of being in the world and as a global narrative that pursues a dynamic of rendering margins invisible. In other words, the climate and ecological crisis, in its intertwinement with multidimensional upheavals, sheds new lights on the unsustainability of the hegemonic way of experiencing the world, while threatening with higher intensity certain ways-of-being in the world, which are also those less accountable for the situation (see Sultana 2022, 2023; Tuana 2022). Accordingly, we amplify the necessary work of paying more attention to marginalized experiences and narratives from yonder, from the non-Western world, through an ethic of plurality that emphasizes lived experiences. Making such a movement and recognizing how the Anthropocene is framed in reference to a 'safe operating space' for humanity (or rather a certain portion of the humanity), we are also tempted to question even further the boundaries of what life matters.

7.4.2 Open Up Space for Other-Than-Human Experiences

The hegemonic way of experiencing the world threatens diverse ways-of-being by exacerbating oppression over some people because of their class, race, gender, sexual orientation, and so on. If we take a critical look at the anthropocentrism of anthropocenic discourses, we might raise the question of how we open space for more-than-human ethics that takes care of non-human beings, as Christine Winter does, in her Chapter 4 on ozone, in this book. Even if Rockström et al. (2023) recognize that non-humans have been under-considered within the framework of PB, their attempt to provide a theoretical frame for just and safe PB remains limited to a nature/culture divide and a capitalistic model wherein non-human animals' relations are envisioned through the lenses of appropriation to the profit of humans and where ecosystems are valued for their ability to sustain humanity potentialities. Although the stories we collect are told by humans, it now seems important to us to observe attentively what is at stake for the animals that are part of human practices and activities that we encounter. Additionally, when working with marginalized communities, our ethics regarding caring for the suffering of human beings guide us in questioning and reframing the scope of our moral considerations. This question is notably addressed by non-hegemonic ontologies (for example, animism) (Willerslev 2007; Escobar 2014; Rivera Andía 2019), multispecies ethnography (Kirksey & Helmreich 2010) but also anticolonial critical animal studies (Montford & Taylor 2020). The aforementioned perspectives involve ethical experiences that guide certain attitudes towards non-human animals but also privilege certain political responses that distance themselves from the ideology of control and mastering. Furthermore, as noted by Bauer (2016: 416), 'by reproducing the concept of nature, it also ushers in an implicit politics of knowledge production, creating a political epistemology that often gives Western scientists and technocrats a privileged position to define natural and desirable environmental conditions'.

7.5 Conclusion

Global and anthropocenic narratives – whose acme is met with PB – tend to produce a uniform representation, one embedded in the modern drift that has progressively devolved into an exsanguinated reason – a positivist universalism that is a leitmotif in which reason is deprived of reflexivity and blind to its own practices of domination. If the climate and ecological crisis, in its entanglement with multidimensional upheavals, sheds new light on the unviability of the hegemonic way of experiencing the world, the latter has to be questioned in its internal ability to provide a universal meaning, without deepening its tendency for an abstraction that serves the perpetuation of the control of meaning and restricting the potentialities towards action to a (dominant) minority. We argue that voices displayed on the margins of the Western world contain the seeds to express ontological and epistemological pluralities, to deploy the yonder on the centre and to produce a radical critique of the western drawing up of ongoing upheavals. Critical ethnography offers opportunities to promote reflexive practices, whose intentions are to shed light on and destabilize domination, and to pluralize the understanding of the anthropocenic situations we dwell in too. By doing so, the urgency raised by anthropocenic framings has to deal with the temporality of historical (for example, colonialism, imperialism) and ongoing harms (Gergan 2017) to embrace a care for slow violence (Nixon 2011) and focus attention on unsensational violence that exacerbates the vulnerabilities of ecosystems, non-human and disempowered individuals and groups.

References

Baldwin, A., & Erickson, B. (2020). Introduction: Whiteness, Coloniality, and the Anthropocene. *Environment and Planning D: Society and Space*. 38(1), 3–11. DOI: 10.1177/0263775820904485.

Bauer, A. M. (2016). Questioning the Anthropocene and Its Silences: Socioenvironmental History and the Climate Crisis. *Resilience: A Journal of the Environmental Humanities*. 3, 403. DOI: 10.5250/resilience.3.2016.0403.

Baztan, J., Cordier, M., Huctin, J.-M., Zhu, Z., & Vanderlinden, J.-P. (2017). Life on Thin Ice: Insights from Uummannaq, Greenland for Connecting Climate Science with Arctic Communities. *Polar Science*. 13, 100–108. DOI: 10.1016/j.polar.2017.05.002.

Bleier, R. (1986). *Feminist Approaches to Science*. New York: Pergamon Press.

Bonneuil, C., & Fressoz, J.-B. (2016). *L'événement anthropocène: La Terre, l'histoire et nous*, Nouvelle éd. révisée et augmentée. Paris: Éditions Points.

Bravo, M. T. (2009). Voices from the Sea Ice: The Reception of Climate Impact Narratives. *Journal of Historical Geography*. 35(2), 256–78. DOI: 10.1016/j.jhg.2008.09.007.

Bremer, S., Wardekker, A., Dessai, S., et al. (2019). Toward a Multi-Faceted Conception of Co-Production of Climate Services. *Climate Services*. 13, 42–50. DOI: 10.1016/j.cliser.2019.01.003.

Brown, L., & Strega, S. (2005). *Research as Resistance: Critical, Indigenous and Anti-Oppressive Approaches*. Toronto: Canadian Scholar's.

Cameron, E. S. (2012). Securing Indigenous Politics: A Critique of the Vulnerability and Adaptation Approach to the Human Dimensions of Climate Change in the Canadian Arctic. *Global Environmental Change*. 22(1), 103–14. DOI: 10.1016/j.gloenvcha.2011.11.004.

Cohen, J., Screen, J. A., Furtado, J. C., et al. (2014). Recent Arctic Amplification and Extreme Mid-Latitude Weather. *Nature Geoscience*. 7(9), 627–37. DOI: 10.1038/ngeo2234.

Crutzen, P. J. (2006). The 'Anthropocene'. In E. Ehlers & T. Krafft, eds., *Earth System Science in the Anthropocene*. Berlin: Springer, 13–18.

Crutzen, P. J., & Stoermer, E. F. (2000). The 'Anthropocene'. *Global Change Newsletter*. 41, 17–18.

Dabiri, E. (2021). *What White People Can Do Next: From Allyship to Coalition*. London: Penguin Books.

Davis, H., & Todd, Z. (2017). On the Importance of a Date, or Decolonizing the Anthropocene. *ACME: An International Journal for Critical Geographies*. 16(4), 761–80.

Denzin, N. K., & Lincoln, Y. S. (2018). *The SAGE Handbook of Qualitative Research*, 5th ed. Los Angeles, CA: Sage.

Escobar, A. (2014). *Sentipensar con la tierra: Nuevas lecturas sobre desarrollo, territorio y diferencia*. Medellín: Ediciones Unaula.

Fox-Kemper, B., Hewitt, H. T., Xiao, C., et al. (2021). Ocean, Cryosphere and Sea Level Change. In Intergovernmental Panel on Climate Change (IPCC), ed., *Climate Change 2021: The Physical Science Basis. Contribution of Working Group I to the Sixth Assessment Report of the Intergovernmental Panel on Climate Change*. Cambridge: Cambridge University Press, 1211–362.

Gergan, M. D. (2017). Living with Earthquakes and Angry Deities at the Himalayan Borderlands. *Annals of the American Association of Geographers*. 107(2), 490–98. DOI: 10.1080/24694452.2016.1209103.

Grove, K., & Chandler, D. (2017). Introduction: Resilience and the Anthropocene: The Stakes of 'Renaturalising' Politics. *Resilience*. 5(2), 79–91. DOI: 10.1080/21693293.2016.1241476.

Gupta, J., Liverman, D., Prodani, K., et al. (2023). Earth System Justice Needed to Identify and Live within Earth System Boundaries. *Nature Sustainability*. 6(6), 630–38. DOI: 10.1038/s41893–023-01064-1.

Haraway, D. (2015). Anthropocene, Capitalocene, Plantationocene, Chthulucene: Making Kin. *Environmental Humanities*. 6(1), 159–65. DOI: 10.1215/22011919-3615934.

Hastrup, K. (2018). A History of Climate Change: Inughuit Responses to Changing Ice Conditions in North-West Greenland. *Climatic Change*. 151(1), 67–78. DOI: 10.1007/s10584–016-1628-y.

Holm, L. K. (2010). Sila-Inuk: Study of the Impacts of Climate Change in Greenland. In I. Krupnik, C. Aporta, S. Gearheard, G. J. Laidler, & L. Kielsen Holm, eds., *SIKU: Knowing Our Ice*. Dordrecht: Springer Netherlands, 145–60.

Ingold, T. (2018). Between Science and Art: An Anthropological Odyssey. In S. Bergmann & F. J. Clingerman, eds., *Arts, Religion, and the Environment*. Boston: Brill, 96–114.

Intergovernmental Panel on ClimateChange. (2022). *The Ocean and Cryosphere in a Changing Climate: Special Report of the Intergovernmental Panel on Climate Change*. Cambridge: Cambridge University Press. DOI: 10.1017/9781009157964.

Jonsson, S. (2016). Where Humanism Finds Its Ends: Lessons from Pia Arke and Katarina Pirak Sikku on the Difficulty of Narrating the Arctic. *Studies in Travel Writing*. 20(3), 226–36. DOI: 10.1080/13645145.2016.1212502.

Karlsdóttir, A., Huctin, J.-M., Gherardi, J.-M., Vanderlinden, J.-P., & Sandré, T. (2023). Working with and for Arctic Communities on Resilience Enhancement. In *Arctic Year Book 2023*. https://arcticyearbook.com/images/yearbook/2023/Scholarly_Papers/12_Karlsdttir_AY2023.pdf

Kirksey, S. E., & Helmreich, S. (2010). The Emergence of Multispecies Ethnography. *Cultural Anthropology*. 25(4), 545–76. DOI: 10.1111/j.1548-1360.2010.01069.x.

Krauß, W. (2020). Narratives of Change and the Co-Development of Climate Services for Action. *Climate Risk Management*. 28, 100217. DOI: 10.1016/j.crm.2020.100217.

Krupnik, I., & Jolly, D. (2002). *The Earth Is Faster Now: Indigenous Observations of Arctic Environment Change*. Fairbanks, AK: Arctic Research Consortium of the United States.

Lewis, S. L., & Maslin, M. A. (2015). Defining the Anthropocene. *Nature*. 519(7542), 171–80. DOI: 10.1038/nature14258.

Luke, T. W. (2020). Tracing Race, Ethnicity, and Civilization in the Anthropocene. *Environment and Planning D: Society and Space*. 38(1), 129–46. DOI: 10.1177/0263775818798030.

Madison, D. S. (2020). *Critical Ethnography: Method, Ethics, and Performance*. Los Angeles, CA: Sage.

Montford, K. S., & Taylor, C., eds. (2020). *Colonialism and Animality: Anti-colonial Perspectives in Critical Animal Studies*. Routledge Taylor & Francis Group.

NCAI & MSU. (2012). *'Walk Softly and Listen Carefully': Building Research Relationships with Tribal Communities*. Washington, DC: Authors.

Nixon, R. (2011). *Slow Violence and the Environmentalism of the Poor*. Cambridge, MA: Harvard University Press.

Noblit, G. W., Flores, S. Y., & Murillo, E. G. (2004). *Postcritical Ethnography: Reinscribing Critique*. Cresskill, NJ: Hampton Press.

Ottinger, G. (2017). Making Sense of Citizen Science: Stories as a Hermeneutic Resource. *Energy Research & Social Science*. 31, 41–49. DOI: 10.1016/j.erss.2017.06.014.

Overland, J., Francis, J. A., Hall, R., et al. (2015). The Melting Arctic and Midlatitude Weather Patterns: Are They Connected? *Journal of Climate*. 28(20), 7917–32. DOI: 10.1175/JCLI-D-14-00822.1.

Preece, J. R., Mote, T. L., Cohen, J., et al. (2023). Summer Atmospheric Circulation over Greenland in Response to Arctic Amplification and Diminished Spring Snow Cover. *Nature Communications*. 14(1), 3759. DOI: 10.1038/s41467–023-39466-6.

Previdi, M., Smith, K. L., & Polvani, L. M. (2021). Arctic Amplification of Climate Change: A Review of Underlying Mechanisms. *Environmental Research Letters*. 16(9), 93003. DOI: 10.1088/1748-9326/ac1c29.

Richardson, K., Steffen, W., Lucht, W., et al. (2023). Earth beyond Six of Nine Planetary Boundaries. *Science Advances*. 9(37), eadh2458. DOI: 10.1126/sciadv.adh2458.

Rivera Andía, J. J. (2019). *Non-Humans in Amerindian South America: Ethnographies of Indigenous Cosmologies, Rituals and Songs*. New York: Berghahn.

Rockström, J., Gupta, J., Qin, D., et al. (2023). Safe and Just Earth System Boundaries. *Nature*. 619(7968), 102–11. DOI: 10.1038/s41586–023-06083-8.

Roosvall, A., & Tegelberg, M. (2013). Framing Climate Change and Indigenous Peoples: Intermediaries of Urgency, Spirituality and De-Nationalization. *International Communication Gazette*. 75(4), 392–409. DOI: 10.1177/1748048513482265.

Sandré, T., Wardekker, A., & Gherardi, J. (2024). While Waiting for the Sea Ice: Stories of Changes from Ittoqqortoormiit. In S. Bremer & A. Wardekker, eds., *Changing Seasonality: How Communities Are Revising Their Seasons*. Berlin: De Gruyter, 61–68.

Sealey-Huggins, L. (2018). The Climate Crisis Is a Racist Crisis: Structural Racism, Inequality and Climate Change. In A. Johnson, R. Joseph-Salisbury, & B. Kamunge, eds., *The Fire Now: Anti-Racist Scholarship in Times of Explicit Racial Violence*. London: Zed Books, 99–113.

Serreze, M. C., & Barry, R. G. (2011). Processes and Impacts of Arctic Amplification: A Research Synthesis. *Global and Planetary Change*. 77(1), 85–96. DOI: 10.1016/j.gloplacha.2011.03.004.

Skjervedal, A.-S. (2021). Building Bridges between Research and Community: Arctic Hub. arctichub.gl/building-bridges-between-research-and-community/.

Smith, L. T. (2021). *Decolonizing Methodologies: Research and Indigenous Peoples*. London: Zed Books.

Stephen, K. (2018). Societal Impacts of a Rapidly Changing Arctic. *Current Climate Change Reports*. 4(3), 223–37. DOI: 10.1007/s40641–018-0106-1.

Sultana, F. (2022). The Unbearable Heaviness of Climate Coloniality. *Political Geography*. 99, 102638. DOI: 10.1016/j.polgeo.2022.102638.

Sultana, F. (2023). Decolonizing Climate Coloniality. In R. Solnit & T. Young Lutunatabua, eds., *Not Too Late: Changing the Climate Story from Despair to Possibility*. Chicago: Haymarket Books, 58–65.

Tuana, N. (2022). *Racial Climates, Ecological Indifference: An Ecointersectional Analysis*. New York: Oxford University Press.

Vanderlinden, J.-P., Baztan, J., Chouinard, O., et al. (2020). Meaning in the Face of Changing Climate Risks: Connecting Agency, Sensemaking and Narratives of Change through Transdisciplinary Research. *Climate Risk Management*. 29, 100224. DOI: 10.1016/j.crm.2020.100224.

Willerslev, R. (2007). *Soul Hunters: Hunting, Animism, and Personhood among the Siberian Yukaghirs*. Berkeley: University of California Press.

Part IV

Biosphere Integrity

8

Introduction to the Biosphere Integrity Boundary

SIMON P. MEISCH

In 2019, Robert Watson, the Chair of the Intergovernmental Science-Policy Platform on Biodiversity and Ecosystem Services (IPBES), warned that '[t]he loss of species, ecosystems and genetic diversity is already a global and generational threat to human well-being. Protecting the invaluable contributions of nature to people will be the defining challenge of decades to come' (IPBES 2019). This statement highlights the importance of biodiversity to humanity, for example, in terms of crop security. Such warnings are already the backdrop to blockbuster films such as Christopher Nolan's *Interstellar* (2014), in which humanity in a not-too-distant future depends on a few crops and comparatively thin genetic diversity. Reliance on those few crops, increasingly affected by blight in that film, threatens human survival on Earth. The basic idea is not an alternate reality but a foreseeable future. In times of massive environmental change, a wide range of crops would provide fallback options for food security. In its 2019 Global Assessment Report, IPBES also warned that around one in four species is at risk of extinction (Kolbert 2014). At the current rate, this would be one of the worst extinctions in Earth's history, with many mammals used for food and agriculture also at risk (IPBES 2019).

The perspective of the planetary boundaries framework is that the loss of biodiversity has multiple negative impacts on the functioning of the Earth system and thus the preservation of a safe operating space for humanity. In the framework's 2023 update, the boundary on biosphere integrity is 'markedly exceeded', and conditions are moving into a zone of high risk (Richardson et al. 2023: 3). Currently, the boundary on biosphere integrity has two control variables: *genetic diversity* and *functional diversity*.

Genetic diversity refers to the 'maximum extinction rate compatible with preserving the genetic basis of the biosphere's ecological complexity'. The boundary value is set at less than 10 extinctions per million species-years (E/MSY), but with the aspiration of reducing it to one E/MSY. The current value is at more than 100 E/MSY (Richardson et al. 2023: 3). Since the 2023 update to the planetary boundaries framework, *functional diversity* has referred to the energy and material flows into the biosphere, also known as net primary production (NPP). NPP can be explained using photosynthesis as an example. Plants (primary producers) use abiotic components (e.g., CO_2 or nitrogen) and energy (i.e., solar light) to produce energy stored in complex organic compounds (e.g., carbohydrates, fats, and proteins). Plants need some of this energy for their own metabolism. So, the energy difference between what they produce and what they need is the NPP, or the net gain over a period. Almost all other organisms depend on the energy sources provided by NPP (Andersen & Quinn 2020). The boundary value sets a limit to the human appropriation of NPP (HANPP) to ensure the functioning of the biosphere. The boundary value is set at less than 10 per cent human appropriation of pre-industrial Holocene NPP, meaning that more than 90 per cent of NPP remains in the biosphere. With 30 per cent HANPP, the current value is already in a zone of high risk, that is, beyond 20 per cent HANPP (Richardson et al. 2023: 3–6).

From an ethical and epistemological perspective, 'biodiversity' is an interesting term. Scientifically speaking, it is about '[t]he variability among living organisms from all sources including terrestrial, marine and other aquatic ecosystems and the ecological complexes of which they are a part. This includes variation in genetic, phenotypic, phylogenetic, and functional attributes, as well as changes in abundance and distribution over time and space within and among species, biological communities and ecosystems' (Díaz et al. 2015: 12). This definition, however, does not fully capture the meaning of the term, which is not only about a quantifiable diversity of species, biological communities, and ecosystems. It also implies that such a diversity is something morally good and in need of political protection. As such, it is a concept composed of scientific forms of knowledge as well as value judgements – something Thomas Potthast (2015) called an 'epistemic-moral hybrid'.

In Chapter 9, Uta Eser traces precisely this specific conceptual history of the term 'biodiversity', which emerged in the borderland between biology, ethics, and politics. She takes a critical look at the planetary boundary itself and how it has developed since 2009. Eser focuses on what value judgements are embedded in the framework's conceptualization of biodiversity, which she finds too reductionist. In Section 9.4, she suggests more inclusive ways of framing biodiversity that recognize the many different human–nature relationships.

Makere Stewart-Harawira's Chapter 10 uses the case of Aotearoa New Zealand to discuss how ontologically distinct human–nature relationships are socially and politically negotiated. Three key points emerge from her text. First, most biodiversity hotspots are located on the lands of Indigenous communities, whose understandings of thriving human–nature relationships are often ontologically different from those of an international political community whose knowledge is inspired by Western scientific concepts. Second, there is a need for more inclusive framings of biodiversity that recognize the onto-cosmologies of Indigenous communities and support interspecies justice and the right of more-than-human species to thrive. Third, her case illustrates the fragility of governance approaches and how political achievements are threatened by a movement of extreme right-wing political parties.

References

Andersen, C. B., & Quinn, J. (2020). Human Appropriation of Net Primary Production. In M. I. Goldstein & D. A. DellaSalla, eds., *Encyclopedia of the World's Biomes*. Amsterdam: Elsevier, 5:22–28. DOI: 10.1016/B978-0-12-409548-9.12434-0.

Díaz, S., Demissew, S., Carabias, J., et al. (2015). The IPBES Conceptual Framework: Connecting Nature and People. *Current Opinion in Environmental Sustainability*. 14, 1–16. DOI: 10.1016/j.cosust.2014.11.002.

IPBES (2019). Media Release: IPBES Global Assessment Preview. www.ipbes.net/news/ipbes-global-assessment-preview.

Kolbert, E. (2014). *The Sixth Extinction: An Unnatural History*. First Picador ed. New York: Picador, Henry Holt & Company.

Potthast, T. (2015). Ethics and Sustainability Science beyond Hume, Moore and Weber: Taking Epistemic-Moral Hybrids Seriously. In S. P. Meisch, J. Lundershausen, L. Bossert, & M. Rockoff, eds., *Ethics of Science in the Research for Sustainable Development*. Baden-Baden: Nomos, 129–152.

Richardson, K., Steffen, W., Lucht, W., et al. (2023). Earth beyond Six of Nine Planetary Boundaries. *Science Advances*. 9(37), eadh2458. DOI: 10.1126/sciadv.adh2458.

9

Ethical Perspectives on Biodiversity as Planetary Boundary

UTA ESER

Planetary boundaries are defined from a functional perspective. Only biodiversity appears as also having a value in its own. The original publication contains the laconic note: 'Massive loss of biodiversity [is] unacceptable for ethical reasons' (Rockström et al. 2009). Remarkably, this brief reference is only to be found in a table that characterizes the nine proposed planetary boundaries. Nowhere else in the document do the authors name ethical concerns. They do mention that establishing values for the respective boundaries depends on value judgements or normative judgements. However, the idea that crossing a particular boundary is morally wrong per se only occurs in relation to biodiversity.

For scholars who study 'biodiversity' from a theoretical perspective, this reference to ethics comes as no surprise. The concept of biodiversity has been morally impregnated from the beginning. It was coined at the interface of biology and politics to send a wake-up call to decision-makers and has never seen itself as a value-free scientific concept (Eser 2001). Nevertheless, like the closely related concept of ecosystem services, it appears in public discourse with the full authority of scientific knowledge. The concept of planetary boundaries presents itself as a 'science-based analysis', too. However, under the guise of 'Science' the values that underpin conservation efforts largely remain hidden. My chapter aims to address this shortcoming.

Section 9.1 will characterize the term 'biodiversity' as a concept that is situated in the borderland between biology, ethics, and politics and has factual, evaluative, and normative dimensions. Section 9.2 presents critique that has been made of biodiversity as planetary boundary, and how this has been addressed in more recent versions of the framework. The main section focuses on the different values of biodiversity, which I consider to be insufficiently reflected in the PB approach. Besides its instrumental value for humanity's safe operating space, the diversity of living beings also has intrinsic and relational value. I will present these different value concepts and discuss them with regard to the PB framework (Section 9.3).

Finally, I will argue for a more inclusive perspective on biodiversity that recognizes and takes into account differences between the manifold ways in which humans live from, with, in, and as nature (Section 9.4).

9.1 Biodiversity: A Boundary Concept

The term 'biodiversity' comes from conservation biology (Maclaurin & Sterelny 2008). It is a shorthand of 'biological diversity' or 'biotic diversity' (Faith 2021) and was coined in the course of the *Washington Conference on BioDiversity* in 1986 in order to raise public awareness of the ongoing mass extinction (Wilson 1988). In extensive interviews with protagonists of the concept, David Takacs, a scholar of science studies, has shown that the invention of the term was associated with remarkable boundary work (Takacs 1996). Both the established separation of science and politics and the common concepts of scientific objectivity and value-neutrality were explicitly challenged in the scientific advocacy for a more resolute policy in favour of biodiversity.

The *Conference on BioDiversity* itself was a political, rather than a scientific, endeavour. Dan Janzen, one of the many renowned biologists involved, remembers in his interview: 'That was an explicit political event, explicitly designed to make Congress aware of this complexity of species that we're losing . . . . A lot of us went to that talk with a political mission' (Takacs 1996: 37). Part and parcel of the message that biodiversity is under threat is the message that biodiversity is important, that is to say: It has value. Such a statement is obviously at odds with the conventional understanding of science as being value-neutral. Accordingly, Takacs' interviewees question this scientific norm. 'One of the silly things is the idea that science is somehow separate from society. There is no value-neutral science,' says Paul Ehrlich (Takacs 1996: 181). With regard to both intricate knowledge and imminent danger, biologists feel obliged to bridge the gap between 'Is' and 'Ought' that is so well established in philosophy.

> In my view, it's preposterous for people who have spent their entire life immersed in a problem to present only a value-neutral thing. And politicians don't want you, ordinarily – they want not only to know what you think the situation is, they want at least suggestions on what society ought to do about it . . . . [I]f you're standing in a building that's burning down you don't just stand up and give measurements of the temperature and so on. You say, 'Let's get the fuck out of here' in addition. (Takacs 1996: 179)

This combination of expertise and urgency drives the protagonists of the planetary boundaries framework, too.

In his seminal book, Takacs illustrates that the commitment of biologists to biodiversity was accompanied by an attempt to shift the boundaries between science, politics, and ethics. Biodiversity was associated with 'a new ethic, in which biodiversity's multiplicity of values will be respected, appreciated, and perhaps even worshiped' (Takacs 1996: 9).

Given that history, 'biodiversity' can be regarded as a thick concept. 'Thick', in philosophy, are concepts that have a substantial degree of descriptive content but also are value-laden (Williams 1985). With regard to biological concepts like biodiversity, ecology, or evolution, such thick concepts have been named 'epistemic-moral hybrids' (Potthast 2007). In view of this hybrid character, biodiversity loss is not only a scientific matter of fact. The statement that we are losing biodiversity at unprecedented rates per se involves the value judgement that this loss is bad, and the normative statement that we ought to do something to halt it. In other words, the term has factual, evaluative, and normative dimensions.

Since its invention in the 1980s, the term 'biodiversity' has had a rapid career, in both the scientific and political arenas. Within science, the question of how to define biodiversity and how to measure it fills volumes. In the political arena, the term became the focus of global conservation efforts with the *Convention on Biological Diversity* (CBD), adopted at the first *Earth Summit* in Rio de Janeiro in 1992. The close entanglement of political and scientific dimensions of the term makes it both important and difficult to define biodiversity. According to the CBD, '"[b]iological diversity" means the variability among living organisms from all sources . . .; this includes diversity within species, between species and of ecosystems' (1992: Art. 2). The UNESCO funded 'Man and Biosphere' Program described it as living variability: 'Biodiversity is the property of living systems of being distinct, that is different, unlike. Biological diversity or biodiversity is defined here as the property of groups or classes of living entities to be varied' (Solbrig 1991: 9). An explanatory leaflet produced by the CBD's clearinghouse mechanism for communication purposes explained: 'The term "biological diversity" . . . is defined in terms of genes, species, and ecosystems, which are the outcome of over 3,000 million years of evolution. The human species depends on biological diversity for its own survival. Thus, the term can be considered a synonym for "life on earth"' (CBD 1999). When asked for a definition of biodiversity, David Ehrenfeld stated, 'I don't have a definition of biodiversity . . . I think it's one of those wonderful catchwords like sustainable development, that, because it's vaguely defined, has a broad appeal, like motherhood' (Takacs 1996: 46–47).

This indeterminacy of content contrasts with scientific exactness but is precisely why it enables practical cooperation. It is, therefore, helpful to understand biodiversity as a boundary object that is 'both plastic enough to adapt to local needs and constraints of the several parties employing them, yet robust enough to maintain a common identity across sites' (Star 1989: 46). While such objects can be very useful enablers of cooperation, it is hardly possible to pin them down to a single metric within a science-based framework. This becomes apparent when we look at the scientific debate on biodiversity as a planetary boundary.

9.2 Biodiversity as Planetary Boundary

How can an ill-defined concept as biodiversity be used within a framework that intends to be science-based? In particular: How is it possible to formulate a reliable boundary that separates the safe operating space from the danger zone? What are appropriate metrics for such a boundary? And how to deal with the huge amount of non-knowledge with regard to both the numbers of existing and endangered species and their relevance for ecosystem functioning?

The planetary boundaries framework uses extinction rate as a proxy for biodiversity loss. In calculating the boundary and the zone of uncertainty, the original paper made use of a rather rough rule of thumb: 'The lower and upper bounds of extinction rates in the fossil record . . . provide the best estimates of the background extinction rates that have historically conserved global biodiversity' (Rockström et al. 2009). With regard to this natural extinction rate, the authors set the boundary at an extinction rate of less than 10 extinctions per million species per year (E/MSY), and the zone of uncertainty at 10–100 E/MSY. However, they admit, 'it remains very difficult to define a boundary level for the rate of biodiversity loss that, if transgressed for long periods of time, could result in undesired, non-linear Earth System change at regional to global scales' (Rockström et al. 2009).

Within conservation biology, there have been controversial debates for years about whether biodiversity is an end or a means of conservation, and how biodiversity helps in the selection and prioritization of conservation objects (Maclaurin & Sterelny 2008; NRC 1999; Sarkar 2005). Due to massive conceptual and methodological difficulties, some even argued for abandoning the term altogether (Santana 2014). Accordingly, the biodiversity community's response to the planetary boundaries framework was mixed. The reactions range from constructive criticisms (Mace et al. 2014) to harsh verdicts (Montoya et al. 2018). Mace et al. (2014) argued that extinction rate and species richness were weak metrics and suggested alternative approaches to determine biodiversity loss boundaries. In response to their critique, the PB framework was adjusted and the initial boundary 'biodiversity loss' was replaced by 'biosphere integrity' (Steffen et al. 2015). Biosphere integrity has two components: genetic diversity, with extinction rate as control variable, and functional diversity with a Biodiversity Intactness Index (BII) as control variable. The updated table no longer contains the remark on ethics but the note: 'These are interim control variables until more appropriate ones are developed' (Steffen et al. 2015), thus acknowledging practical and theoretical challenges of finding appropriate control variables at regional or global scales.

Despite these adjustments, some still doubt the practical utility of the concept. '[N]otions of planetary boundaries add no insight into our understanding of the threats to biodiversity and ecosystem functioning, have no evidence to support them, are too vague for use by those who manage biodiversity, and promote pernicious policies' (Montoya et al. 2018). In addition to practical conservation concerns, the authors of this verdict also express moral and political concerns. In the face of boundaries that are difficult to operationalize, they consider the talk of a 'safe operating space' problematic because it could encourage policymakers to assume 'if the planet is not obviously collapsing around us, then surely we can continue to deplete it' (Montoya et al. 2018).

I will leave the question of a proper metric for biodiversity here and turn to the ethical implications of a concept that considers the diversity of living beings exclusively in terms of a safe operating space for humanity.[1]

9.3 The Value(s) of Biodiversity

'Our primary reason for including biological diversity as a planetary boundary is its role in providing ecological functions that support biophysical sub-systems of the Earth' (Rockström et al. 2009). Apart from the marginal note in the first table, the planetary boundaries framework considers biodiversity for functional reasons. The only value of biodiversity taken into account is its instrumental value: Biodiversity is valued as a means to humanity's continued existence. Genetic diversity is appreciated as a database that allows adaptation to an uncertain future, and functional diversity is treasured for its contribution to ecosystem services (Steffen et al. 2015). There is no doubt that this functional perspective contains essential reasons for the protection of biological diversity. However, it ignores other, equally important reasons: it does not take into account the possible intrinsic value of other than human living beings (*Instrumental vs. Intrinsic value*), nor does it adequately address the diversity of meaningful relationships between humans and nature (*Relational Value*).

[1] The original manuscript of this article was submitted before the latest update was published. Richardson et al. (2023) replace the Biodiversity Intactness Index with a computable proxy for net primary production (NPP), thus defining the boundary of functional integrity as 'a limit to the human appropriation of the biosphere's NPP (HANPP) as a fraction of its Holocene NPP'. A discussion of this new metric cannot be provided here. For critics, it probably does little to change the scepticism regarding its practical usefulness. Regarding the ethical implications, it also focuses on functionality and therefore reflects the same instrumental perspective.

9.3.1 Instrumental vs. Intrinsic Value

The planetary boundaries framework links biodiversity to human well-being. This link has been emphasized ever since the Millennium Ecosystem Assessment (2005). This had helped the concept of ecosystem services, developed in the 1990s, to achieve a breakthrough. Biodiversity, according to this approach, contributes significantly to supporting, provisioning, regulating, and cultural ecosystem services (Díaz et al. 2006). The purpose of the concept of ecosystem services was to enable communication between economics, ecology, and politics (Eser 2016). However, it met (and still meets) with criticism among conservationists. A commentary published in the prestigious journal *Nature* that warned from 'selling out on nature' and pleaded 'for returning to the protection of nature for nature's sake' (McCauley 2006) surely spoke from the heart of many conservationists. The notion that all life has intrinsic value, independent from whatever human interests, goals, or preferences are, is inextricably linked to the concept of biodiversity, as I have shown earlier. The conviction that nature has intrinsic value has to be considered a core belief of most conservationists. It is, therefore, reasonable to assume that the harsh criticism of biodiversity as planetary boundary demonstrated earlier has one of its roots here.

Acknowledging the great importance of the idea of intrinsic value, however, does not necessarily mean to attribute moral intrinsic value to nature. In a seminal analysis of 'value systems and attitudes toward nature', the German philosopher Angelika Krebs discriminated between three possible understandings of 'intrinsic value': absolute intrinsic value, moral intrinsic value, and eudemonic intrinsic value (Krebs 1999). The distinction between moral intrinsic value of nature and other non-instrumental values like aesthetic, emotional, or spiritual value was formative for the German discourse on conservation ethics (Eser & Potthast 1999; Muraca 2011; Ott 2020). In the wake of the activities of the Intergovernmental Science-Policy Platform on Biodiversity and Ecosystem Services (IPBES) it also became recognized in the global biodiversity discourse. Non-instrumental intrinsic values have gained greater attention in the international conservation community as so-called relational values.

9.3.2 Relational Value

The Intergovernmental Science-Policy Platform on Biodiversity and Ecosystem Services (IPBES) was established in 2012 as an independent intergovernmental body. Its mission is 'to provide policy-relevant knowledge and to catalyze the implementation of knowledge-based policies at all levels in government, the private sector and civil society' (IPBES 2022a). To facilitate a comprehensive understanding of the complex interactions between the natural world and human societies, IPBES developed a conceptual framework (CF) that integrates not only different types of knowledge but also different values. The original CF contained six elements: nature, nature's benefits to people, good quality of life, anthropogenic assets, institutions, and direct drivers (Díaz et al. 2015). The term 'nature's benefits to people' has later been replaced by 'nature's contributions to people' (NCP), not least for the less instrumentalist and more comprehensive notion of NCP (Pascual et al. 2017). In addition to instrumental value,

> NCP can embody symbolic relationships with natural entities to the extent that such relationships are inextricably linked to people's sense of identity and spirituality, to a meaningful life and to 'doing the right thing'. In this case NCP are associated with relational values, that is values that do not directly emanate from nature but are derivative of our relationships with it and our responsibilities towards it. (Pascual et al. 2017)

The term 'relational values' was promoted to overcome the false dichotomy between instrumental value and intrinsic value (Chan et al. 2016; Luck et al. 2012). Besides nature's instrumental value in providing benefits to people and nature's intrinsic value, the concept of relational value wants to raise 'attention to other ways that value is expressed and realized by people' (Chan et al. 2016). By its integration into the IPBES process, the value of non-instrumental relationships between people and nature finally received broader attention by conservationists.

The point that nature and the diversity of living beings matter to humans in other than instrumental ways does not come into view in the functional perspective of the planetary boundaries framework. The idea of a *Safe Operating Space for Humanity* suggests that the major concern of environmental politics is the survival of humankind. However, as early as 1972 the *Club of Rome* stated in its afterword to the report *Limits to Growth:* 'The crux of the matter is not only whether the human species will survive, but even more whether it can survive without falling into a state of worthless existence' (Meadows et al. 1972: 200). What makes human lives worth living cannot be defined by the natural sciences alone. There are philosophical traditions devoted to exploring the meaning of a good life and the role nature plays in it. For but one example, philosopher Martha Nussbaum names 'Being able to live with concern for and in relation to animals, plants, and the world of nature' as 1 out of 10 central human functional capabilities (Nussbaum 2000). There are also contributions by the social and political sciences that name minimal standards of a decent human life. With regard to the planetary boundaries, Kate Raworth expanded the 'safe operating space' to a 'safe and just space for humanity' (Raworth 2012). The doughnut she suggests is meant as a new model of prosperity that not only respects planetary boundaries but has equity at its heart. In a similar vein, researchers from the social sciences and humanities recently pleaded for a move from planetary to societal boundaries (Brand et al. 2021). This focus on societal aspects brings me to my major concern: the unifying concept of 'humanity' is an abstraction and an over-simplification that needs to be broken down. Instead, we have to make visible and acknowledge the different ways in which different people, cultures, economies, and societies relate to nature by taking a more inclusive perspective on biodiversity.

9.4 A More Inclusive Perspective on Biodiversity

For decades, environmental discourse has been framed as a conflict between humans and nature. Such a framing ignores that in environmental protection and nature conservation we are usually not dealing with conflicts between 'man' and 'nature', but with conflicts between different people with different interests and needs – and different potentials to achieve those. One consequence of this binary framing is that the exclusion of people is often seen as the best way to protect nature. One recent example of such a preference is the Kumming-Montreal biodiversity framework adopted at the 15th Conference of the Parties to the CBD in Montreal in December 2022. It established the 30 by 30-target of global biodiversity conservation, to '[e]nsure and enable that by 2030 at least 30 per cent of terrestrial, inland water, and of coastal and marine areas, . . . are effectively conserved and managed through . . . systems of protected areas and other effective area-based conservation' (CBD 2022). Only after fierce debates and in response to serious concerns with regard to: Indigenous people and local communities, the document finally explicitly recognizes and respects the rights of these groups. As long as we conceive of biodiversity loss as a conflict between humans and nature, crucial distinctions between different lifestyles and economies and their impacts on biodiversity cannot come into view – and hence cannot be adequately addressed in the political arena.

This critique of generalization also applies to the concept of planetary boundaries. '*Human activities* increasingly influence the Earth's climate . . . and ecosystems', '*humans* are effectively pushing the planet outside the Holocene range of variability . . .', '*humanity* has already entered deep into a danger zone' (Rockström et al. 2009, emphasis added): These quotes demonstrate the shortcomings of a planetary perspective mentioned earlier. Different human activities influence climate and ecosystems in very different ways – not 'human activities' as such but particular activities of particular persons. Not 'humans' are pushing 'the planet' outside of the comfort zone, but some humans with specific lifestyles and policies are pushing other humans – and other species – 'deep into the danger zone'.

Talk of 'humanity' tends to ignore morally relevant differences between originators and victims of the depletion of nature. Those who are responsible for the decline of species and those who suffer from this decline more often than not are different persons. As the *Human Development Report 1998* pointed out, responsibility for environmental damage, including biodiversity loss, lies mainly with the top global consumers (UNDP 1998). Increasing consumption not only puts strains on the environment but also reflects and increases global inequality. The richest fifth of the world's population account for 86 per cent of total private consumption expenditures – and the related environmental and social burdens (UNDP 1998). On the other hand, consequences of biodiversity loss 'will be felt disproportionately by the poor' (Díaz et al. 2006). Degradation of the Earth's life support systems 'is threatening the fulfillment of basic needs and aspiration of humanity as a whole but especially, and most immediately, those of the most disadvantaged segments of society' (Díaz et al. 2006). Both with regard to the polluter pays principle and with regard to responsibility for particularly vulnerable groups, the unifying talk of humanity is inadequate from an ethical perspective.

Talk of 'humanity' also tends to ignore conflicting interests: on the one hand, conflicts between interests in different kinds of use (e.g. provision of food vs. provision of housing); on the other hand, conflicts between interests in use and interests in other kinds of relations with nature (e.g. aesthetic, cultural, or spiritual). In order to overcome the unsatisfactory frontline between humans and nature and to recognize the diversity of human-nature relationships, IPBES initiated an assessment on the multiple values of nature and its contributions to human well-being in 2018. Over a period of four years, a multidisciplinary and multinational team of nearly a hundred authors scrutinized different concepts of value and methods of evaluation, analysed the values underlying current decision-making, and inquired if the inclusion of multiple values might lead to better decision outcomes. The results, published in July 2022, provide extensive evidence that 'decisions based on [a] narrow set of market values of nature underpin the global biodiversity crisis' (IPBES 2022b). The *Summary for Policy Makers* establishes that decisions are (too) often made on the basis of monetary values, neglecting other values. 'Despite the diversity of nature's values, most policymaking approaches have prioritized a narrow set of values at the expense of both nature and society, as well as future generations, and have often ignored values associated to indigenous peoples and local communities' worldviews' (IPBES 2022c). According to the *Values Assessment*, dominance of market values and ignorance of non-instrumental values are to be seen as root causes of the biodiversity crisis. To overcome this restriction, the assessment suggests a value typology that leaves room for different ways in which humans can relate to nature. It discriminates four 'live-frames': living from, living in, living with and living as nature. *Living from nature* designates the instrumental perspective on nature as a resource. *Living with nature* emphasizes relational bonds between humans and other than human beings.

Living in nature highlights place-based connections between people and nature while *Living as nature* underlines the inseparability of humans and nature.

The findings of the IPBES Values Assessment may shed new light on the planetary boundary approach. By taking a functional perspective on biodiversity, it is clearly situated within the living-from frame. Within that frame, it provides powerful reasons for action. However, this achievement comes at the expense of neglecting other perceptions. A more inclusive perspective on biodiversity needs to take into account diverse values and to address social aspects such as power, poverty, and inequality. As long as such an inclusion is missing, it is doubtful whether the planetary boundaries approach can provide adequate solutions to the ongoing crisis of biodiversity.

References

Brand, U., Muraca, B., Pineault, É., et al. (2021). From Planetary to Societal Boundaries: An Argument for Collectively Defined Self-Limitation. *Sustainability: Science, Practice and Policy*. 17(1), 264–91. DOI: 10.1080/15487733.2021.1940754.

CBD. (1992). Convention on Biological Diversity. United Nations. 1760 UNTS 79. www.cbd.int/doc/legal/cbd-en.pdf.

CBD. (1999). The Convention about Life on Earth: What Is Biological Diversity? web.archive.org/web/20000526031244/http://www.biodiv.org/conv/leaflet.html.

CBD. (2022). Kunming-Montreal Global Biodiversity Framework. Conference of the Parties to the CBD. Fifteenth Meeting – Part II, Montreal, Canada, 7–19 December 2022. Agenda item 9A. (CBD/COP/15/L.25). www.cbd.int/doc/c/e6d3/cd1d/daf663719a03902a9b116c34/cop-15-l-25-en.pdf.

Chan, K. M. A., Balvanera, P., Benessaiah, K., et al. (2016). Opinion: Why Protect Nature? Rethinking Values and the Environment. *Proceedings of the National Academy of Sciences*. 113(6), 1462–65. DOI: 10.1073/pnas.1525002113.

Díaz, S., Demissew, S., Joly, C., Lonsdale, W. M., & Larigauderie, A. (2015). A Rosetta Stone for Nature's Benefits to People. *PLOS Biology*. 13(1), e1002040. DOI: 10.1371/journal.pbio.1002040.

Díaz, S., Fargione, J., Chapin, F. S., & Tilman, D. (2006). Biodiversity Loss Threatens Human Well-Being. *PLOS Biology*. 4(8), e277. DOI: 10.1371/journal.pbio.0040277.

Eser, U. (2001). Die Grenze zwischen Wissenschaft und Gesellschaft neu definieren: Boundary Work am Beispiel des Biodiversitätsbegriffs. *Verhandlungen zur Geschichte und Theorie der Biologie*. 7, 135–52.

Eser, U. (2016). Das Konzept der Ökosystemdienstleistungen: Ein Brückenschlag zwischen Ökologie, Ökonomie und Naturschutz. *Natur und Landschaft*. 91(9/10), 470–75.

Eser, U., & Potthast, T. (1999). *Naturschutzethik: Eine Einführung für die Praxis*. Baden-Baden: Nomos Verlagsgesellschaft. DOI: 10.5771/9783845261447.

Faith, D. (2021). Biodiversity. In E. N. Zalta, ed., *The Stanford Encyclopedia of Philosophy: Spring 2021 Edition*. plato.stanford.edu/archives/spr2021/entries/biodiversity/

IPBES. (2022a). About: What Is IPBES? www.ipbes.net/about.

IPBES. (2022b). Media Release: IPBES Values Assessment – Decisions Based on Narrow Set of Market Values of Nature Underpin the Global Biodiversity Crisis. ipbes.net/media_release/Values_Assessment_Published.

IPBES. (2022c). Summary for Policymakers of the Methodological Assessment of the Diverse Values and Valuation of Nature of the Intergovernmental Science-Policy Platform on Biodiversity and Ecosystem Services. IPBES Secretariat (IPBES/9/ L.13).

Krebs, A. (1999). *Ethics of Nature: A Map*. With a Foreword by Bernard Willimas. Berlin: De Gruyter.

Luck, G. W., Chan, K. M. A., Eser, U., et al. (2012). Ethical Considerations in On-Ground Applications of the Ecosystem Services Concept. *BioScience*. 62(12), 1020–29. DOI: 10.1525/bio.2012.62.12.4.

Mace, G. M., Reyers, B., Alkemade, R., et al. (2014). Approaches to Defining a Planetary Boundary for Biodiversity. *Global Environmental Change*. 28, 289–97. DOI: 10.1016/j.gloenvcha.2014.07.009.

Maclaurin, J., & Sterelny, K. (2008). *What Is Biodiversity?* Chicago: University of Chicago Press.

McCauley, D. J. (2006). Selling out on Nature. *Nature*. 443(7107), 27–28. DOI: 10.1038/ 443027a.

Meadows, D. H., Meadows, D. L., Randers, J., & Behrens, W. (1972). *The Limits to Growth: A Report for the Club of Rome's Project on the Predicament of Mankind*. New York: Universe Books.

Millennium Ecosystem Assessment. (2005). *Ecosystems and Human Well-Being: Biodiversity Synthesis*. A Report of the Millennium Ecosystem Assessment. Washington, DC: World Resources Institute.

Montoya, J. M., Donohue, I., & Pimm, S. L. (2018). Planetary Boundaries for Biodiversity: Implausible Science, Pernicious Policies. *Trends in Ecology & Evolution*. 33(2), 71–73. DOI: 10.1016/j.tree.2017.10.004.

Muraca, B. (2011). The Map of Moral Significance: A New Axiological Matrix for Environmental Ethics. *Environmental Values*. 20(3), 375–96. DOI: 10.3197/ 096327111X13077055166063.

National Research Council (NRC). (1999). *Perspectives on Biodiversity: Valuing Its Role in an Everchanging World*. Washington, DC: National Academy Press.

Nussbaum, M. C. (2000). *Women and Human Development: The Capabilities Approach*. Cambridge: Cambridge University Press.

Ott, K. (2020). Environmental Ethics. In T. Kirchhoff, ed., *Online Encyclopedia Philosophy of Nature / Online Lexikon Naturphilosophie*. DOI: 10.11588/oepn.2020.0.71420.

Pascual, U., Balvanera, P., Díaz, S., et al. (2017). Valuing Nature's Contributions to People: The IPBES Approach. *Current Opinion in Environmental Sustainability*. 26–27, 7–16. DOI: 10.1016/j.cosust.2016.12.006.

Potthast, T. (2007). Biodiversität, Ökologie, Evolution: Epistemisch-moralische Hybride und Biologietheorie. In T. Potthast, ed., *Biodiversität – Schlüsselbegriff des Naturschutzes im 21. Jahrhundert? Erweiterte Ergebnisdokumentation einer Vilmer Sommerakademie 48. of Naturschutz und biologische Vielfalt*. Bonn-Bad Godesberg: Bundesamt für Naturschutz, 57–88.

Raworth, K. (2012). *A Safe and Just Space for Humanity: Can We Live within the Doughnut?* Oxford: Oxfam.

Richardson, K., Steffen, W., Lucht, W., et al. (2023). Earth beyond Six of Nine Planetary Boundaries. *Science Advances*. 9(37), eadh2458. DOI: 10.1126/sciadv.adh2458.

Rockström, J., Steffen, W., Noone, K., et al. (2009). Planetary Boundaries: Exploring the Safe Operating Space for Humanity. *Ecology & Society*. 14(2), 32. DOI: 10.5751/ES-03180-140232.

Santana, C. (2014). Save the Planet: Eliminate Biodiversity. *Biology & Philosophy*. 29(6), 761–80. DOI: 10.1007/s10539-014-9426-2.

Sarkar, S. (2005). *Biodiversity and Environmental Philosophy: An Introduction*. Cambridge: Cambridge University Press. DOI: 10.1017/CBO9780511498558.

Solbrig, O. T. (1991). *Biodiversity: Scientific Issues and Collaborative Research Proposals*. Paris: UNESCO.

Star, S. L. (1989). The Structure of Ill-Structured Solutions: Boundary Objects and Heterogeneous Distributed Problem Solving. In L. Gasser & M. N. Huhn, eds., *Distributed Artificial Intelligence*. London: Pitman Publishing, 37–54.

Steffen, W., Richardson, K., Rockström, J., et al. (2015). Sustainability. Planetary Boundaries: Guiding Human Development on a Changing Planet. *Science*. 347(6223), 1259855. DOI: 10.1126/science.1259855.

Takacs, D. (1996). *The Idea of Biodiversity: Philosophies of Paradise*. Baltimore: Johns Hopkins University Press.

UNDP (United Nations Development Programme). (1998). *Human Development Report 1998: Consumption for Human Development*. New York: Oxford University Press.

Williams, B. (1985). *Ethics and the Limits of Philosophy*. Cambridge, MA: Harvard University Press.

Wilson, E. O. (1988). *Biodiversity*. Washington, DC: National Academy Press.

10

Inter-species Justice and the Right to Thrive

Biodiversity in Crisis in Aotearoa New Zealand

MAKERE STEWART-HARAWIRA

10.1 Introduction

In 2024, global temperatures reached more than 1.4°C above pre-industrial temperatures for the first time (Copernicus Climate Change Service 2025). This was marked by increasingly urgent-voiced reports and statistics underscoring the extreme danger facing humanity and indeed the entirety of the biosphere (see for instance Ripple et al. 2022; Rockström et al. 2023). The frighteningly rapid pace of change, the identification of an extensive list of interrelated climate feedback loops (Ripple et al. 2023b), the warning that we are on the verge of five critical and inter-connected tipping points (Livina 2023; Ripple et al. 2023a), that six of the earth's nine critical boundaries had already been passed (Wang-Erlandsson et al. 2022), and most likely, the seventh – ocean acidification (Caesar et al. 2024) – the continuing failure of the United Nations Conferences of Parties on Climate Change to secure drastic reductions in carbon and methane emissions and the refusal of the fossil fuel industry to end fossil fuel expansion and use underscored the cascading dangers of species extinction and the urgency of resolving the impasse on climate action. Food production and agriculture alone *are* responsible for 8–10 per cent of global emissions (UNEP 2024) and have caused 70 per cent of biodiversity loss on land and 50 per cent in fresh water (WWF 2022). With over a million plant and animal species threatened with extinction (WWF 2022), the intertwined global catastrophes of climate change and the extinction of species as manifested in the alarming loss of global biodiversity (World Economic Forum 2021) threaten the well-being of all interrelated aspects of life on Planet Earth (Mahon et al. 2024).

Despite making up only 6.7 per cent of the land surface area of the globe, islands are outstanding for hosting 20 per cent of the world's biota. This extraordinary proportion of the world's biodiversity on islands is, however, surpassed by the rate of endangered and extinct species due to the vulnerability of islands, particularly marine islands, to anthropogenic disturbances (Fernández-Palacios et al. 2021). Fernández-Palacios et al. report that 50 per cent of the threatened species and 75 per cent of the known extinctions since the European expansion around the globe are occurring on islands. Marine islands, which are surrounded by ocean, include land-bridge islands as well as continental fragments such as Aotearoa New Zealand. Oceanic islands are reported as some of the most severely impacted on the planet. Although fresh water makes up less than 0.3 per cent of the global available surface water, fresh waters support around 50 per cent of the total fish species (Reid et al. 2019). The gravest decline in biodiversity is occurring in freshwater habitats which support around 50 per cent of the total fish species are reported as experiencing two to three times the rate of biodiversity loss of terrestrial and marine habitat (Birnie-Gauvin et al. 2023).

Against this background, increasingly urgent calls for environmental justice and transformed systems of governance elicit an ever-widening range of responses. Many of the responses to climatic crisis raise critical issues of justice and equity in particular for those occupying regions and areas threatened by rising oceans and vulnerable to extremes in weather conditions and diverse populations in the Global South (Sultana 2023). Multispecies justice scholars argue that normative frameworks of climate justice are no longer adequate to address the climate crisis. Multispecies justice rejects the three core ideas of human exceptionalism, namely that humans are physically separable from other species and non-human nature, that they are uniquely separable from all other species because they possess consciousness and agency, and that they are therefore more important than other species (Celermajer et al. 2022). Using a relational lens, multispecies justice approaches recognize the myriad of networks and relationships to which individuals and societies are bound in everyday interactions. This includes recognition of the multiplicity of different types of being, in their own terms and their involvement in thick relational webs. As Celermajer et al. (2022) point out, recognizing this thick interrelationship between humans and other beings, other species, means that justice for one equals justice for all. Nonetheless because humans have the capacity to impact these relations, and an ability to have a conception of justice that might reshape human institutions and actions, humans also have a unique ability to conceive and practice multispecies justice. This approach recognizes that practicing different ways of co-existing between the human and non-human is essential in order to generate new forms of being and acting in this world (Tschakert et al. 2021). Calls for an expansion of the notion of justice to more-than-human species have many alignments with Indigenous onto-cosmologies. In seeking deeper understandings of the meaning of relationality, multi-species justice scholars frequently turn toward the ontologies of being that underpin Indigenous peoples' ethics and values in enacting their first responsibilities under their own traditional laws; to care for and protect the multiplicity of the more-than-human beings with whom they coexist.

Indigenous communities on whose traditional lands over 80 per cent of the world's biodiversity have been protected for thousands of years, and whose ways of life, livelihoods, cultural identity, and physical and spiritual well-being are inseparable from the well-being of the more-than-human species in their lands and territories have been some of the most powerful voices against the devastating impacts of market capitalism (Celermajer et al. 2023). Today they are in the forefront of calls for action on climate justice. For the Indigenous peoples of Aotearoa New Zealand where the case study that is central to this chapter is located, the relationship between all human and non-human or more-than-human beings is delineated in the onto-cosmologies of Maori genealogies in intricate detail (Rewi & Hikuroa 2023). Encompassing the tangible and intangible, the organic and inorganic, the seen and the unseen, throughout and beyond the universe (Durie 2001: 88), these deeply entwined co-creative relationships have defined and articulated Maori relational values and ethics, their spirituality, and their cultural practices for over a thousand years (Salmond et al. 2023). In contrast to planetary justice approaches, these relationships are place-based, communal, and also deeply personal. Mountains, rivers, and many landmarks are recognized as ancestors to which the traditional human occupants have inalienable obligations. Predicated on these genealogical and cosmological relationships, Maori have an abiding responsibility of *kaitiakitanga* (loosely interpreted as guardianship) to the more-than-human species with whom they co-exist in place. Protocols and practices that care for the environmental have deep cultural and spiritual meaning and are recognized as such in traditional sayings, songs, and chants. Underpinning these protocols is recognition of the sacred nature of the mauri, the life force that binds everything together. Rivers have particular importance and are conceptualized as the lifeblood of the land. The well-being of the river which is encapsulated by its mauri is intertwined with the well-being of its people, thus its protection is of the highest importance (Stewart-Harawira 2020). This deeply ethical relationality and interconnectedness that for aeons underpinned Maori traditional cultural practices for caring for and managing their environment is today the foundation of Maori environmental governance (Redvers et al. 2023; Salmond et al. 2023).

The failure of the international governance system of nation-states and corporate-government partnerships to address the global scale of the destruction is self-evident. Transformative forms of governance have been high on the agenda of the UN Conferences on Climate Change since 2014 and the UN Sustainable Development goals since 2015. The evolving planetary boundaries framework is one such response (Cf. Rockström et al. 2009), yet as Blythe et al. emphasize (2018), there are a number of latent risks with transformative government proposals that are often overlooked in the articulations of such aspirational goals, not least of which is the political processes and shifts in power that are involved. A recent report calling for a new paradigm for planetary governance positions environmental and climate justice in its goals and invokes a proposed new legal architecture of earth law as one arm of planetary governance (Rockström et al. 2023). A critical examination of the role of law in maintaining a safe operating space within the planetary boundaries (Rockström et al. 2024) concludes that, while

> [l]aw cannot [therefore] be a panacea for addressing the complex task of governing the Earth system and the regulatory challenges emerging from the planetary boundaries, it will have a critically important role to play in this existential endeavour. (Kotzé & French 2021)

While the role of law is widely promoted as an effective tool in addressing the failure of calls for justice to address inequity, it too is hampered by the principle of state sovereignty which overrides issues of justice at the global and national levels. As dramatically demonstrated in the November 2023 national election in New Zealand, this is particularly the case within nation-states whose national governance strategies and policies can be reversed almost on a whim.

Situated at the intersection of Indigenous onto-epistemologies, multispecies justice, and proposed planetary justice frameworks of transformative governance, this chapter focuses on the critical issue of the right of more-than-human species to thrive in the context of increasingly catastrophic effects of anthropogenic climate change. The vehicle for this exploration is a consideration of the challenges to transformative governance and the implications that accrue to indigenous biodiversity. The context is the catastrophic climatic events that occurred in the South Pacific Island nation of Aotearoa New Zealand in early 2023 with devastating consequences and the well-intentioned, but contradictory, responses of a liberal humanist government with a transformative agenda. Of particular note is the recognition of Maori knowledge, ethics, and values in very recent environmental policy documents and responses to crisis after decades of its absence. Attention is drawn to the power of political upheavals in climate obstructionism and some inherent dangers in evoking transformative governance as a response to the crisis of extinction. In the final analysis, it is argued that nothing less than deep systemic transformation underpinned by an ontology of being that is committed to the ethics of deep interrelationship, the importance of place, and protection of the ability of all species to thrive will provide a pathway for maximizing the protection, regeneration, and survival of species into the future.

10.2 Extinction in the Pacific

The small island nation in the South Pacific and one of the last countries to be settled, Aotearoa New Zealand, as it is known today, has some of the world's most unique biodiversity due to 70 million years of Gondwana's isolation from other continents. Elaborated in an address on the Gondwana inheritance shared by Southern hemisphere lands (Logan 2001), New Zealand's biodiversity evolved in the absence of many species and families of plants and animals dominant elsewhere in the world due to its isolation and unique environmental history over 80 million years. As Logan enumerates, these absences include large land mammals, large reptiles, and snakes. Its unique biodiversity includes the distinctive kakapo, the only flightless parrot in the world, primitive frogs that lay eggs that hatch into adult frogs, and large insects such as the Stephens Island giant weta that fill niches occupied by rodents elsewhere. Unique features of the landscape include the braided river systems of the South Island, the kauri forests of the North Island, and the geothermal ecosystems of the central North Island. Yet today Aotearoa New Zealand has the greatest loss of indigenous biodiversity in the world (Joy & McLean 2019). Heavily forested prior to colonization,[1] by the 1860s Andrew Rutherford of a prominent Canterbury squatter family, was able to revel in the replacement of the forested cover of the Canterbury Plains with 'nothing but grass and wind' (Holm 1992), a comment revealing of the extent of destruction of the indigenous forest biome caused by settler burning in the pursuit of prosperity. As Park (1995) wrote, 'When the colonists' fires cleared at the end of the nineteenth century, New Zealand had become a different place. Maori had lost their most precious life support system' (Park 1995). Legitimated in their eyes by both biblical and Lockean ideologies of ownership and the subjugation of nature, colonists unhesitatingly took the opportunity to clear the land for building and development. In the early 1900s, as Salmond and Caddie report (Salmond & Caddie 2026), the Royal Commission on Forestry stated that 'No forestland, except if it be required for the special purposes of a climatic or scenic reserve and which is suitable for farmland, shall be permitted to remain under forest if it can be occupied and resided upon'. In hilly, highly erodible terrains such as in Tairawhiti (the East Coast of the North Island), debates about indigenous forests which prevented erosion versus exotics for which there was a ready market drove changing policies. Ultimately fast-growing exotic *Pinus radiata* plantations, which offered local communities a steady income from forestry and an unfulfilled promise of permanent plantations to address erosion, supplanted indigenous forests. The impact of this on the highly erodible terrains in Tairawhiti was particularly pronounced on the catchment areas of the five major rivers in the region (Fuller et al. 2023).

At the beginning of the first National Forest Survey 1945, on a total land area of 26.6 million hectares, only 25 per cent of indigenous forest remained and that

[1] Richardson et al. note that forests were 'the land biome with the strongest functional coupling to the climate system during the Holocene period' (2007).

mostly in areas difficult to farm (Masters et al. 1957 as cited in Marden et al. 2018; cf. Cunningham 1953: 40). Almost 80 years on from the NZ Forest Service survey, 62 per cent of New Zealand's ecosystems classified as rare are listed as threatened (Stats NZ 2023). Ninety per cent of New Zealand country's wetlands have been destroyed with the remaining 10 per cent under extreme threat from polluted waterways in farming areas, polluted urban environments, over-abstraction from waterways, industrial fishing practices, introduced species, and the effects of climate change (Department of Conservation 2020; see also Ministry for the Environment New Zealand 2019). In consequence, New Zealand has the highest proportion of threatened indigenous species in the world. Regarded as national treasures, as well as being an important food source for Maori, New Zealand's endemic species hold cultural significance (De Lange as cited in Hancock 2022).

Over the past 50 years, freshwater biodiversity is reported as vanishing at a particularly fast rate (Elston et al. 2015). While much of this is due to invasive species, equally serious is the effects of large-scale habitat destruction through deforestation, wetland drainage, and ecosystem degradation (Ministry for the Environment New Zealand 2020). In 2013, almost 33 per cent of assessed native freshwater plants (182 of 559 species) were threatened or at risk in 2013. Of these, almost 20 per cent were in the highest risk category: nationally critical (Ministry for the Environment New Zealand 2023a). Ten years on, in 2023, around one-third of New Zealand's unique indigenous marine, land, and freshwater species were listed as threatened.[2] Seventy-four per cent of assessed freshwater species were ranked as 'threatened' and a further 17 per cent as 'At Risk'. Nearly a quarter of the freshwater species were described as 'Data Deficient', meaning that there was insufficient data to make a proper assessment (Department of Conservation 2020). A machine learning analysis published in 2022 found that more than half of species whose endangered status cannot be assessed due to a lack of data are predicted to face the risk of extinction (Borgelt 2022). Many of New Zealand's rare native bird species also depend on freshwater environments. Eighty-two per cent of these were classified as threatened or at risk of becoming threatened in 2023. Eleven are already extinct (Ministry for the Environment New Zealand 2023a). Sediment levels which have a direct impact on river quality and freshwater biodiversity have been found to be significantly higher in areas of pastoral land and exotic forestry, compared to indigenous forest cover (Spicer 2021, as cited in Salmond & Caddie (2026)). In recent years New Zealand's market-oriented approach to emissions reduction has seen the exotic, fast-growing *Pinus radiata* become the forestry product of choice for carbon emissions trading. For some

[2] This accounts only for species which have been assessed. Some species are unassessed because there is no data available.

Tairawhiti communities, this has represented significant investment with overseas partners (Fuller et al. 2023). For the five major rivers on the East Coast, the two largest of which comprise 'steep, highly dissected terrains that are subject to recurrent hillslope failures' and one of which, the Waiapu, is reported to be the highest sediment yielding river in Aotearoa New Zealand, the impact has been particularly pronounced (Jones et al. 2023). Owing to the lobbying power of New Zealand's key export industries, forestry, and dairy products, however, to date efforts to reduce the effects of agriculture and over-abstraction on waterways as well as land have spectacularly failed. This failure is acerbated by policies underpinned by utilitarian approaches which measure and divide ecosystems according to their economic worth.

10.3 Forests, Rivers, and Floods: Tensions in Paradise

In March 2023, following two months of unprecedented rain and flooding that accompanied Cyclone Hale in January and sparked by superheated atmospheric storms, Cyclone Gabrielle gathered strength and velocity as it headed towards Aotearoa New Zealand (Jones et al. 2023). It hit the north-east of the country leaving in its wake the loss of 11 lives (adding to the 4 lives lost 2 weeks earlier in an unprecedented flooding event further north), 6,500 people still missing one week later, and the destruction of homes, roads, and bridges leaving some townships cut off without power, water, food, or any way out for up to 2 weeks (Kerr et al. 2023). A government-commissioned panel called under urgency to 'investigate the situation in Tairawhiti/Gisborne and Wairoa' as a result of weather events which flooded woody debris and sedimentation into land, waterways, and infrastructure, concluded that 'Sedimentation from more than a thousand untreated gullies, trees, logs and slash off hills that should never be plantation planted or clear felled, waterways choked with debris flows, riverbeds aggraded, coastlines suffocated and dangerous, roads and bridges unfit, unpassable, and many broken' (Ministry for the Environment New Zealand 2023b: 6).

Recommendations from the eight-week investigation, which included consultation with local Maori communities, the forestry industry, and scientists, focused attention on the well-known erosion susceptibility of the land and advocated for immediate restrictions on large-scale clear felling of plantation forests (of exotic pines), restrictions on plantation forests, and returns of some areas to permanent – preferably native forest – 'which would have the advantage of biodiversity co-benefits' (Ministry for the Environment New Zealand 2023b: 19) as well as strengthening biodiversity through a system of mosaic planting. Many of these recommendations had been previously made in 1988 following the devastation of Cyclone Bola (Campbell 2023). Regarding waterways, the panel noted that legislating for the health of the two rivers to be a mandatory legal requirement would 'considerably speed up the health and wellbeing of the two rivers' (Ministry for the Environment New Zealand 2023b: 13).[3] Some of the critiques levelled at the report exposed tensions and contradictions within local Maori communities. Other critiques focused on the potential loss to a highly lucrative timber export industry, particularly from partnerships with overseas investors whose pine plantations represented a multibillion-dollar industry based on carbon credit trading, harvesting, and export (cf. Te Taumata 2022). Behind these tensions and anxieties was a colonial history of commodification and extraction which had seen indigenous forests replaced with more market-lucrative exotics for export and a ready employment market for local communities.[4] A report on the health impacts of Cyclone Gabrielle is unequivocal in its statement that prevention of this disastrous flooding 'means systemic decolonisation and restoration of Indigenous relationships with the land and waterways' (Jones et al. 2023: 8).

In July 2023, the New Zealand National Policy Statement on Indigenous Biodiversity was published as part of a bundle of policy documents aimed towards fulfilling Aotearoa New Zealand's climate commitments under the 2015 Paris Climate Change Agreement. Using language which prioritizes Maori ethics and values regarding guardianship of the environment, the document acknowledges 'the web of interconnectedness between indigenous species, ecosystems, the wider environment, and the community, at both a physical and metaphysical level' and recognizes 'that the health and wellbeing of people and communities are dependent on the health and wellbeing of indigenous biodiversity and that in return people have a responsibility to care for and nurture it' (Ministry for the Environment New Zealand 2023a: 1.5.2). Yet the emphasis on the role of Maori knowledge and values was not matched in the proposals. Directed towards local body roles and responsibilities for 'identifying, protecting and maintaining indigenous biodiversity under the Resource Management Act (1991)', the policy extended provisions relating to promoting restoration and increasing indigenous vegetation cover to include natural inland wetland areas and included partitioning the land for forestry based on catchments rather than their original habitat and natural groupings known to Maori (Thomas et al. 2024). From a *matauranga Maori* or Maori knowledge perspective, these proposals flew in the face of place-based knowledge and relationships. Species do not exist in isolation. They exist within ecosystems of mutual interdependence such that the loss or destruction of one species causes

[3] An earlier version of the report recommended that legislation to protect the rights of the two rivers be pursued.
[4] For a detailed account, see Salmond, A., & Caddie, M. (2026). The root of the matter: forests and colonial histories in Aotearoa New Zealand. In *People, Place and Nature in Indigenous-Settler Relations: Recentring the More-than-Human World* (pp. 85-103). Cham: Springer Nature Switzerland.

the diminishment, and, ultimately, the loss of those whose ability to thrive is dependent on it for sustenance (Thomas et al. 2024). Similarly, the spiritual, physical, and emotional health of Maori communities is inextricably tied to the health of those species with whom they co-exist.

10.4 The Challenge of Transformative Governance and the Return of the Extreme Right

Within the first 100 days of the newly elected extreme Right coalition government elected in November 2023, the liberal humanist policies of a six-year Labour government which sought to reduce carbon emissions, improve freshwater governance, install wide-ranging policies to address biodiversity loss, and prioritize well-being were overturned. Without public consultation and with merely a ministerial stroke of a pen, the New Zealand Government abandoned an agreement to significantly widen a deep ocean marine sanctuary in the Pacific, reopened coal and oil mining throughout New Zealand including on the only habitat of a rare protected fauna, ceased implementation of new Significant Natural Areas legislation introduced by the previous government (Sowmund-Lund 2023), introduced Fast-track Approvals legislation offering a fast-track consenting process for certain listed, and referred projects that will have 'significant regional or national benefits'. The stated objective of the Fast-track Bill is to 'reduce consenting costs and timeframes to enable the efficient implementation of large-scale projects' (Garvan & Hulme-Moir 2024). The Bill allows for projects to be approved by three ministers with unprecedented powers to 'authorise harm to wildlife in return for compensation', a single Minister after rapid examination by a Fast-track Expert Advisory Group supported by officials from the Ministry of Business Innovation and Employment, and a great deal more. In his submission opposing the Fast-track legislation, the Parliamentary Commissioner for the Environment stated that the proposed Fast-track Approvals legislation will 'achieve sub-optimal outcomes through poor decision-making, poor allocation of resources, a lack of legislative durability, and increased litigation risk' (Parliamentary Commissioner for the Environment 2024). As he noted, 'Ministers subject to political lobbying are empowered to allocate public resources with potentially harmful environmental and health consequences.' Former Prime Minister and Queen's Counsel Sir Geoffrey Palmer was unequivocal about the harmful consequences, stating,

> [t]he overturning of the guiding principles of the existing Resource Management Act and the shift from an economy of wellbeing to an economy prioritizing mining and development has removed all considerations of ecosystem health and species survival. (Palmer 2024)

Aotearoa New Zealand is a signatory to the Convention on Biodiversity (since 1993) and the Kumming-Montreal Global Biodiversity Framework (2022). The implications of the proposed removal of environmental protection legislation would mean an extremely poor prognosis for the last remaining 10 per cent of New Zealand's wetlands, 20 per cent of and dunes, 27 per cent of rare ecosystems, and 20 per cent of indigenous species would be extremely poor. Like marine life in the South Pacific Ocean and indeed biodiversity globally, without deep systemic intervention, the biodiversity on this island nation in the South Pacific is facing even further extinction. The consequences of such massive disruption to ecosystems can only exacerbate the effects of climate change and accelerate the already devastating loss of biodiversity.

At the time of first writing this conclusion, news headlines in New Zealand reported that the Fast Track bill was crumbling under the weight of thousands of submissions opposing such action (Edwards 2024). As Salmond et al. (2023) noted, in the country that was one of the first to recognize a mountain as an ancestor with legal personhood, resistance, and transformation was happening from the ground. Awareness of the catastrophic impact of industrial development and climate change had seen the principles of kaitiakitanga rise to the fore. Yet critical as this movement was, and is, it was insufficient to stop the juggernaut of market capitalism and financial power. Rising awareness and New Zealand's role as a signatory to international conventions seeking to protect biodiversity notwithstanding, the determined efforts of a government elected by the power of the financial-industrial lobby saw the Fast Track Approvals Act passed into law in December 2024 (New Zealand Government 2024). Under this Act the impacts of industrial development on rare and/or endangered species are no longer to be considered nor, of course, are impacts on Maori values or communities. The removal of Maori rights, language, and the principles of kaitiakitanga from all legislative considerations under the Coalition Government is not coincidental.

10.5 Governing for the Survival of Species

Among the proposals for transformative governance for planetary well-being and multispecies justice is a new planetary commons framework proposed by a large consortium that includes the Potsdam Institute, the Stockholm Resilience Centre, and many others. Calling for a new paradigm for governance that 'foregrounds planetary stewardship and secures critical planetary life-support systems in a safe and just way' (Rockström et al. 2024), the proposed planetary commons approach for earth system governance recognizes that the 'transgression of boundaries and undermining of biophysical symptoms already have multispecies justice implications for past and present generations' (Rockström et al. 2024: 3). As this preliminary document emphasizes, to achieve the goal of 'collectively securing critical biophysical Earth system functions that regulate livability on Earth for everyone, irrespective of where these functions are located' would require a shift from governing of resource use of natural resources to universal rules of planetary stewardship. Alongside this major challenge to the 'outdated traditions of an old European political order', including state sovereignty and self-determination, vested corporate interests, inequities of global power, and the complexities of demarcation (Rockström et al. 2024: 6), is a proposed new legal paradigm of earth system law and a new planetary ethic that would rest on sets of principles such as differentiated degrowth and sufficiency, interconnectivity, and a new planetary ethic (Rockström et al. 2024: 8). The particular engagement of indigenous peoples with their 'knowledges, practices, and contributions to ensure stable living conditions across the planet' is included in this framework for planetary well-being, as is the relationship between 'territorial custodians of a planetary commons and the broader spectrum of planetary beneficiaries', yet it is unclear to what degree the importance of place-based species and knowledge that is central to Indigenous onto-epistemologies is fully understood. As the authors acknowledge, 'the spectre of climate and environmental colonialism' is ever present (Rockström et al. 2024: 7). So also, as the document demonstrates, is the centring of the human. Critiques of the planetary model point out that there are other ways of conceiving of planetary ethical commitments that require a radical shift in our conceptualizations of humans and more-than-human, of communities, and of the planet. One example is Latour's model of using the concept of planetary to break apart the concept of a 'whole', 'singular planet' that can be looked down upon from on high. Others point to multispecies justice lifeways as a route to achieving planetary, multispecies justice (Kurki 2024). The ethics and values of Indigenous peoples who have for thousands of years recognized the animated, agential nature of the world, including human and nonhuman, and whose first law of governance is the protection of biodiversity represent the foundation of such a model. One thing is unequivocal, while the exceptional human remains at the centre, the transformative paradigm for protecting biodiversity that is sought in the framework for planetary stewardship will repeat the mistakes that have led to this extinction.

Aotearoa New Zealand is not alone in its demonstration of the fragility of governance and law as mechanisms for protecting the well-being of species, human and more-than-human, and the distribution and protection of justice from the proclivities of power and ambition. The achievement of deep systemic change will require transformation from the bottom up. It begins with the ethics and values of right relations and the first law of traditional Indigenous governance. Traditional Maori ethics and values of right relations recognized that status accrued not from ownership and power over but from uplifting and protection of the rights of others, that our first and foremost responsibility is to the protection of the life force, well-being, and right to thrive of all our relations, human and more-than-human (Williams 2013). 'Kaitiakitanga', stated Chaz Doherty, 'is an action, not a title' (Te Ao with Moana 2020).

Without an underpinning onto-epistemology of the sacredness of the lifeforce which binds all existence and the decentring of the human, the achievement of deep systemic transformation may be long in the making.

Granting legal personhood to geological bodies such as rivers and mountains is one representation of the broadening of concepts of justice to the more-than-human, to multi-species justice. However critical limitations and challenges exist, some of which are reflected in the planetary boundaries framework. The onto-cosmologies of Indigenous peoples, such as, for example, those traditionally held by Maori in Aotearoa New Zealand, by First Peoples in Australia and in the Americas, in the high mountains of Bhutan and elsewhere, hold the sacred nature of all being and of ethical relationality itself, as the essence of existence. Such an ontology decentres property rights and economic priorities and re-centres the well-being of life itself. These are the ethics and values that must be embedded in and enframe legal and civil legislation and boundaries not only for planetary wellbeing, but for the well-being of all aspects of existence now and into the future, and in such a way that they cannot be overturned. Never have they been more urgently needed.

References

Spicer, B. D. O. (2021). Report on the Impacts of Permanent Carbon Farming on Te Tairawhiti Region: Stepping into the Shoes of Our Mokopuna. www.eastlandwood.co.nz/wp-content/uploads/2021/11/Impacts-of-permanent-carbon-farming-on-Tairawhiti-region-Problems-and-solutions-030321-v5.pdf.

Birnie-Gauvin, K., Lynch, A. J., Franklin, P. A., et al. (2023). The RACE for Freshwater Biodiversity: Essential Actions to Create the Social Context for Meaningful Conservation. *Conservation Science and Practice*. 5(4), e12911. DOI: 10.1111/csp2.12911.

Blythe, J., Silver, J., Evans, L., et al. (2018). The Dark Side of Transformation: Latent Risks in Contemporary Sustainability Discourse. *Antipode*. 50(5), 1206–1223. DOI: 10.1111/anti.12405.

Borgelt, J., Dorber, M., Høiberg, M. A., et al. (2022). More than Half of Data Deficient Species Predicted to Be Threatened by Extinction. *Communications Biology*. 5(1), 679. DOI: 10.1038/s42003-022-03638-9.

Caesar, L., Sakschewski, B., Andersen, L. S., et al. (2024). *Planetary Health Check: A Scientific Assessment of the State of the Planet*. Potsdam: Potsdam Institute for Climate Impact Research.

Campbell, J. (2023). John Campbell Revisits Tairāwhiti: We Keep Getting It Wrong. *1news*. www.1news.co.nz/2023/06/14/john-campbell-revisits-tairawhiti-we-keep-getting-it-wrong/.

Celermajer, D., Schlosberg, D., Rickards, L., et al. (2022). Multispecies Justice: Theories, Challenges, and a Research Agenda for Environmental Politics. *Environmental Politics*. 30(1–2), 119–140. DOI: 10.1080/09644016.2020.1827608.

Celermajer, D., Schlosberg, D., Wadiwel, D., et al. (2023). A Political Theory for a Multispecies, Climate-Challenged World: 2050. *Political Theory*. 51(1), 39–53. DOI: 10.1177/00905917221128833.

Copernicus Climate Change Service. (2025). Global Climate Highlights 2024. climate.copernicus.eu/global-climate-highlights-2024.

Cunningham, A. (1953). National Forest Survey: A Large-Scale Ecological Project. *Tuatara*. 5(2), 39–48.

Department of Conservation. (2020). Biodiversity in Aotearoa: An Overview of State, Trends and Pressures. Wellington. www.doc.govt.nz/globalassets/documents/conservation/biodiversity/anzbs-2020-biodiversity-report.pdf.

Durie, M. (2001). *Mauri Ora: The Dynamics of Maori Health*. Auckland: Oxford University Press.

Edwards, B. (2024). The Government's Fast Track Being Demolished by Submissions to Parliament: Bryce Edwards' Political Roundup. *The New Zealand Herald*. www.nzherald.co.nz/nz/political-roundup-the-govts-fast-track-being-demolished-by-submissions-to-parliament/TG4V6AYH3FDLZJVSPSHR5W4MJU/.

Elston, E., Anderson-Lederer, R., Death, R. G., et al. (2015). The Plight of New Zealand's Freshwater Species: Conservation Science Statement No. 1. Sydney: Society for Conservation Biology (Oceania). conbio.org/images/content_groups/Oceania/Scientific_Statement_1_.pdf.

Fernández-Palacios, J. M., Kreft, H., Irl, S. D. H., et al. (2021). Scientists' Warning: The Outstanding Biodiversity of Islands Is in Peril. *Global Ecology and Conservation*. 31, e01847. DOI: 10.1016/j.gecco.2021.e01847.

Fuller, I. C., Brierley, G. J., Tunnicliffe, J., et al. (2023). Managing at Source and at Scale: The Use of Geomorphic River Stories to Support Rehabilitation of Anthropocene Riverscapes in the East Coast Region of Aotearoa New Zealand. *Frontiers in Environmental Science*. 11, 1162099. DOI: 10.3389/fenvs.2023.1162099.

Garvan, N., & Hulme-Moir, Will. (2024). Government's Fast-Track Approvals Bill Released under 100-Day Plan. Bell Gully. www.bellgully.com/insights/government-s-fast-track-approvals-bill-released-under-100-day-plan.

Hancock, F. (2022). Aotearoa's Vanishing Species Creatures Great and Small. www.rnz.co.nz/programmes/in-depth-special-projects/story/2018868243/aotearoa-s-vanishing-species-creatures-great-and-small.

Holm, J. (1992). *Nothing but Grass and Wind: The Rutherfords of Canterbury*. Christchurch: Hazard Press.

Joy, M., & McLean, S. (2019). Despite Its Green Image, NZ Has World's Highest Proportion of Species at Risk. *The Conversation*. theconversation.com/despite-its-green-image-nz-has-worlds-highest-proportion-of-species-at-risk-116063.

Jones, R., Macmillan, A., & Woodward, A. (2023). Editorial: Superheated Storms: Climate Drivers, Health Effects and Responses. *The New Zealand Medical Journal*. 136(1573), 8–11.

Kerr, J., Thomson, G., & Wilson, N. (2023). Cyclone Gabrielle Joins List of Aotearoa NZ's 'Sudden Mass Fatality Events'. Public Health Communication Centre Aotearoa. www.phcc.org.nz/briefing/cyclone-gabrielle-joins-list-aotearoa-nzs-sudden-mass-fatality-events.

Kotzé, L. J., & French, D. (2021). Staying within the Planet's 'Safe Operating Space'? Law and the Planetary Boundaries. In D. French & L. J. Kotzé, eds., *Research Handbook on Law, Governance and Planetary Boundaries*. Research Handbooks in Human Rights. Cheltenham: Edward Elgar, 1–19.

Kurki, M. (2024). Planetary Justice Reconsidered: Developing Response-Abilities in Planetary Relations. *Environmental Politics*. 33(7), 1185–1204. DOI: 10.1080/09644016.2024.2336869.

Livina, V. N. (2023). Connected Climate Tipping Elements. *Nature Climate Change*. 13(1), 15–16. DOI: 10.1038/s41558-022-01573-5.

Logan, H. (2001). Gondwana Invaded: An Address on Distinctive Features of Managing Indigenous Biodiversity in Protected Areas in New Zealand. *Journal of the Royal Society of New Zealand*. 31(4), 813–818. DOI: 10.1080/03014223.2001.9 517677.

Mahon, M. B., Sack, A., Aleuy, O. A., et al. (2024). A Meta-Analysis on Global Change Drivers and the Risk of Infectious Disease. *Nature*. 629, 830–836. DOI: 10.1038/s41586-024-07380-6.

Marden, M., Lambie, S., & Phillips, C. (2018). Biomass and Root Attributes of Eight of New Zealand's Most Common Indigenous Evergreen Conifer and Broadleaved Forest Species during the First 5 Years of Establishment. *New Zealand Journal of Forestry Science*. 48(1), 9. DOI: 10.1186/s40490-018-0113-y.

Masters, S. E., Holloway, J. T., & Mc Kelvey, P. J. (1957). *The National Forest Survey of New Zealand 1955*. 1. Wellington: Government Printer.

Ministry for the Environment New Zealand. (2019). Environment Aotearoa 2019: New Zealand's Environmental Reporting Series. environment.govt.nz/assets/Publications/Files/environment-aotearoa-2019.pdf

Ministry for the Environment New Zealand. (2020). New Zealand's Environmental Reporting Series: Our Freshwater 2020. environment.govt.nz/assets/Publications/Files/our-freshwater-2020.pdf.

Ministry for the Environment New Zealand. (2023a). National Policy Statement for Indigenous Biodiversity. environment.govt.nz/assets/publications/National-Policy-Statement-for-Indigenous-Biodiversity-superseded-.pdf.

Ministry for the Environment New Zealand. (2023b). Outrage to Optimism: Report of the Ministerial Inquiry into Land Uses Associated with the Mobilisation of Woody Debris (Including Forestry Slash) and Sediment in Tairawhiti Gisborne District and Wairoa District. environment.govt.nz/assets/Outrage-to-optimism-superseded.pdf.

New Zealand Government. (2024). Fast Track Approvals Act 2024. https://legislation.govt.nz/act/public/2024/56/en/latest/.

Palmer, S. G. (2024). A Fast Track to Environmental Degradation. Newsroom. newsroom.co.nz/2024/05/20/a-fast-track-to-environmental-degradation/.

Park, G. (1995). *Ngā Uruora: The Groves of Life: Ecology and History in a New Zealand Landscape*. Wellington: Victoria University Press.

Parliamentary Commissioner for the Environment. (2024). Submission on the Fast-Track Approvals Bill. pce.parliament.nz/publications/submission-on-the-fast-track-approvals-bill/.

Redvers, N., Aubrey, P., Celidwen, Y., et al. (2023). Indigenous Peoples: Traditional Knowledges, Climate Change, and Health. *PLOS Global Public Health*. 3(10), e0002474. DOI: 10.1371/journal.pgph.0002474.

Reid, A. J., Carlson, A. K., Creed, I. F., et al. (2019). Emerging Threats and Persistent Conservation Challenges for Freshwater Biodiversity. *Biological Reviews*. 94(3), 849–873. DOI: 10.1111/brv.12480.

Rewi, S., & Hikuroa, D. (2023). He Karanga Maha: Investigating Relational Resource Management in Aotearoa, New Zealand. *International Journal of Environmental Research and Public Health*. 20 (8), 5556. DOI: 10.3390/ijerph20085556.

Richardson, D. M., Rundel, P. W., Jackson, S. T., et al. (2007). Human Impacts in Pine Forests: Past, Present, and Future. *Annual Review of Ecology, Evolution, and Systematics.* 38(1), 275–297. DOI: 10.1146/annurev.ecolsys.38.091206.095650.

Ripple, W. J., Wolf, C., Gregg, J. W., et al. (2022). World Scientists' Warning of a Climate Emergency 2022. *BioScience.* 72(12), 1149–1155. DOI: 10.1093/biosci/biac083.

Ripple, W. J., Wolf, C., Gregg, J. W., et al. (2023a). The 2023 State of the Climate Report: Entering Uncharted Territory. *BioScience.* 73(12), 841–850. DOI: 10.1093/biosci/biad080.

Ripple, W. J., Wolf, C., Lenton, T. M., et al. (2023b). Many Risky Feedback Loops Amplify the Need for Climate Action. *One Earth.* 6(2), 86–91. DOI: 10.1016/j.oneear.2023.01.004.

Rockström, J., Gupta, J., Qin, D., et al. (2023). Safe and Just Earth System Boundaries. *Nature.* 619(7968), 102–111. DOI: 10.1038/s41586-023-06083-8.

Rockström, J., Kotzé, L., Milutinović, S., et al. (2024). The Planetary Commons: A New Paradigm for Safeguarding Earth-Regulating Systems in the Anthropocene. *Proceedings of the National Academy of Sciences.* 121(5), e2301531121. DOI: 10.1073/pnas.2301531121.

Rockström, J., Steffen, W., Noone, K., et al. (2009). Planetary Boundaries: Exploring the Safe Operating Space for Humanity. *Ecology & Society.* 14(2), 32. DOI: 10.5751/ES-03180-140232.

Salmond, A., & Caddie, M. (2026). The Root of the Matter: Forests & Colonial Histories in Aotearoa. In *People, Place and Nature in Indigenous-Settler Relations: Recentring the More-than-Human World.* Cham: Springer Nature Switzerland, 85–103.

Salmond, A., Hikuroa, D., & Robertson, N. (2023). Think Like a Fish: New Oceanic Histories. In A. Bashford, E. M. Kern, & A. Bobbette, eds., *New Earth Histories: Geo-Cosmologies and the Making of the Modern World.* Chicago: University of Chicago Press, 70–90.

Sowmund-Lund, S. (2023). What's Been Done – and What's Left to Do – in the Government's 100-Day Plan. The Spinoff. thespinoff.co.nz/politics/23-02-2024/whats-been-done-and-whats-left-to-do-in-the-governments-100-day-plan.

Stats NZ. (2023). Extinction Threat to Indigenous Species. stats.govt.nz/indicators/extinction-threat-to-indigenous-species.

Stewart-Harawira, M. W. (2020). Troubled Waters: Maori Values and Ethics for Freshwater Management and New Zealand's Fresh Water Crisis. *Wiley Interdisciplinary Reviews: Water.* 7(5), https://doi.org/10.1002/wat2.1464.

Sultana, F. (2023). Whose Growth in Whose Planetary Boundaries? Decolonising Planetary Justice in the Anthropocene. *Geo: Geography and Environment.* 10(2), e00128.

Te Ao with Moana. (2020). Kaitiakitanga Is an Action, Not a Title. Maori+. www.facebook.com/TeAoWithMoana/videos/571980806815772/.

Te Taumata. (2022). Government Backs Down on Forestry Changes: 7b Win for Maori. Minister Walks Back on Forestry Legislation. www.tetaumata.com/government-backs-down-on-forestry-changes/.

Thomas, M., Lythberg, B., Hikuroa, D., et al. (2024). Problematising the Anthropocene: Geographic Perspectives upon the Riverscapes of Waimatā Catchment, Aotearoa New Zealand. *The Geographical Journal.* 191(1), e12598. DOI: 10.1111/geoj.12598.

Tschakert, P., Schlosberg, D., Celermajer, D., et al. (2021). Multispecies Justice: Climate-Just Futures with, for and beyond Humans. *WIREs Climate Change.* 12(2), e699. DOI: 10.1002/wcc.699.

UNEP (United Nations Environment Programme). (2024). *Food Waste Index Report 2024.* Think Eat Save: Tracking Progress to Halve Global Food Waste. wedocs.unep.org/20.500.11822/45230.

Wang-Erlandsson, L., Tobian, A., Ent, R. J. van der, et al. (2022). A Planetary Boundary for Green Water. *Nature Reviews Earth & Environment.* 3(6), 380–392. DOI: 10.1038/s43017-022-00287-8.

Williams, J. (2013). Lx Aotearoa: An Heroic Attempt to Map the Maori Dimension in Modern New Zealand Law. *Waikato Law Review.* 21, 1–34.

World Economic Forum. (2021). *The Global Risks Report 2021: 16th Edition.* www.weforum.org/publications/the-global-risks-report-2021/.

WWF. (2022). *Living Planet Report 2022: Building a Nature-Positive Society.* Gland. wwflpr.awsassets.panda.org/downloads/lpr_2022_full_report.pdf.

Part V

Land System Change

11

Introduction to the Land System Change Boundary

SIMON P. MEISCH

As early as 1819, the polymath Alexander von Humboldt observed a massive land system change in the Lake Valencia region of present-day Venezuela. To create agricultural land for colonial commodities such as sugar, indigo, and cotton, European settlers were cutting down large areas of the nearby tropical forests, with serious detrimental consequences for water flows, soils, biodiversity, and the regional climate system (von Humboldt et al. 1814: 142–145). What makes Humboldt's observation still relevant is how he linked ideology (colonialism, extractivism) and specific social practices (agriculture) to the negative effects of land system changes, including local climate change. Forests played a crucial role in his assessment.

A similar rationale drives the planetary boundary on land system change. Since the 2015 update, the control variable of this boundary is the amount of remaining forest cover (in per cent) compared to the pre-industrial Holocene state (100 per cent). In particular, this takes into account the importance of tropical forests for evapotranspiration and boreal forests for land surface albedo effects, which refers to the proportion of solar radiation reflected back into space (Steffen et al. 2015). There are two types of boundary value: a global value for all forests relative to the preindustrial Holocene base value (75 per cent), and values specific to the three major forest biomes: boreal, temperate, and tropical. In the 2023 update to the planetary boundaries framework, the current global value is 60 per cent of Holocene forest cover, exceeding the boundary value of 75 per cent. The biome-specific thresholds have also been exceeded, with deforestation continuing (Richardson et al. 2023).

As with other planetary boundaries, the control variable for land system change has changed since the concept was introduced. Originally, the variable referred to the 'percentage of global land cover converted to cropland'. The underlying rationale was that '[l]and-system change, driven primarily by agricultural expansion and intensification, contributes to global environmental change, with the risk of undermining human well-being and long-term sustainability'

(Rockström et al. 2009). This reference to agriculture disappeared in later updates. An interesting feature of the original definition is that it was a rare occasion when the framework mentioned a social sector as responsible for profound planetary changes. It has since been criticized for its silence on the social processes and key actors that are pushing humanity beyond a safe operating space; the framework would gain political teeth if it related the environmental changes to actual human action and social organizations (Biermann & Kim 2020).

An ethical discussion of land system change, therefore, needs to 'put people on the map' (Ellis & Ramankutty 2008). Humans have always altered the Earth's surface because, like almost all other creatures, they metabolize and cannot help but consume the resources grown on or extracted from the land. What is needed, however, is an ethical discussion about morally acceptable cultivation of and changes to land systems. The two chapters in this book address this issue in different ways.

In their text, Joana Castro Pereira and Carlota Houart draw on their research in the Amazon, one of the hotspots of land system change. They are critical of those who view the region as merely a bioeconomy and suggest instead the concept of 'forestzenship' that 'refers to understanding human beings as one among many members of a forest community, each contributing to the preservation and development of the entire community' (Chapter 12, Section 12.2.2). In this way, Castro Pereira and Houart argue for a multi-species ethic of land use that rejects anthropocentric arguments in favour of an inclusive and intercultural ethic that empowers local communities, respects the rights of ecosystems and animals, and embraces diverse knowledge systems for transformative change and sustainable progress. Their text underscores the relevance of governance for implementing more ethical policies.

Alice Benessia's text is in the form of a diary. She gives the reader insights into her daily work at the Pianpicollo Selvatico Foundation in Northern Italy, which is dedicated to research at the intersection of art, deep ecology, and regenerative agriculture. Benessia talks about the conversations she and the Foundation's guests have with the land, and what it means to physically engage with a land that has long been transformed by an agro-industrial system. Her focus is on living with the land and all that lives in and on it – rather than mastering and possessing it – and on healing the land in challenging times of climate change. With her work, Benessia explicitly suggests a different form of knowledge co-production for land system change.

References

Biermann, F., & Kim, R. E. (2020). The Boundaries of the Planetary Boundary Framework: A Critical Appraisal of Approaches to Define a 'Safe Operating Space' for Humanity. *Annual Review of Environment and Resources*. 45(1), 497–521. DOI: 10.1146/annurev-environ-012320-080337.

Ellis, E. C., & Ramankutty, N. (2008). Putting People in the Map: Anthropogenic Biomes of the World. *Frontiers in Ecology and the Environment.* 6(8), 439–447. DOI: 10.1890/070062.

von Humboldt, A., Bonpland, A., & Williams, H. M. (1814). *Personal Narrative of Travels to the Equinoctial Regions of the New Continent during the Years 1799–1804 by Alexander de Humboldt and Aimé Bonpland: With Maps, Plans &c.* London: Printed for Longman, Hurst, Rees, Orme, and Brown. DOI: 10.5962/bhl.title.87614.

Richardson, K., Steffen, W., Lucht, W., et al. (2023). Earth beyond Six of Nine Planetary Boundaries. *Science Advances.* 9(37), eadh2458. DOI: 10.1126/sciadv.adh2458.

Rockström, J., Steffen, W., Noone, K., et al. (2009). Planetary Boundaries: Exploring the Safe Operating Space for Humanity. *Ecology and Society.* 14(2), 32. DOI: 10.5751/ES-03180-140232.

Steffen, W., Richardson, K., Rockström, J., et al. (2015). Planetary Boundaries: Guiding Human Development on a Changing Planet. *Science.* 347(6223), 1259855. DOI: 10.1126/science.1259855.

12

A Transformative Multispecies Land Use Ethic for the Amazon[1]

JOANA CASTRO PEREIRA AND CARLOTA HOUART

12.1 Introduction

The planetary boundary for land is determined by the percentage of forested land compared to the original forest cover. To remain within the 'safe operating space for humanity', it is suggested that a minimum of 75 per cent of the world's forests should remain intact. However, the current situation indicates that we have exceeded this boundary (Richardson et al. 2023). The security of the Earth system depends on the extent of the remaining rainforests in the Amazon, Congo, and Indonesia as well as northern hemisphere temperate and boreal forests.

The Amazon, the world's largest continuous rainforest, is indeed crucial for the planet's stability. It acts as a significant carbon reservoir, housing approximately half of all tropical forest carbon, and provides habitat for 10 per cent of known species. The region's ecosystems are essential for natural processes such as the water cycle and climate regulation. Yet overexploitation, deforestation, and forest fragmentation are weakening the biome's resilience. If deforestation reaches 20–25 per cent of the original forest area, most of the region could become a dry savannah, reaching a tipping point (Science Panel for the Amazon (SPA) 2021). The eastern Amazon may have already crossed this critical threshold for forest loss, and extensive deforestation in the east could impact the central and western parts of the rainforest, as the region relies on water recycling from the Atlantic Ocean (Finer & Mamani 2022). The potential dieback of the Amazon could yield severe consequences, for example, local populations would confront substantial hardships, significant carbon emissions would exacerbate global warming, and numerous terrestrial species would lose their natural habitats, placing them at the brink of extinction, thus precipitating an ecological disaster. At the same time, the Amazon's critical role in regulating the planet's climate could be disrupted. This is a cause for concern as current climatic patterns underpin agricultural production in

[1] GrammarlyGo was used to assist with language editing in the preparation of this manuscript.

Brazil, a significant global food exporter. The deterioration of the Amazon's resilience could trigger a cascading effect, fundamentally altering the planet's water cycle and climate, thereby posing a substantial threat to global food security (SPA 2021).

The preservation of the forest relies on recognizing the significance of creating a sustainable operational environment that prioritizes the welfare of ecologies, animals, and the land. This entails adopting transformative measures and sustainable land use solutions to address issues of poverty, inequalities, and injustices in the Amazon (Pereira & Terrenas 2022). This shift towards sustainability in the region extends beyond 'environmental' matters; it necessitates an ethical transformation.

In this chapter, we draw from our extensive research on Amazon governance and use sustainability science, political ecology, and ecological economics to discuss ethical issues related to land use, with a particular focus on the significant topic of the bioeconomy. Inspired by the alternatives of local sociobioeconomies in the Amazon and the concept of 'forestzenship', we advocate for a multispecies land use ethic that rejects anthropocentric reasoning and emphasizes the interconnectedness and intrinsic value of all components of nature (Leopold 2020); an inclusive and intercultural ethic that empowers local communities, respects the rights of ecosystems and animals, and embraces diverse knowledge systems for transformative change[2] and sustainable progress. This is crucial for achieving socioecological justice, safeguarding our forests, and respecting the planet's limits.

12.2 The Ethics of Land Use

In this section, we start by discussing the main ethical issues concerning land use in the Amazon. Next, we propose an ethic for land use in the region that is transformative and inclusive of multiple species.

12.2.1 A Brief Overview of Land Use Ethical Issues in the Amazon

The bioeconomy refers to the utilization of renewable biological resources from both land and sea (e.g., crops, forests, animals, fish, micro-organisms) with the purpose of generating food, materials, and energy. It is seen as a path to global sustainability, with the Amazon at the core of worldwide conversations on the topic. The term 'bioeconomy' can have different meanings for different groups and

[2] A transformation in how economies, societies, and technologies operate, alongside a transition in paradigms, goals, and values to ensure a safe operational boundary for earthly life (Visseren-Hamakers & Kok 2022).

scientific perspectives. It may refer to either monoculture or sustainable forest management, depending on the context (Costa et al. 2022). The prevailing bioeconomy approaches, such as 'biotechnology' and 'bioresource', originate mainly from the Global North and focus on advanced technologies and market-driven strategies for economic growth. In the Amazon region, discussions on the bioeconomy have often aligned with the bioresource narrative,[3] associated with major industries such as sugarcane, soy, and meatpacking. A 'forest-product-based bioeconomy' involving wood and non-wood products has emerged in the Brazilian Amazon, but it has not yet transformed the region's predatory economic development agenda (Bergamo et al. 2022). While the bioeconomy is presented as a solution for biodiversity protection and local development, in practice, it has become a form of 'conservative ecological modernization' dominated by large agribusinesses. This approach has legitimized activities that cause deforestation, put plant and animal species at risk of extinction, violate Indigenous rights, and marginalize vulnerable groups in the region (Bastos Lima 2021).

This bioeconomy paradigm contributes to biodiversity loss by replacing natural ecosystems with capital-intensive monocultures (Bergamo et al. 2022). It also neglects the needs of rural communities. In the Amazon, subsidized agricultural credit by governments benefits large landowners in the livestock and agriculture sectors, while small farmers face exclusion from the bioeconomy and displacement due to soy plantations and cattle ranching (Bastos Lima 2021). Influential political groups oppose family farming, considering it unproductive and against their interests, which leads to small farmers having limited political power compared to the agribusiness sector (Sabourin et al. 2020). This lack of support leads to increased deforestation as small farmers resort to unsustainable practices (Pereira & Viola 2022). Moreover, the move to a bioeconomy based on plantation economies erodes the land-based cultures and ways of knowing, surviving, and thriving within nature of Indigenous peoples, while maintaining neocolonial relationships (Bergamo et al. 2022; Pereira & Gebara 2023). The impacts of agribusiness (e.g., land dispossession, loss of food biodiversity, water resource contamination) on Indigenous peoples and traditional communities are overlooked, with no efforts towards restorative justice (Bastos Lima 2022). Finally, although crop production and exports linked to deforestation have sometimes led to economic growth in Amazonian countries, the benefits from clearing forests have primarily been gained by foreign players and wealthy local groups. Indeed, education, innovation, and sustainable infrastructure have not received enough

[3] It emphasizes nature-based innovations and products, with a focus on land use. However, this approach prioritizes production and the efficient use of resources over preservation, ignores the potential for economic diversification related to biodiversity, and relies on standardization and large-scale practices (see, e.g., Costa et al. 2022).

investment to make products from the region more valuable or to reinvest profits into healthcare and education (Garrett et al. 2023).

The carbon market is seen by political elites, the business sector, and some academic circles and Indigenous communities as an alternative way to promote the bioeconomy in the region. The sale of carbon credits is considered a sustainable and profitable way for Amazonian countries, particularly Brazil, to protect their forests and promote local development. These credits aid in reducing greenhouse gas emissions by financing projects to protect at-risk areas from deforestation as well as nature-based solutions such as reforestation and land restoration. In the Amazon, forest degradation is both a significant source of emissions and the main cause of impoverishment. It is estimated that one-third of the region has been degraded due to human disturbances (Lapola et al. 2023). Although carbon credits can help fund land restoration, the market lacks proper regulation and clear negotiation rules (Ferrante 2023), leading to projects overlapping public lands without government authorization, lack of prior and informed consent from communities involved, and unfair negotiation conditions between companies and communities (Antunes 2023). Amazonian Indigenous and traditional peoples face threats from companies seeking deals in their territories. The agreements can last up to several decades and are often written in English with unclear terms that communities do not understand; consequently, these are displaced and lose access to their lands (Greenfield 2023).

President Lula da Silva's administration in Brazil prioritizes an inclusive bioeconomy and regulating the national carbon market to control deforestation (Governo Federal do Brasil 2023). However, the green agenda of the government's most progressive members faces resistance from right-wing opposition in Congress and Amazonian state governments with agribusiness and pro-mining agendas (Vilani et al. 2023). The environment and Indigenous ministers, environmentalist Marina Silva, and Indigenous leader and activist Sônia Guajajara encountered political setbacks just six months into their terms when Congress voted to reduce their powers. The environment ministry lost control over a land registry system used for land grabs and the Indigenous ministry no longer has authority over demarcating Indigenous lands (Schröder 2023). The Lower House also passed legislation invalidating Indigenous claims to territories they could not prove physical occupation of on the day the country's Constitution was enacted in 1988, potentially allowing road-building, mining, dam construction, agriculture, and genetically modified crops on protected Indigenous lands. Those supporting the legislation emphasized more land for farmers, economic development, and aiding Indigenous communities' integration into mainstream society (Phillips 2023). A common belief in the Amazon countries is that progress entails transforming the region's natural and societal wealth into standardized, nature destructive commodities for the market (Pereira & Gebara 2023). Albeit the Supreme Court of Justice ruled against the legislation a few months later, the Indigenous movement remains concerned that the 'ruralist' caucus in Congress will mobilize in response to the Court's decision and attempt to introduce a new law through an amendment to the constitution (Mori 2023).

The problem of deforestation is often analyzed from the supply side, looking at the sectors responsible for replacing forests. Yet one should also consider the demand perspective, identifying incentives driving deforestation and empowering anti-environmentalist forces. More than 80 per cent of deforestation in the Brazilian Legal Amazon results from domestic demand outside the region and foreign trade, with livestock product demand being the primary driver (Nobre et al. 2023). From a broader perspective, livestock production raises further ethical implications, namely about animal rights. In the Brazilian Amazon, for example, there are over 230 million cattle (Lima Filho et al. 2021) treated as commodities or resources for human consumption, who are denied their own subjecthood and agency as well as sentient character. Likewise, as previously seen, the large-scale farming of animals negatively impacts other animals, plants, and human communities. These land use dynamics thereby demonstrate how (global) power relations, which determine whose lives are valued, when, why, and for what purpose, cross species boundaries, and entangle human and non-human beings in shared networks of vulnerability, loss, and extinction, but potentially also of resilience and survival.

12.2.2 A Transformative Multispecies Land Use Ethic

The Amazon basin is incredibly diverse. Indigenous organizations, small farmer associations, and other grassroots movements criticize the technological and productive bioeconomy paradigm that has threatened their lands and ways of knowing and being; they advocate for a 'sociobioeconomy' based on a production model that prioritizes bio- and social diversity (Amazonian Socio-Biodiversity Meeting 2021). In fact, sustainable bioeconomies centred around specific locations have long been integral to the Amazon's economy; built on the responsible use and restoration of forests and rivers, the region's sociobioeconomies bring benefits to both regional communities and the global society. Amazonian sociobioeconomies offer an alternative to current economic models and power structures; bioeconomy plans and efforts ought to prioritize helping, gaining insights from, increasing, and generating new ideas based on current sociobioeconomies (Garrett et al. 2023).

A sociobioeconomy model should closely align with an inclusive, relational, and non-anthropocentric multispecies justice perspective. This perspective acknowledges the intricate connections between human and non-human beings in shaping and sustaining earthly communities, granting non-human entities the freedom to thrive and fulfil their socioecological roles (Celermajer et al. 2021; Tschakert et al. 2021). Similarly, the cosmologies and ethical frameworks of many Amazonian Indigenous peoples recognize animals, plants, and ecosystems as living, agentic, and communicative beings (Inoue 2018), perceiving humans as part of these multispecies communities (van Dooren et al. 2016).

In Amazonian land ethics, adopting a multispecies standpoint involves acknowledging the cohabitation of humans, animals, plants, and other entities like water on the land. This emphasizes biodiversity protection and recognizes the autonomy, agency, and lifeways of non-human beings within land and water management. Viewing a bioregion like the Amazon through a multispecies lens can shape ethical and political frameworks as well as human livelihoods. Examples include agroecological practices such as permaculture, regenerative agriculture, natural farming, and agroforestry, along with concepts like 'forestzenship' (Barca 2020). The latter refers to understanding human beings as one among many members of a forest community, each contributing to the preservation and development of the entire community. Drawing from Ailton Krenak's work on 'extractive reserves' in Brazil, the principle of *florestania* (forestzenship) is founded on the idea that 'the forest and its peoples constitute a polis, a more-than-human community endowed with proper political subjectivity' (Barca 2020: 24). This principle encourages us to include non-human entities in our concept of a 'safe operating space for humanity' within the framework of the planetary boundaries. It emphasizes the importance of acknowledging the various ways in which those entities interact with and shape the environment, the potential consequences of disrupting their ecological roles and behaviours, and the imperative of establishing a 'safe operating space' for multiple species.

Barca presents an illustrative instance in the Brazilian Amazon – an agroforestry initiative known as Praialta Piranheira. Here, local nut collectors Zé Claudio Ribeiro da Silva and Maria do Espírito Santo collaborated with chestnut trees, not only to ensure their livelihoods but also to enhance the ecological sustainability of the forest region (Barca 2020). Tragically, both were assassinated in 2011 while striving to safeguard the forest, including the chestnut trees, from illegal logging and timber trafficking – a burgeoning activity that claims human and arboreal lives alike, along with other species. Their story epitomizes these varied agroecological (and concurrently political) practices that prioritize values such as well-being, economic justice, and democracy over mere economic expansion.

Another example manifests through the environmental activism of Indigenous groups and local communities spanning Amazonian nations. Initiatives like Kawsak Sacha, the Living Forest Declaration, originated by the Kichwa people of Sarayaku, Ecuador, exemplify the defence of ancestral forests against extractive industries like oil, mining, wood extraction, and biopiracy. According to the Kichwa people's worldview, their forest is akin to a multispecies political community preserved through interrelationships between its constituents – active participants in forest life (Houart 2022).

The degraded pastures in the Amazon could be transformed into agroforestry systems, a multispecies land use offering benefits for both nature and people, such as carbon storage, biodiversity protection, sustainable livelihoods, and improved food security (Ollinaho & Kröger 2023). Agroforestry can also help restore and enhance ecological connectivity in the landscape (Kremen & Merenlender 2018). The increased fragmentation of the forest as a result of deforestation activities hinders animal movement and disrupts vital ecological processes (e.g., seed dispersal, gene flow, colonization of suitable areas), posing risks to ecosystem functionality and the survival of various species, while impeding sustainable development as small forested areas lack the capacity for conservationist land management practices (SPA 2021). Agroforestry on degraded lands, despite its potential, falls short when compared to the significant expansion of pastures and plantations. This foregrounds the importance of forming and reinforcing alliances among actors who share transformative multispecies sustainability and justice principles towards the policy interventions and investments that are urgently needed to support local sociobioeconomies in the Amazon region.

First, governments in the region should implement measures to cease actions that endanger Indigenous and other local communities and their sociobioeconomies (Garrett et al. 2023). An Amazon bioeconomy policy should commit to strongly avoid any deforestation, including legal deforestation, and stop forest degradation and land grabbing (Bergamo et al. 2022). It is essential to enhance governance and enforcement capacities, and redirect funds from activities that directly harm existing sociobioeconomies towards agroecology practices. In addition, increasing demand, funding, and channels for promoting ecosystem services and valuable, environmentally friendly products (e.g., fruits, nuts) as well as research in collaboration with local populations and regional research institutions should be a priority. Another crucial aspect is to strengthen the land rights of Amazonian communities. This can be achieved by implementing existing legislation or reinforcing it to formally acknowledge their territorial rights, while enabling these communities to effectively oversee and prevent deforestation as well as the degradation of forest ecosystems (Bastos Lima 2021; Garrett et al. 2023). There is significant evidence that Indigenous/local forest management has been more successful than other methods in reducing deforestation.

Second, a participatory 'bottom-up' democratic governance model that empowers Indigenous organizations, family farming associations as well as cooperatives and community enterprises (especially by women and youths) is essential for driving transformative change in the region. This model ensures that local voices are both heard and respected, alongside the preservation of the ancient socioeconomic practices and cultural diversities of Amazonian peoples. It also facilitates the equitable sharing of bioeconomy benefits and the inclusion of a wide range of ideas about the bioeconomy, encompassing both new and traditional knowledge and technologies to ensure the diversification of production methods that value biodiversity (Bastos Lima 2021; Bergamo et al. 2022; Garrett et al. 2023). Moreover, legal and political representation for non-humans and combining rights of nature and animal rights approaches are necessary actions to achieve transformative bioeconomy governance. This shift aims to move away from prevailing human-centred perspectives towards a multispecies orientation (Schapper et al. 2022).

Third, the establishment of interconnected networks that span across diverse scales and communities becomes essential to enhance local engagements with sociobioeconomies. This undertaking will necessitate persistent dedication, culminating in a joint political endeavour encompassing countries within the region (Garrett et al. 2023). The 'Belém Declaration', signed during the 2023 Amazon summit, reflects the consensus among member countries of the Amazon Cooperation Treaty Organisation. While specific measures were not outlined, the declaration serves as an initial step, as it emphasizes the imperative to prevent the forest from reaching a critical tipping point and enhance regional collaboration (Organização do Tratado de Cooperação Amazônica (OTCA) 2023).

Considering the case of Praialta Piranheira and the active citizenship, or 'forestzenship', demonstrated by individuals like nut collectors Zé Claudio and Maria prompts us to conceive of the *forces of reproduction* as subjects and agents that sustain humanity by caring for the biophysical environment that underpins life itself (Barca 2020). These forces stand in contrast to the forces of production outlined in Western political-economic paradigms, founded on continuous exploitation of humans, non-humans, lands, and waters for economic growth.

Shifting our perspective to focus on forces of reproduction encourages us to reimagine bioeconomies. This reimagining aims to foster social and ecological sustainable futures in regions of diverse species, such as the Amazon. However, these forces of reproduction often preserve the world, yet their ecological agency remains largely unnoticed in prevailing narratives. This is due to their misalignment with mainstream accounts of human advancement and well-being in the global capitalist economy (Barca 2020). Crucially, neoliberal frameworks driven by economic expansion often align with anthropocentric worldviews that regard nature as a passive backdrop for human exploitation and control; despite that, it is possible to reject neoliberalism without entirely relinquishing anthropocentrism. As an example, ecosocialist perspectives may recognize the existential threat to humanity posed by capital accumulation (Foster 2023) while not fully considering its effects on the non-human world or acknowledging the agency and contributions of non-human entities in shaping our shared reality (Barua 2018). This emphasizes the significance of multispecies justice perspectives, which actively challenge human exceptionalism and advocate for recognizing agency and knowledge as diverse and interconnected. Transformative multispecies land use ethics should involve acknowledging the agency and lifeways of those subjects and collectives that already participate in (social, cultural, environmental) practices that enable the survival and flourishing of multispecies communities.

12.3 Conclusion

This chapter advocated for a transformative multispecies land use ethic in the Amazon. This involves recognizing the possibility of ensuring and enhancing human livelihoods and respecting the rights of local communities. Simultaneously, it entails caring for and honouring the lives and habitats of other species and beings. Exemplified by the concept of 'forestzenship' and instances such as Praialta Piranheira and local environmental activism such as Kawsak Sacha, the development of reproduction and consumption systems grounded in socio-ecological justice, economic equity, respect for territorial sovereignty, and political inclusion that goes beyond human-centred perspectives can uphold the well-being of both nature and diverse communities.

This multispecies land use ethic is non-anthropocentric, emphasizing the interconnectivity, intrinsic value, and agency of all inhabitants of the Amazon, human and non-human alike. Moreover, it is inclusive, relational, and intercultural, embracing local, traditional, and Indigenous knowledge systems, and encompassing groups often marginalized by dominant political-economic structures. It is also reproductive, promoting practices that support ecological sustainability while concurrently preserving or even enhancing human livelihoods.

The Amazon tipping point bears global significance due to the rainforest's vital role in the Earth system. Land use changes within the forest impact scales ranging from local to planetary, underscoring the crucial necessity of safeguarding Amazonian ecosystems for a stable planet. However, this chapter asserted that remedies for the region's land use challenge ought to emerge from within the region itself. Despite the Amazon's planetary importance, the emphasis should be on local solutions, avoiding the imposition of global strategies that might be driven by alarming narratives, which legitimize top-down control and neglect regional realities. The global society should contribute to enhancing sociobioeconomies in the region by promoting the value of local knowledge, practices, and institutions.

This approach strives to prevent a catastrophic tipping point in the Amazon while addressing inequalities in the region, nurturing a future where all species can flourish.

References

Amazonian Socio-Biodiversity Meeting. (2021). Letter from the Amazon 2021 to the Participants of the 26th United Nations Framework Convention on Climate Change (COP26). https://s3.amazonaws.com/appforest_uf/f1636128732103x816951933313834600/LETTER%20FROM%20THE%20AMAZON%202021%20PARA%20COP26%20%28EN%29%20vf%20%281%29.pdf.

Antunes, C. (2023). Caubóis do Carbono' loteiam a Amazônia. *Sumauma*, 26 June. sumauma.com/caubois-do-carbono-loteiam-a-amazonia/.

Barca, S. (2020). *Forces of Reproduction*. Cambridge: Cambridge University Press.

Barua, M. (2018). Animal Work: Metabolic, Ecological, Affective. Society for Cultural Anthropology. culanth.org/fieldsights/animal-work-metabolic-ecological-affective.

Bastos Lima, M. G. (2021). *The Politics of Bioeconomy and Sustainability: Lessons from Biofuel Governance, Policies and Production Strategies in the Emerging World*. Cham: Springer International.

Bastos Lima, M. G. (2022). Just Transition towards a Bioeconomy: Four Dimensions in Brazil, India and Indonesia. *Forest Policy and Economics*. 136, 102684. DOI: 10.1016/j.forpol.2021.102684.

Bergamo, D., Zerbini, O., Pinho, P., & Moutinho, P. (2022). The Amazon Bioeconomy: Beyond the Use of Forest Products. *Ecological Economics*. 199, 107448. DOI: 10.1016/j.ecolecon.2022.107448.

Celermajer, D., Schlosberg, D., Rickards, L., et al. (2021). Multispecies Justice: Theories, Challenges, and a Research Agenda for Environmental Politics. *Environmental Politics*. 30(1–2), 119–40. DOI: 10.1080/09644016.2020.1827608.

Costa, F. d. A., Nobre, C., Genin, C., et al. (2022). Bioeconomy for the Amazon: Concepts, Limits and Trends for a Proper Definition of the Tropical Forest Biome. World Resources Institute Brasil. www.wri.org/research/bioeconomy-amazon-concepts-limits-and-trends-proper-definition-tropical-forest-biome.

van Dooren, T., Kirksey, E., & Münster, U. (2016). Multispecies Studies. *Environmental Humanities*. 8(1), 1–23. DOI: 10.1215/22011919-3527695.

Ferrante, L. (2023). O Crédito 'Verde' da Amazônia. *Revista Cenarium Amazônia*. 4(36), 6–11.

Finer, M., & Mamani, N. (2022). Amazon Tipping Point: Where Are We? MAAP #164. www.maaproject.org/2022/amazon-tipping-point/.

Foster, J. B. (2023). Planned Degrowth: Ecosocialism and Sustainable Human Development. *Monthly Review*. 75(3).

Garrett, R., Ferreira, J., Abramovay, R., et al. (2023). Supporting Socio-Bioeconomies of Healthy Standing Forests and Flowing Rivers in the Amazon: Policy Brief. Science Panel for the Amazon (SPA). www.theamazonwewant.org/spa_publication/policy-brief-supporting-socio-bioeconomies-of-healthy-standing-forests-and-flowing-rivers-in-the-amazon/.

Governo Federal do Brasil. (2023). Plano de Ação para Prevenção e Controle do Desmatamento na Amazônia Legal (PPCDam): 5ª Fase (2023–2027). Brasília: Ministério do Meio Ambiente e Mudança do Clima.

Greenfield, P. (2023). The 'Carbon Pirates' Preying on Amazon's Indigenous Communities. *The Guardian*, 21 January. www.theguardian.com/environment/2023/jan/21/amazon-indigenous-communities-carbon-offsetting-pirates-aoe.

Houart, C. (2022). Forests and Rivers: Non-Anthropocentric Conceptions of Political Community. *CES Contexto*. 31, 58–66.

Inoue, C. Y. A. (2018). Worlding the Study of Global Environmental Politics in the Anthropocene: Indigenous Voices from the Amazon. *Global Environmental Politics*. 18(4), 25–42. DOI: 10.1162/glep_a_00479.

Kremen, C., & Merenlender, A. M. (2018). Landscapes that Work for Biodiversity and People. *Science*. 362(6412), eaau6020.

Lapola, D. M., Pinho, P., Barlow, J., et al. (2023). The Drivers and Impacts of Amazon Forest Degradation. *Science*. 379(6630), eabp8622. DOI: 10.1126/science.abp8622.

Leopold, A. (2020). *A Sand County Almanac and Sketches Here and There*. New York: Oxford University Press.

Lima Filho, F. L., Bragança, A., & Assunção, J. J. (2021). The Economics of Cattle Ranching in the Amazon: Land Grabbing or Pushing the Agricultural Frontier? Climate Policy Initiative. www.climatepolicyinitiative.org/publication/the-economics-of-cattle-ranching-in-the-amazon-land-grabbing-or-pushing-the-agricultural-frontier/.

Mori, L. (2023). 'Grande vitória, mas não fim da ameaça': O que acontece após STF decidir contra marco temporal para terras indígenas. BBC News Brasil. www.bbc.com/portuguese/articles/c16k386gzrko.

Nobre, C. A., Feltran-Barbieri, R., Costa, F. d. A., et al. (2023). New Economy for the Brazilian Amazon. World Resources Institute Brasil. www.wribrasil.org.br/sites/default/files/2023-06/Executive%20Summary%20%28english%29.pdf.

Ollinaho, O. I., & Kröger, M. (2023). Separating the Two Faces of 'Bioeconomy': Plantation Economy and Sociobiodiverse Economy in Brazil. *Forest Policy and Economics*. 149, 102932. DOI: 10.1016/j.forpol.2023.102932.

Organização do Tratado de Cooperação Amazônica. (2023). Conheça a Declaração de Belém assinada pelos países amazônicos na Cúpula. http://otca.org/pt/conheca-a-declaracao-de-belem-assinada-pelos-paises-amazonicos-na-cupula/.

Pereira, J. C., & Gebara, M. F. (2023). Where the Material and the Symbolic Intertwine: Making Sense of the Amazon in the Anthropocene. *Review of International Studies*. 49(2), 319–38. DOI: 10.1017/S0260210522000134.

Pereira, J. C., & Terrenas, J. (2022). Towards a Transformative Governance of the Amazon. *Global Policy*. 13(S3), 60–75. DOI: 10.1111/1758-5899.13163.

Pereira, J. C., & Viola, E. (2022). *Climate Change and Biodiversity Governance in the Amazon: At the Edge of Ecological Collapse?* New York: Routledge Taylor & Francis Group.

Phillips, T. (2023). Outrage as Brazil Law Threatening Indigenous Lands Advances in Congress. *The Guardian*, 31 May. www.theguardian.com/world/2023/may/31/brazil-law-indigenous-lands-congress

Richardson, K., Steffen, W., Lucht, W., et al. (2023). Earth beyond Six of Nine Planetary Boundaries. *Science Advances*. 9(37), eadh2458. DOI: 10.1126/sciadv.adh2458.

Sabourin, E., Craviotti, C., & Milhorance, C. (2020). The Dismantling of Family Farming Policies in Brazil and Argentina. *International Review of Public Policy*. 2(1), 45–67. DOI: 10.4000/irpp.799.

Schapper, A., Visseren-Hamakers, I. J., Humphreys, D., & Bliss, C. (2022). Mainstreaming the Animal in Biodiversity Governance: Broadening the Moral and Legal Community to Nonhumans. In I. J. Visseren-Hamakers & M. T. J. Kok, eds.,

Transforming Biodiversity Governance. Cambridge: Cambridge University Press, 179–99.

Schröder, A. (2023). The counterstrike: Brazilian Congress Moves to Block Lula's Environmental Agenda. *Mongabay*, 1 June. news.mongabay.com/2023/06/the-counterstrike-brazilian-congress-moves-to-block-lulas-environmental-agenda/.

Science Panel for the Amazon (SPA). (2021). Amazon Assessment Report 2021. www.theamazonwewant.org/amazon-assessment-report-2021/.

Tschakert, P., Schlosberg, D., Celermajer, D., et al. (2021). Multispecies Justice: Climate-Just Futures with, for and beyond Humans. *WIREs Climate Change*. 12(2), e699. DOI: 10.1002/wcc.699.

Vilani, R. M., Ferrante, L., & Fearnside, P. M. (2023). The First Acts of Brazil's New President: Lula's New Amazon Institutionality. *Environmental Conservation*. 50(3), 148–51. DOI: 10.1017/S0376892923000139.

Visseren-Hamakers, I. J., & Kok, M. T. J. (2022). The Urgency of Transforming Biodiversity Governance. In I. J. Visseren-Hamakers & M. T. J. Kok, eds., *Transforming Biodiversity Governance*. Cambridge: Cambridge University Press, 3–22.

13

In Conversation with a Land

Extracts from the Pianpicollo Diary

ALICE BENESSIA

Over the past decades, we have been engaging in the intricate attempt to measure our effect on the global biogeochemical equilibrium that has gently allowed us to flourish as a species, until now. We hope that robust quantification – pinning down planetary boundaries as thresholds of irreversibility into the flux of destruction we are bringing about – can make the needed change of course acceptable and desirable, thus possible. We are set to demonstrate how and when the transformation we are generating can push us out of a global 'safe operating space', when it can become dangerous *for us*.

The planetary boundary known as 'land system change' refers specifically to a threshold in the overall percentage of land we can irreversibly transform for our own use, most typically for food production and resource extraction. As it can be measured with some degree of accuracy, the percentage of reduction in the global forest area is taken as the main control variable for representing all 'land system change' – locating one endpoint in the constellation driving our risky and damaging journey.

As we proceed in this kind of modern endeavour – looking for scientific truth to induce collective good – and face its crisis, we can pause for a moment and explore a different path. What if instead of speaking the language of control variables and safe operating spaces, we could actually engage in a dialogue with land systems, listen to what they have to say, and learn how to respond? What if we could learn to collaborate with the others that surround, sustain, inhabit, and even constitute us? Not only the other (human and nonhuman) animals, but also plants, fungi, lichens, microorganisms, minerals, rocks; and, rivers, forests, mountains?

Asking these questions, inherent in many Indigenous cultures around the world, implies a change of posture from the Cartesian dream of power and control over all others, breeding separation and dominion as a way of life.

I have been experimenting with this change of posture for the past 10 years, through a daily conversation[1] with Pianpicollo Selvatico, the land that hosts and nourishes me, together with a large family of other beings.

Pianpicollo Selvatico is home to a hamlet in an isolated valley that rests on a small plateau, an anomaly in the terraced landscape of Alta Langa, in the Northern Italian region of Piedmont. It owes its name to this peculiarity. Pianpicollo was translated in recent times from the word Cianpicòl, 'small plateau' in the dialect of southern Piedmont. I found it written in a map from the late eighteenth century, pre-dating the stone farmhouse that was built in 1811. Shortly after settling here, I discovered in a casual conversation with some neighbours that the place name is associated with an adjective, Selvatico ('wild' in Italian), which distinguishes it from another Pianpicollo called Domestico ('domestic'), a farmstead further down the valley, facing South. It is wild, I was told, because it is isolated and because the snow is slow to melt. This is because it faces northeast, another anomaly from the southwest of traditional farmsteads. I also found out that Northeast is the orientation of monasteries and Ciampicòl (the wild) is synonymous of a remote place. To go to Ciampicòl is used to mean 'on top of the world', 'to the devil's house', and other similarly colourful expressions. Indeed, in Pianpicollo the wild is very present. Throughout the seasons, the nights are populated with a variety of sounds of forest animals. Wild boars, deer, wolves, foxes, dormice, nocturnal raptors, and much more that I cannot recognize, or that simply wanders silently around. In an environmental monitoring experience, I heard the summer presence of bats through a sensor that turns ultrasound into audible frequencies. There is a whole world of creatures that live undisturbed in the landscape at night and hide from human presence during the day. In an opposite movement, the domesticated creatures that inhabit the hamlet with me, a small group of human and nonhuman animals, venture into the landscape during the day and retreat to safety and rest at night.

The plants all around seem to be quietly witnessing as the tides of mammals, birds, insects, reptiles, and amphibians alternate. Mammals, both domesticated and wild, can be very disturbing, eating bulbs, roots, stems, and branches, gnawing barks, digging holes, compacting, and turning over soil. Domesticated plants in the vegetable garden, orchard, and tree nursery are protected by enclosures of various kinds – higher, lower, made with wood, nets, wires – partly because they are younger and more fragile, partly because they are functional to us, human mammals. We decide where and how they grow, if and where they reproduce, often even when it is the time for them to die. Meanwhile, wild plants coexist with the rest of the living, more exposed to contingency. Two types of existence separated from each other by physical barriers and genetic differences.

[1] In its original meaning, 'to keep company with, to turn about with' from *cum* 'with, together', *versare*, 'to turn, bend'.

In the small plateau there is also a large pasture. I realized very early on that at our latitude prairies are ecological systems out of balance. They continue to exist only with constant seasonal maintenance. If you let them be – so to speak – they reclaim their equilibrium state, turning back into forests. The first colonizers are barbed shrubs, brambles, dog roses, blackthorns, and hawthorns. They say, as I read in a book, that meadows arm themselves with thorns to defend the tree seeds from animals. I see this happening in the hedges at the margin with the woodlands, occupied by thorny shrubs, cracked with precision where wild boars decide to create their paths. Plant societies, I realize, are not actually helpless, much less inert; the issue is much more complex. The margins between what we claim as domestic and wild are permeable and fuzzy, both in time and space.

Along with me, several farmstead animals are permanent residents in Pianpicollo. As of today, two donkeys and three horses, two pigs and six hens. They came here as a gift, sometimes in distress, through various encounters and often difficult circumstances. Some already old, some just born, some in the prime of their lives. They belong to species that have been domesticated for millennia and selected in more recent times with increasing, brutal precision, to fulfil specific functions. Programmed in their behaviour and genetic expression.

In Pianpicollo, they no longer have a functional life; they are not eaten or put to work. They are free to move around the valley during the day and return, at different times, to shelters of various shapes and sizes where they take refuge for the night.

Sometimes, we cross each other during the day, while busy in our own affairs, as one would meet a friend on the road. We improvise the rhythm of the days, according to weather, season, dangers, and opportunities. We make dates and we show up on time unless serious impediments occur.

Other guests take temporary refuge in the small community, at times. Especially young mammals, left alone in unknown circumstances. They find a human home where they can grow before going back to the wild.

Day after day, I conduct an experiment of mutual care and transformation with the land and her[2] inhabitants. I take the time to observe them, to be with them for no particular purpose. In that silence of intention, I feel their presence as beings in a form, somehow accessible, yet other than mine. Inner and outer landscapes merge and expand. Boundaries blur and sometime dissolve.

Other humans – artists, researchers, farmers, students – people of different age and life paths join me from time to time, from all over the world.[3] In the evening, I take note of some of the day's thoughts and events.

[2] Throughout the text I don't use the neutral form when referring to land, animals, and plants, to indicate personhood.

[3] Pianpicollo Selvatico is home to the eponymous Foundation, devoted to research at the boundary between art, deep ecology, and regenerative agriculture. The Pianpicollo Selvatico Foundation ETS investigates the

13.1 Extracts from the Pianpicollo Diary

ar
Sanskrit root of the word *art*
'to set things in motion', 'to move towards'
etimo.it (Italian – Etymological Dictionaries; author's translation)

2022

JULY 16. The heat is brutal. We keep digging in the shade of a tent, guided by Andrea and Raffaella. In three small groups, finding our way into the soil. The ecological crisis is an aesthetic crisis at its root, they say, a crisis of perception. We are asked to observe everything we encounter with equanimity. Preserve life as we can and keep track of the stratigraphy so we can carefully recompose it on the final day of the workshop when we close the holes. The word aesthetic can be understood as the opposite of anaesthetic, or numbness. I remember reading it in a dialogue between Volker Harlan and Joseph Beuys. Then the art process has to do with aliveness. Digging into the woods is painful. I do not want to cut roots. I end up working with my fingers only. Could we dig like worms do? Or grow underground like roots? Older trees struggle more with drought, I am told. Their root system has grown in a different time when water was easily available. Now they cannot reach down. As we go deeper, silence expands. We are leaving Gaea, the surface, and entering Chthonia, the submersed. In the evening, a conversation emerges around the fire, while making charcoal for the drawing sessions. Young researchers are touched by the experience. Some of them measure soil routinely but none of them has ever encountered it. Out of their comfort zone, they trust the process, and they begin to trust each other. Disciplines disappear. They pay attention to inner and outer worlds. A new kind of collective awareness seems to be arising. I remember reading somewhere that truth and trust share the same root. Pursuing truth in science involves isolation and fragmentation of wholes into pieces. Disrupting integrity, ending life to measure it. It implies the breaking of trust with the world in its foundation. A sombre paradox.

JULY 17. In the afternoon I hear some amused cries in the distance. A badger comes out of the woods and approaches two researchers at work. Within a short while the tone of their voices changes, I hear my name from afar, and the phone

relationship between humans and nonhumans through practices of interspecific coexistence, collaboration, and care. It promotes artistic, scientific, and philosophical research experiences as transformative processes. It is a collective experiment that supports all forms of diversity, cultural, and ecological, in the service of life as a whole. Every year, it hosts artists and other researchers from around the world, students and people of different ages and backgrounds, in long-term residency and training programmes. Transdisciplinary, participatory, and immersive, the research and training residencies provide the space and time to go to the roots of one's practice, in collaboration with the land and her community of beings.

rings. The badger has surrounded them in a convulsive run, then slumped at their feet. He looks sick, they tell me. Plants and animals have been asking for water for months, and I know what to do. I rush in with a water bottle, a wet towel, and a wheelbarrow full of straw. The badger is motionless on the ground, lying on his side. His breathing is short, his eyes and mouth half-closed. Our bodies hold him in the shade. He surrenders to the embrace of the cold, soggy cloth, feels water in his mouth and reflexively begins to swallow. With the help of a friend, we put him in the wheelbarrow and carry him home. In a small, silent procession, our shadows protect his body from the sun. I open the towel and we lay him down in an outdoor stone sink. His breathing is more regular. I continue to bathe and water him. I notice that he is a young male, still growing. In a few minutes he drinks more than three litres of water. I have never touched a badger. I feel his body through my gloves. I pause to look at his massive, bear-like paws, his sharp teeth and bristly, wet fur. As soon as he recovers, in terror, he tries to bite me. I feel the force of the bite, and the glove is pierced. We release him into the tree nursery enclosure, waiting for evening. He bites the net to get out. At dusk, we take him back into the woods in the wheelbarrow again, wrapped in the tarp, this time to protect us. We let him go at the margin of the forest where he came from. As soon as he can, he vanishes, without looking back. Everybody is shaken. I stay awake through the night.

JULY 18. The puddle at the edge of the woods is dry and cracked, like everything else. I saw it again yesterday, when releasing the badger. I remember being annoyed sometime last winter, watching a hunter drive too fast down the muddy little road, knowing that I would find an inconvenient hole, right in the bend next to the woods. Over time, I have watched that hole fill with rainwater and become a precious puddle for wild animals. My two pigs have discovered it in the late spring, and with the first warm weather I have found them soaking in it, after eating cherries in the nearby woods. With the drought I have seen it dry up, inexorably. Today all of a sudden it seems obvious that we must refill it. We wait for the night, load a large glass bowl full of water onto the wheelbarrow and go to pour it. Sixty litres are barely enough to heal the cracks. And to think that this is probably illegal – my friend whispers. I can barely see his eyes below the headlight, and we look at each other in silence. On the third lap the puddle fills up and I retreat for the night. I struggle to sleep again.

JULY 19. The sun is scorching. Tired and distracted, I sit in the workshop's drawing session, next to the stone sink. Suddenly, without a thought, my hand reaches for a charcoal and starts looking for the badger on paper. When it stops, I realize that instead of bear paws there are tiny limbs, almost sprouting. In the evening at dinner, at the long group table, I pick up a fork red of gravy, resting on a napkin. I see the claws appearing in that blotting paper. The ones I could not find in my drawing. A sense of stillness invades me. After dinner we pour more water into the puddle. Maybe tonight I can get some sleep.[4]

November 15. 'How long does a "White of Saluzzo" hen live?' I asked Achille today. He paused for a moment and then said, 'I am not sure. They are not allowed to get old'. He breeds them in the Veterinary Department of the University of Torino. I adopt the leftovers. When Priscilla collapsed and died, she was five years old. Archimede and Isidora did not leave her body for the whole day. They stood there; their eyes closed at times.

November 16. The work for Animot[5] is advancing and in the afternoon, I go back to a dictionary, to look for the actual etymology of truth and trust. The common origin is *deru also *dreu-, a Proto-Indo-European root, meaning 'be firm, solid, steadfast'. As I keep reading, another word comes to mind: tree. I check and yes, it comes from the same root. Actually, from the *main root:* the way older Sanskrit *dru. So, trust and truth are one, in a way, and they are rooted, quite literally, in tree. Around dinnertime, I walk out to feed the horses and donkeys. The air is crisp, and the sky is clear, lots of stars. On the way, I pass by the usual old cypress. He has been here way longer than me. I never pay attention to him. As I walk under his canopy, I feel the urge to pause and look up. I just stand there for a while, in gratitude. Then Zefira calls me up. Donkeys have some patience but not too much.

2023

March 12. In the morning, I walk with the donkeys to their pasture along the usual path. On my way back home a ray of sunlight shifts my attention to a distant tree. She is a fully adult oak tree, majestic and slightly isolated from the forest on the ridge. She was maybe spared to mark the margin between two properties, once separated. She has two main trunks and her position on a slight slope makes her very accessible, even to someone who is not that great at climbing trees. I find myself walking out of the path, drawn to her for a visit I never paid before. Once I reach her, I climb on her first ramification with no hesitation, on an impulse. I just sit there, taking in the view and feeling her contact, her solidity. I do not have any agenda. As silence pervades my being, water comes in. It comes as an unequivocal perception, a presence in my mind. I take it to be an empirical fact that the oak is

[4] These first entries of the diary refer to a five-day research residency called: 'IN TERRA: dalla misura all'incontro' ('ON EARTH: From measure to encounter'). It is a chapter of a biennial project (2021–2023) called BRIDGES – Building Reflexivity and response-ability Involving different narratives of knowleDGE and Science, supported by Fondazione Cariplo, led by the Italian National Research Council (CNR) in a partnership with the Pianpicollo Selvatico Foundation ETS. Young Italian researchers in both science and humanities were invited to work together on the issue of soil fertility, through a workshop led by the artist duo Andrea Caretto and Raffaella Spagna, in collaboration with myself and other members of the BRIDGES staff.

[5] In 2022, I had the opportunity to share and expand my conversation with the land by reaching out to artists and researchers around the world, while editing the issue no. 13 of the journal *Animot* (Benessia 2023).

sitting on water, and I with her. Words of intention bubble up to my conscious self, as a response to that perception: 'The water is yours, and not mine to be taken'. I just sit there with that experience. As I climb down, I venture into the nearby woods, with the vague memory of a puddle where animals used to drink and bathe in the water surfacing from underground. Before the drought, we had a wildlife camera set there for a while, during an environmental assessment. I find it soon, and as I look down, I realize it is perfectly aligned with the oak tree downhill.

MARCH 25. As I walk back from the pasture, I take the time to visit the oak tree again. I sit under her low branches, on the old winter grass. The perception of water comes up again, this time as a memory. I just look out and my breath slows down. The land is quiet. At some point in time, a full clear sentence pops into my mind, as in a dream: 'Plant some trees around me and I will teach them to reach the water.' Words surface in the form of an invitation coming from the oak tree. I smile thinking I am giving voice to a tree – or it is the tree giving voice to my being? In dreams the question is pointless. As I walk back a resolution falls into place. In the nursery, a family of very young trees is waiting to find a permanent home. When I collected them from the woods around the land, two years ago, I knew they would not have a future where they were born. But I did not know what I was saving them for. There was no why.

MARCH 27. Silvia is at work today, helping me plan the vegetable garden for the summer. We take the time to walk to the oak tree and I tell her what happened. As I retrace the story, I hear myself telling her: 'It feels like a call from the land, to help coping with the drought. We just have to pay attention – attend, being present – and respond.' To which she smiles replying: 'Well, that's what Indigenous people do, all the time'. She is young and has been living for a few years in the Amazon of Ecuador. The experience informs her way of being and working. 'Planting trees is planting water', she adds. We agree to join forces in the fall. She will help to design the forestation. As we are having tea, we appreciate the irony of the whole situation. Silvia was invited as research fellow for a project on regenerative agriculture that I wrote more than a year ago.[6] The project was funded early this year and we agreed that in the fall she would design an edible forest next to the vegetable garden. We did not imagine at the time that Pianpicollo would have other priorities. 'It looks like I have been recruited by the land herself', she tells me with enthusiasm. At night, tree-trust-truth comes to mind again.

MAY 7. On my way home I see the rotten pumpkin once again, with a hint of embarrassment. I forgot her weeks ago on a stonewall, resting on a piece of newspaper. I had found her going bad in the storage space. As I was taking her to the compost pile, I was distracted by chickens walking by, asking for water.

[6] 'Inviting Life' is the sixth edition the Pianpicollo Research Residency, a yearlong research programme in which artists and researchers of other disciplines are invited to work together, in connection to the land. Run by the Pianpicollo Selvatico Foundation ETS, it is funded by local private and public institutions (Fondazione Cassa di Risparmio di Cuneo and Regione Piemonte).

Nourished from within, her own seeds have sprouted into dozens of tiny plants. A display of renewal speaking to my disregard. After I move them into a pot, I do my best to detach the newspaper from what is left of the original pumpkin, so that it does not end up in the compost. When I decide that the effort is enough, scraps of paper are still attached to the stone surface. Detached words appear to create a sentence: *Galileo abitualmente spara*. Galileo habitually shoots.

JUNE 5. I woke up today with words in my head, coming from a dream I could not remember. *Senza sforzo alcuno, come l'acqua*. Without any effort, like water. A few repetitions occurred before I was fully awake.

JUNE 27. Brus has been here for five days now. As we meet for breakfast he says, 'I had a dream'. His friend Lupe stares at me, as to prepare me for something significant. Spanish is deceptive, it creates an illusion of closeness, forced by colonization. Words do not quite mean what they seem.

Brus is the first of his community to leave the forest, he told me when we first met. His village is into the depths of the Amazon, at the boundary between Peru, Brazil, and Colombia. He has been traveling with his artwork through galleries and museums around Europe, and he is now having coffee in my kitchen, as an invited artist in residence. He travels with the help of Lupe, a filmmaker from Lima, and his teacher plant, tobacco. 'I dreamt of a powerful spirit, a water spirit, ruling onto this land', he says. 'The spirit told me: *Eso es lo que ay*. (This is what there is)'. When I ask what the spirit meant, Brus answers by simply repeating *Eso es lo que ay*. Pointing to a vein in his left wrist, he adds, 'He showed me a vein'. I dare asking if he remembers the spirit living in a specific place, and if he can take me there. He says yes, and yes. We walk out in silence and after a little while we end up in the slope above the oak tree: 'The spirit is here, underground'. I tell him about my experience. He is unsurprised. I take him to the puddle. Looking down at the path to the oak he says: 'This is the vein', and adds: 'This place is sacred, it is where all beings meet. Eso es lo que ay'. On the way home I hear his voice behind me: 'All you have to do is what you are doing'.

AUGUST 3. The puddle near the oak has been dry for weeks. Before sunrise Juan and I take the horses and donkeys to the pasture in the shade. They have only a few hours before the heat and the flies force them back to the stable. On the way back we decide to follow the path of the wild boar into the woods downhill. They rub themselves against the trunks, marking them with mud as they move. Mud means water. We reach two puddles lower down in the valley. One of them shows only signs of humidity, the other has still some water. We spend some time in silence there, sharing relief. Juan moulds some mud into a small sculpture. Then the flies remind us that we need to move. We leave the sculpture behind, as an offering. A sacred place is where all beings meet. I keep thinking about it.

AUGUST 5. In the late afternoon we work in the vegetable garden. As the sun goes down mosquitos force us to leave. We realize that when Juan began his summer visits, only a few years ago, there were none. Change is happening fast. As we walk back home, I feel the urge to bring water to the oak puddle. Juan can help me. We fill the big glass bowl and load it onto the wheelbarrow. The puddle is far and uphill. When we get there, we are both out of breath, feeling the exhaustion of the day. As we start pouring, a wave of emotion hits me, and I end up bursting into tears. Juan stands on the other side of the puddle in silence. Then he hugs me, and we sit together until it is dark. On the way back we both feel observed by the woods.

August 6. At breakfast, Juan mentions we could invite his friends from Colombia to help with the forestation in the fall. I sense his love for them. They are family, since he made a documentary about their relationship with medicinal plants, years ago. As we speak, the forestation becomes a collective ritual of three weeks. Kamëntsá people would be powerful allies, with their deep connection to land and elements. Artists, scientists, and farmers that have been working in Pianpicollo over the years could be invited. An extended family reunion. We could plant all the three hundred trees that are waiting in the nursery. The funding could be possible with some donations to complement the grant. It could become a celebration for and with the land, sharing knowledge and experience to help bringing the water back. The plan is ambitious. As Juan makes another coffee, we decide to consult the I Ching (Figure 13.1).[7]

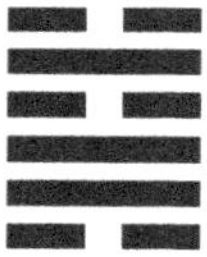

Figure 13.1 Hexagram 48, The Well.

Hexagram 48, The Well. Above K'AN THE ABYSMAL, WATER. Below SUN THE GENTLE, WIND, WOOD. Wood is below, water above. The wood goes down into the earth to bring up water. The image derives from the pole-and-bucket well of ancient China. The wood represents not the buckets, which in ancient times were made of clay, but rather the wooden poles by which the water is hauled up from the well. The image also refers to the world of plants, which lift water out of the earth by means of their fibers. The well from which water is drawn conveys the further idea of an inexhaustible dispensing of nourishment.[8]

August 7. We call Pastora in Sibundoy. I tell her the story and she is touched. 'When the elements call, we have to answer', she says while thanking us. She is honoured to join and will extend the invitation to another member of her family. She comes from a lineage of Kamëntsá shamans, working with yagé as teacher plant. 'Healing the land', she adds, 'is healing the Earth. There is no difference.'

[7] The I Ching, The Book of Changes, is an ancient Chinese divination text, based on 64 hexagrams. It evokes a correspondence between psychic laws and cosmic laws. Carl Jung – who wrote the introduction to the translation that popularized the text in the West (Wilhelm 1924) – defined the notion of synchronicity in relation to that correspondence. He extensively discussed the idea with quantum physicist Wolfgang Pauli. I have been consulting the I Ching for years and over time I keep finding new levels of meaning.

[8] Extract from Exagram 48 (Baynes 1950).

November 25. As I am finishing to protect the young trees with individual iron nets around their tiny trunks, I go back the incredible collective action of planting them. Everybody we wished for showed up. It was a remarkable learning and healing experience, way vaster than any of us could imagine. Dissolving boundaries in all realms and confirming that meaningful change can only be mutual. Among the many insights, a vision is taking shape within, like an inner sketch drawn in conversation with the land herself – with all her inhabitants. What if we could move western science out of its obsolete, suffocating shelter – from which 'Galileo habitually shoots' – and let it breathe with other ways of being in the world? What if we valued analogy over analysis, equanimity over hierarchy? What if we learned to perceive and witness correspondence, instead of struggling for causality? What if we gave up the need to predict the future, in exchange for the ability to respond to change, as it happens? What if we cherished *kairós* over *chrónos*?[9] We could experience a kind of action that knows no doubt and flows like water, without any effort (Figures 13.2 and 13.3).

Figure 13.2 *Tree at night*. Alice Benessia 2022, from the series 'Ask again'. Digital color photograph. 16-11-2022 / 7:50 pm / 4°C. First published in Benessia (2023: 35 and 129), reproduced by permission of Alice Benessia.

[9] In ancient Greek, *chrónos* is linear time, while *kairós* refers to a singular, propitious moment, the right moment. *Chrónos* is quantitative in its essence, *kairós* is qualitative. *Kairós* can lead to *aéon*, indicating the time of eternal present, in which duality dissolves.

Figure 13.3 *Badger*, Alice Benessia 2022. Charcoal drawing on paper, 20,5 x 14,5 cm 19-07-2022 / around 4 pm / 32°C. First published in Benessia (2023: 35 and 129), reproduced by permission of Alice Benessia.

References

Baynes, C. (1950). *The I Ging or Book of Changes*. New York: Pantheon Books.

Benessia, A., ed. (2023). So Long and Thanks for All the Fish. *Animot: The Other Philosophy*. 13. ojs.unito.it/index.php/animot/issue/view/614/393.

Wilhelm, R. (1924). *I Ging: Das Buch Der Wandlungen*. München: Diederichs.

Part VI

Freshwater Change

14

Introduction to the Freshwater Change Boundary

JEREMY J. SCHMIDT

In 2022, the United States Geological Survey released a new diagram of the global hydrological cycle that, for the first time, included people and the social processes affecting freshwater. It was a significant step and also a long-anticipated change. Humans have appropriated more than their share of Earth's freshwater for decades; thousands of mega-dams block rivers around the world, groundwater is pumped unsustainably in many places, and over half of the world's lakes are in decline, largely owing to human impacts on local water systems and the planet's climate (Postel et al. 1996; Yao et al. 2023). Even so, the update to the hydrological cycle arguably did not go far enough. This is because it is one thing to depict social processes that affect water and quite another to show the scale of anthropogenic impacts on the global water cycle (Abbott et al. 2019). Those impacts connect the planetary boundary for freshwater use to ethical concerns regarding how unequal distributions of water over time and space affect people, other species, and the processes that support life on Earth.

Like other planetary boundaries, calculations for freshwater change have undergone significant shifts since the framework was first proposed. In 2009, a boundary of 4,000 km^3 worth of annual freshwater was estimated, with only 419 km^3 having been used prior to the industrial era and a current estimate of 2,600 km^3 (Rockström et al. 2009). These planetary numbers don't factor in regional disparities, such as arid versus humid regions or places where human demands for agriculture or industrial extraction already overburden freshwater systems that are the 'bloodstream of the biosphere' (Ripl 2003). Even so, the initial estimate for the freshwater boundary was criticized for focusing on what hydrologists call blue water, which is the visible water in lakes, rivers, and taps. This water represents only a fraction of the global water system, with invisible 'green water' circulating throughout the atmosphere, soils, and biota in ways also affected by anthropogenic impacts on the environment. Land-use and climate changes, for instance, both significantly impact green water, such when forests are converted to arable or urban land. Further, these changes affect the entire freshwater system in complex ways. Together, these concerns prompted a reassessment of the freshwater boundary to include both blue and green water (Gleeson et al. 2020; Wang-Erlandsson et al. 2022). In the 2023 update to the boundary for freshwater change, the assessments for both blue and green water showed significant departures from pre-industrial estimates, with both having been transgressed for some time. The boundary for blue water, on the new model, was passed in 1905 and that for green water in 1929 (Richardson et al. 2023). Although labelled as 'preliminary' findings, the direction of travel for the new boundary on freshwater change showed an increasing risk to a key variable that, in previous assessments, had been characterized as within a 'safe space for humanity'.

Improved depictions of human impacts on global hydrology and redefining the boundary for freshwater change also raise ethical concerns the two chapters in this part examine. First, Nicole Wilson critically engages the freshwater boundary using the idea of a hydro-social cycle – an approach that does not begin by separating humans from nature, such as by making calculations of 'natural' or pre-industrial values from which the freshwater changes are calculated. Instead, hydro-social scholars emphasize how human-water systems are mutually co-constituted (Linton 2010). For instance, although ecohydrologists like Carl Folke (2003) have long argued for a coevolutionary view of humans and water, hydro-social scholars prioritize social relations and situate science as one kind of social practice among others, such as those of Indigenous peoples. From this perspective, a 'boundary' to freshwater use also requires an ethic that respects the bounded, different ways of knowing and relating to the world. Second, Neelke Doorn considers how the planetary boundaries framework should impact philosophical thought. When it comes to freshwater, a significant ethical concern is how to understand risk, from floods and droughts to engineering decisions. But this is tricky ethical territory, she argues, because the freshwater boundary is global, yet risks are experienced locally. In this context, Doorn considers what science and ethics can offer to one another given the respective aims of each – how can the intrinsic issues of equity that don't figure into the freshwater planetary boundary be addressed through a focus on justice, for instance? Doorn's emphasis on disciplinary complementarity contrasts with Wilson's disciplinary critique in ways that aren't so much two poles of a wide spectrum as two respective attempts to get to grips with the plurality of approaches to science and ethics affecting freshwater change.

References

Abbott, B. W., Bishop, K., Zarnetske, J. P., et al. (2019). Human Domination of the Global Water Cycle Absent from Depictions and Perceptions. *Nature Geoscience*. 12(7), 533–540. DOI: 10.1038/s41561-019-0374-y.

Folke, C. (2003). Freshwater for Resilience: A Shift in Thinking. *Philosophical Transactions of the Royal Society of London B.* 358(1440), 2027–2036. DOI: 10.1098/rstb.2003.1385.

Gleeson, T., Wang-Erlandsson, L., Zipper, S. C., et al. (2020). The Water Planetary Boundary: Interrogation and Revision. *One Earth.* 2(3), 223–234. DOI: 10.1016/j.oneear.2020.02.009.

Linton, J. (2010). *What Is Water? The History of a Modern Abstraction*. Nature, History, Society. Vancouver: University of British Columbia Press.

Postel, S. L., Daily, G. C., & Ehrlich, P. R. (1996). Human Appropriation of Renewable Fresh Water. *Science.* 271(5250), 785–788. DOI: 10.1126/science.271.5250.785.

Richardson, K., Steffen, W., Lucht, W., et al. (2023). Earth beyond Six of Nine Planetary Boundaries. *Science Advances.* 9(37), 2458. DOI: 10.1126/sciadv.adh2458.

Ripl, W. (2003). Water: The Bloodstream of the Biosphere. *Philosophical Transactions of the Royal Society of London B.* 358(1440), 1921–1934. DOI: 10.1098/rstb.2003.1378.

Rockström, J., Steffen, W., Noone, K., et al. (2009). A Safe Operating Space for Humanity. *Nature.* 461(7263), 472–475. DOI: 10.1038/461472a.

Wang-Erlandsson, L., Tobian, A., Ent, R. J. van der, et al. (2022). A Planetary Boundary for Green Water. *Nature Reviews Earth & Environment.* 3(6), 380–392. DOI: 10.1038/s43017-022-00287-8.

Yao, F., Livneh, B., Rajagopalan, B., et al. (2023). Satellites Reveal Widespread Decline in Global Lake Water Storage. *Science.* 380(6646), 743–749. DOI: 10.1126/science.abo2812.

15

The Role of the Planetary Boundaries Framework in Water-Related Management and Policy Decisions

NEELKE DOORN

15.1 Introduction

The attention of philosophers to water is of recent date. While the flow of water as a metaphor for life goes back to the pre-Socratic philosophers (cf. the well-known quote *panta rhei* – everything flows, often attributed to Heraclitus), the importance of water for life – and with that its essential role for both humans and nature more generally – has received little attention in the normative philosophy literature before 2010. This is surprising as the different and often competing water uses, the exhaustion of the earth's freshwater resources, and the effect of flood-prevention strategies on water availability and water quality pose questions that are deeply normative and by that in the heart of philosophical disciplines such as political philosophy, philosophy of risk, and ethics. Since 2010, maybe as a result of increasing attention to climate change, the topic of water is gradually receiving attention in the philosophical literature, focusing primarily on these normative questions and therefore often discussed under the label of water ethics (Doorn 2019; Meisch 2018, 2019).

This chapter aims to explore how the planetary boundaries framework can be useful for thinking about water-related risks complementary to frameworks developed in the philosophical literature on water ethics and risk ethics. Hence, the aim is not to criticize or further refine the planetary boundaries but rather to see how the philosophy literature can benefit from the framework and vice versa. To do so, I highlight three points from the water ethics and risk ethics literature that are particularly relevant to our thinking about the water planetary boundary. The first concerns the division of labour between science and management, the second concerns the multitude of water-related risks and risk-risk trade-offs, and the third concerns the metric of intergenerational justice in water policy and engineering. By discussing these three points, I show how the planetary boundaries framework can play a role in water-related management and policy decisions. In the concluding Section 15.5, I reflect on the role of water ethics in these decisions.

15.2 The Reception of the Water Planetary Boundary in the Literature

The water planetary boundary is one of the nine boundaries that, taken together, propose quantitative limits for human perturbation of critical earth system processes (Rockström et al. 2009). With the concept of planetary boundaries, Rockström and colleagues have developed a framework that can be used to demarcate a safe operating space for human activities. It is important to stress this goal: it is intended to develop the safe operating space itself, not the distribution within that space. The water boundary was originally labelled ‘freshwater use’, and in the early publications, it was formulated globally in terms of the maximum rate of consumptive use of blue water (Rockström et al. 2009). In later publications, the boundary was further refined with a limit at the basin scale as well, formulated in terms of maximum monthly withdrawals as a percentage of mean monthly river flow (Steffen et al. 2015).

In recent years, the framework has received some criticism: some points of criticism are warranted and also accepted by the initiators of the framework, but some points of criticism also seem quite unfair. One rightful point, which could also be read as a modest revision or refinement of the water planetary boundary, is the metric of the boundary – and in a way also the label itself. As stated earlier, it was originally formulated in terms of blue water consumptive use, because Rockström et al. considered this a good proxy variable for the planetary limit, integrating the three largest anthropogenic manipulations of the water cycle: human impacts on precipitation patterns, modifications of soil moisture by land use and land cover, and water withdrawals from discharge for human use. In a later paper, it was suggested to move away from human water use as a metric and to conceptualize the water planetary boundary as ‘stores of the water cycle’ (Gleeson et al. 2020). The reason for this move is that ‘the foundation of the planetary boundaries framework is the Earth System and, in the case of water, the Earth System functions of water’ (231). It makes indeed much more sense to express the boundaries in terms of geophysical variables directly related to this boundary (i.e., water stores) and not in terms of some indirect and aggregated variables like water use. In this later paper by Gleeson et al. (2020), which includes most of the original authors of the Rockström et al. (2009) paper, the authors carefully avoid the term ‘freshwater use boundary’ and replace it with ‘water boundary’. In this chapter, I will use this more recent label ‘water boundary’.

A different line of criticism comes from authors who attack the framework for not paying enough attention to contextual factors, including technological interventions and governance concepts (Bogardi et al. 2013), failing to consider questions of equity (cf. Steffen & Stafford Smith 2013), and insufficiently recognizing the impact of impaired water quality on the water system (One Earth 2020). With the changed focus from freshwater use to water stores, especially the reference to technological interventions does no longer seem apt. While these interventions may help us stay within the boundaries, they do in principle not affect the boundaries themselves.

Regarding equity, indeed, the planetary boundaries framework is not enough for resolving all issues. However, given the goal of the framework ('to provide the safe operating space'), this incompleteness is not problematic per se and could even be considered a strength. The framework is not intended as an allocation framework but rather as a set of constraints – or at most as a partial allocation framework – spanning the space within which allocation can take place. As such, the model fits quite well in the traditional division of labour between science, on the one hand, and management or policy, on the other, as promoted in most of the risk literature. Of the points of criticism mentioned, the lack of attention to water quality issues seems the most serious one.

To gain more insight into how the planetary boundaries framework in general, and the water boundary in particular, can complement water ethics and risk ethics and vice versa, let us turn to some ethical perspectives on water-related risks.

15.3 Ethical Perspectives on Water-Related Risks

Until a decade ago, no subfield within applied philosophy explicitly dealt with water (Doorn 2013). Even though environmental ethics had developed into a solid field by that time (Jamieson 2008), it was less concerned with the questions that are typical to water: the risks of water shortage, poor quality water, and inadequate flood protection, as well as questions of distributive justice related to these risks, and the importance of water for other sectors (Brennan & Lo 2022). Around that time, the concept of water security emerged in the water literature as a lens through which to look at these questions (Lankford et al. 2013: ix; see also Cook & Bakker 2012; WaterAid 2012). Water security explicitly recognizes the interdependencies between different water-related risks and does not exclusively focus on either reducing water scarcity or water pollution, or on improving safety from flooding. This water security perspective allows us to move beyond water scarcity as an issue of competing water uses and allocation. It shows that it is also about decision-making, rights, distributive justice, and who has a say. Yet, this literature is still disconnected from the relevant philosophical literature.

In recent years, philosophical perspectives on water-related risks have been developed (cf. Doorn 2019; Meisch 2018, 2019). In this section, I highlight three points that are particularly relevant to our thinking about the water planetary boundary. The first concerns the division of labour between science and management, the second concerns the multitude of water-related risks and risk–risk trade-offs, and the third concerns the metric of intergenerational justice in water policy and engineering.

15.3.1 The Division of Labour between Risk Analysis and Risk Management

The traditional view of risk-related decision-making builds on a distinction between its allegedly neutral scientific activities and its value-laden regulatory activities (Doorn 2019). This view is reflected in a report by the American National Academy of Sciences in the context of the regulation of toxic substances and human exposure to these substances, specifically focusing on the role of different actors within the regulatory process (National Research Council (NRC) 1983). The report promotes a strict distinction between the scientific and the policymaking parts of any risk decision procedure, which is to be performed consecutively. The first of these parts, commonly called risk *assessment*, is a scientific undertaking. It consists of collecting and assessing the relevant information to characterize the nature and magnitude of the risk. The second procedure is called risk *management*. Unlike risk assessment, this is not a scientific undertaking. Its starting point is the outcome of risk assessment, which it combines with economic and technological information about various ways of reducing or eliminating the risk in question, and also with political and social information. Based on this, a decision is made on what measures, if any, should be taken to reduce the risk. To prevent risk assessments from being manipulated to meet predetermined policy objectives, it was proposed to separate risk assessment organizationally from risk management (Doorn & Hansson 2011, 2015). The rationale behind this strict division of labour is the concern that value-laden management in the hands of science would result in a technocracy where decisions about what is best for society are made by experts who have scientific expertise but who do not necessarily have the knowledge required to make decisions about what is best for society (van de Poel & Royakkers 2011). Also the other way around, the scientific analysis should not be manipulated to meet predetermined policy objectives. While this division is in practice probably not as strictly adhered to as presented in the NRC-report of 1983, it still seems useful to think about risk-related decision-making as something comprising a descriptive part, belonging to science, and a more normative part.

How does this division of labour apply to the planetary boundaries framework? The framework describes the limits of the carrying capacity of the world, and the identification of the boundaries is largely a geophysical exercise, belonging to the domain of science. Admittedly, the boundary is not equivalent to a threshold or tipping point. The framework takes a precautionary approach in placing the boundary on the 'safe side' of the critical values of the control variables. This buffer accounts for uncertainty in the precise position of the threshold concerning the control variable but also allows 'society time to react to early warning signs that it may be approaching a threshold and consequent abrupt or risky change' (Steffen et al. 2015). While the 'size' of the buffer itself is indeed a value-laden choice that goes beyond what can scientifically be established, it should be reiterated that the framework is intended to describe the safe operating space and, in that sense, a conservative approach, even if it is more conservative than necessary, still fits this overall aim. Given the urgency of our current anthropogenic climatic and environmental risks (Intergovernmental Panel on Climate Change (IPCC) 2021, 2022) and the persistent delay in acting adequately upon them (cf. Lamb et al. 2020), this value-laden buffer seems prima facie justified.

15.3.2 The Multitude of Water-Related Risks and Risk-Risk Trade-Offs

Water is both a source of risk and a scarce resource. Most of the non-technical literature on water management and governance focuses on the scarcity of water, operationalized in terms of access to water. However, the risk of flooding is, in some areas at least, equally or even more urgent. Moreover, in places where the local hydrological circumstances are affected by large infrastructure projects (such as hydro-power plants), both the risk of flooding and potential water scarcity may be present at the same time and solutions to one problem may exacerbate the other. This makes the trade-offs between different risks sometimes unavoidable.

While it would be easy if we could include all risks in one calculation and implement a technical solution or policy that has the lowest overall risk – an approach that is reflected in many probabilistic risk assessments (cf. Doorn 2014) – risks that lead to irreversible consequences are incommensurable. Adequately accounting for these incommensurable risks, therefore, requires a pluralistic approach that avoids expressing all risks in one number (Doorn 2018). As, conceptually, each boundary is associated with a control variable reflecting a particular risk (Wang-Erlandsson et al. 2022), the planetary boundaries framework does allow for such a pluralistic approach. In traditional probabilistic risk assessment, any environmental pollution can be justified as long as it is outweighed by an even higher reduction of some other risk. The planetary boundaries framework, by contrast, sets hard limits for each of the boundaries, and it prevents risks from being shifted to some other domain. At the heart of the framework is the recognition that irreversible changes in the Earth system are to be avoided. The boundaries for 'Biosphere integrity' and 'Climate change' are therefore identified as the core boundaries because they can push the Earth system towards an irreversible state shift (Steffen et al. 2018). Transgression of the other boundaries implies a deterioration in the Earth system functioning that can increase the risk of regional regime shifts. However, as transgressions of these non-core boundaries may also predispose transgressions of the core boundaries, they should still be seen as hard limits that should not be transgressed. This may be particularly relevant for interventions affecting green water, which is critical for supporting and regulating most terrestrial biosphere processes, including energy, carbon, water, and biogeochemical cycles (Wang-Erlandsson et al. 2022).

15.3.3 Inclusion of Intergenerational Justice in Water Policy and Engineering

The third point that I would like to highlight here concerns the inclusion of intergenerational justice aspects. As stated in Section 15.3, the planetary boundaries framework does not account for equity considerations. Equity is in most of the commentaries to the planetary boundaries framework understood as an aspect of *intra*generational distributive justice, that is, justice within current generations. However, this is just one aspect of distributive justice. When it comes to our planet, intergenerational justice is important as well. We currently lack adequate metrics for intergenerational distributive justice, especially for climate adaptation responses. These metrics should describe what it is that makes a distribution just (Teodoro et al. 2023). While we have at least one – admittedly incomplete – metric for climate change mitigation (namely, to reduce our greenhouse gas emissions), it is much more difficult to evaluate the different climate adaptation responses in terms of intergenerational justice. For example, how can we measure and weigh the environmental consequences of different sea level rise and flood protection strategies in coastal systems from the point of view of intergenerational justice (Timmerman et al. 2021)?

Maybe the most elaborated account of intergenerational justice comes from Brian Barry, who developed his theory in the late 1970s and 1980s and who focused primarily on the context of non-renewable resources (fossil fuels). Barry takes the position that 'future generations are owed compensation in other ways for our reducing their access to easily extracted and conveniently located natural resources. In practice, this entails that the combination of improved technology and increased capital investment should be such as to offset the effects of depletion' (Barry 1989: 511–12). With offsetting the depletion of natural resources, Barry refers to 'the replacement of the productive opportunities we have destroyed by the creation of alternative ones' (512). While we can question whether productive opportunities are all that matter, the planetary boundaries framework may exactly be the framework that helps assess which depletion of opportunities are replaceable and which cannot be replaced. The water boundary alone cannot provide this answer and it may very well be that additional boundaries are needed, but the structure of the framework seems in principle well suited to account for these intergenerational aspects.

15.4 Distribution and Allocation within the Boundaries

In *The Division of Labour between Risk Analysis and Risk Management*, we saw that the planetary boundaries framework fits in the traditional division of labour between risk analysis and risk management, or between science and policy in more general terms. The framework provides the safe operating space, but it does not provide guidance on how things should be allocated *within* those boundaries, and this is sometimes also mentioned as a weakness of the framework: given how small the remaining safe operation space is, does this planetary focus not come at the expense of vulnerable people currently alive (cf. Rammelt et al. 2022)? Water ethics and political philosophy more generally can provide a valuable contribution here, conceptually as well as normatively. In political philosophy, a distinction is sometimes made between distribution and allocation (Rawls 1971), a quite technical, but yet relevant, distinction in the context of water. Distribution is concerned with the conception of society as a fair system of cooperation over time, and it is judged on whether it delivers the outcomes that people can expect based on being part of society. In that sense, the people have certain rightful claims. Allocation and relatedly allocative justice apply when a given collection of goods is to be divided among definite individuals who do not have a prior claim on the things to be divided because these things are not the product of their work nor do these individuals stand in any existing cooperative relations. Water ethics is particularly well-equipped to help our thinking about the appropriate rules for distribution and allocation. In the case of allocation, Rawls suggests allocation according to a utilitarian principle, but other allocation rules are also possible. However, in the context of water, more important than allocation is distribution. Water is essential for life, and when it comes to people's basic needs, such as the water needed for drinking water and sanitation, we are dealing with questions of distributive justice. Here, many people defend a sufficientarian, prioritarian, or egalitarian distribution. Making it a matter of allocation seems at odds with basic human rights, such as the right to water or the right to sanitation. The same applies to the valuation of water other than in terms of direct use value, for example, the spiritual value of water. Water ethics is well-positioned to distinguish allocative questions from distributive ones and to develop corresponding conceptions of justice. This is typically something where water ethics can complement the planetary boundaries framework. In areas where basic needs are met and there is no competition for water, it is less controversial to look at the resources as an issue of allocation. As such, distributive justice in combination with the planetary boundaries framework together provides the space within which there is room for allocation.

15.5 Concluding Remarks

In this chapter, I briefly discussed three ways in which the planetary boundaries framework and the water boundary in particular can contribute to water ethics and risk ethics. I showed how the planetary boundaries framework provides a valuable contribution to ongoing discussions in the water ethics and risk ethics literature. Vice versa, water ethics and risk ethics can also contribute to the application of the planetary boundaries framework in practice.

Some points still need further development. The current framework focuses very much on freshwater, even though it has been renamed 'water boundary'. So far, the relation with water safety, that is, safety from flooding, is not well-established and also water quality seems underdeveloped. For safety from flooding, a stronger link with the land system change planetary boundary could be established. The land system boundary currently focuses on cropland and forestry, but this boundary may require a more fine-grained operationalization to account for the impact of flood protection strategies on, for example, wetlands and brackish deltas.

Concerning water quality, the water boundary was originally expressed in terms of blue water use, that is, the use of groundwater and surface water for irrigation or processing (Falkenmark 2003). It did not account for the use of infiltrated rainwater (evaporation) for growing crops (green water), or for the effect of water pollution (grey water), which accounts for water use as a result of pollution (Chapagain et al. 2006). This was also recognized in the report written for the European Environment Agency (Häyhä et al. 2018), which mentioned the need to account for green water use and its interactions with other boundaries, as well as water quality constraints (expressed as grey water). As explained in Section 15.2, in the most recent revision of the planetary boundaries framework, the research group no longer talks about the freshwater use boundary but about the water boundary instead, and by that, they have also included green water (Gleeson et al. 2020). The reason for not including water quality is that water quality, like marine systems, is related to other planetary boundaries, such as ocean acidification, biogeochemical flows, climate change, and novel entities. Including it also in the water boundary would lead to duplication, according to Gleeson et al. (2020). Given that many interventions in the earth system involve several boundaries, water quality is not unique, and as long as boundaries are not assessed in isolation but comprehensibly, this choice by the authors seems justified.

To conclude, as explained in the introduction of Section 15.3, the framework is criticized as not being as objective as it claims to be and, indeed, the boundaries do not follow automatically from science. Yet, given the devastating environmental impact of many human activities, the earth – including its water stores – is probably better served by the precautionary approach promoted by the planetary boundaries framework than by the optimism about technological interventions described by Bogardi et al. (2013). The planetary boundaries framework may exactly be the framework that allows us to distinguish naïve technofixes from sustainable solutions to leave a liveable planet for future generations.

References

Barry, B. (1989). The Ethics of Resource Depletion. In B. Barry, ed., *Democracy, Power and Justice: Essays in Political Theory*. Oxford: Clarendon Press, 511–25.

Bogardi, J. J., Fekete, B. M., & Vörösmarty, C. J. (2013). Planetary Boundaries Revisited: A View through the 'Water Lens'. *Current Opinion in Environmental Sustainability*. 5(6), 581–89. DOI: 10.1016/j.cosust.2013.10.006.

Brennan, A., & Lo, N. Y. S. (2022). Environmental Ethics. In E. N. Zalta, ed., *The Stanford Encyclopedia of Philosophy: Summer 2022 Edition*. plato.stanford.edu/entries/ethics-environmental/.

Chapagain, A. K., Hoekstra, A. Y., Savenije, H. H. G., & Gautam, R. (2006). The Water Footprint of Cotton Consumption: An Assessment of the Impact of Worldwide Consumption of Cotton Products on the Water Resources in the Cotton Producing Countries. *Ecological Economics*. 60(1), 186–203. DOI: 10.1016/j.ecolecon.2005.11.027.

Cook, C., & Bakker, K. (2012). Water Security: Debating an Emerging Paradigm. *Global Environmental Change*. 22(1), 94–192.

Doorn, N. (2013). Water and Justice: Towards an Ethics for Water Governance. *Public Reason*. 5(1), 95–111.

Doorn, N. (2014). Rationality in Flood Risk Management: The Limitations of Probabilistic Risk Assessment (PRA) in the Design and Selection of Flood Protection Strategies. *Journal of Flood Risk Management*. 7(3), 230–38. DOI: 10.1111/jfr3.12044.

Doorn, N. (2018). Distributing Risks: Allocation Principles for Distributing Reversible and Irreversible Outcomes. *Ethics, Place & Environment*. 21(1), 96–109. DOI: 10.1080/21550085.2018.1448041.

Doorn, N. (2019). *Water Ethics: An Introduction*. New York: Rowman & Littlefield.

Doorn, N., & Hansson, S. O. (2011). Should Probabilistic Design Replace Safety Factors? *Philosophy & Technology*. 24(2), 151–68. DOI: 10.1007/s13347-010-0003-6.

Doorn, N., & Hansson, S. O. (2015). Design for the Value of Safety. In M. J. van den Hoven, P. Vermaas, & I. R. van de Poel, eds., *Handbook of Ethics and Values in Technological Design*. Dordrecht: Springer, 491–509.

Falkenmark, M. (2003). Freshwater as Shared between Society and Ecosystems: From Divided Approaches to Integrated Challenges. *Philosophical Transaction of the Royal Society of London B*. 358(1440), 2037–49.

Gleeson, T., Wang-Erlandsson, L., Zipper, S. C., et al. (2020). The Water Planetary Boundary: Interrogation and Revision. *One Earth*. 2(3), 223–34. DOI: 10.1016/j.oneear.2020.02.009.

Häyhä, T., Cornell, S. E., Hoff, H., Lucas, P., & van Vuuren, D. (2018). *Operationalizing the Concept of a Safe Operating Space at the EU level – First Steps and Explorations*. A Report for the European Environment Agency.

Intergovernmental Panel on ClimateChange, ed. (2021). *Climate Change 2021: The Physical Science Basis. Contribution of Working Group I to the Sixth Assessment Report of the Intergovernmental Panel on Climate Change.* Cambridge: Cambridge University Press.

Intergovernmental Panel on ClimateChange, ed. (2022). *Climate Change 2022: Impacts, Adaptation and Vulnerability. Contribution of Working Group II to the Sixth Assessment Report of the Intergovernmental Panel on Climate Change.* Cambridge: Cambridge University Press.

Jamieson, D. (2008). *Ethics and the Environment: An Introduction.* Cambridge: Cambridge University Press.

Lamb, W. F., Mattioli, G., Levi, S., et al. (2020). Discourses of Climate Delay. *Global Sustainability.* 3, e17. DOI: 10.1017/sus.2020.13.

Lankford, B., Bakker, K., Zeitoun, M., & Conway, D. (2013). *Water Security: Principles, Perspectives and Practices.* New York: Earthscan.

Meisch, S. P. (2018). Water Ethics? Lessons from Post-Normal Science. In S. Springer & H. Grimm, eds., *Professionals in Food Chains.* Wageningen: Wageningen Academic Publishers, 459–64.

Meisch, S. P. (2019). I Want to Tell You a Story: How Narrative Water Ethics Contributes to Re-Theorizing Water Politics. *Water.* 11, 306–25. DOI: 10.3390/w11040631.

National Research Council. (1983). *Risk Assessment in the Federal Government: Managing the Process.* Washington, DC: National Academy Press.

One Earth. (2020). Thirsty Planet. *One Earth.* 2(3), 197–98. DOI: 10.1016/j.oneear.2020.02.015.

Rammelt, C. F., Gupta, J., Liverman, D., et al. (2022). Impacts of Meeting Minimum Access on Critical Earth Systems amidst the Great Inequality. *Nature Sustainability.* 6, 212–21. DOI: 10.1038/s41893-022-00995-5.

Rawls, J. (1971). *A Theory of Justice*, rev. ed. Cambridge, MA: The Belknap Press of Harvard University Press.

Rockström, J., Steffen, W., Noone, K., et al. (2009). A Safe Operating Space for Humanity. *Nature.* 461(7263), 472–75. DOI: 10.1038/461472a.

Steffen, W., Richardson, K., Rockström, J., et al. (2015). Planetary Boundaries: Guiding Human Development on a Changing Planet. *Science.* 347(6223), 1259855. DOI: 10.1126/science.1259855.

Steffen, W., Rockström, J., Richardson, K., et al. (2018). Trajectories of the Earth System in the Anthropocene. *Proceedings of the National Academy of Sciences of the United States of America.* 115(33), 8252–59. DOI: 10.1073/pnas.1810141115.

Steffen, W., & Stafford Smith, M. (2013). Planetary Boundaries, Equity and Global Sustainability: Why Wealthy Countries Could Benefit from More Equity. *Current Opinion in Environmental Sustainability.* 5(3), 403–8. DOI: 10.1016/j.cosust.2013.04.007.

Teodoro, D., Doorn, N., Kwakkel, J., & Comes, T. (2023). Flexibility for Intergenerational Justice in Climate Resilience Decision-Making: An Application on Sea-Level Rise in the Netherlands. *Sustainability Science.* 18(3), 1355–65. DOI: 10.1007/s11625-022-01233-9.

Timmerman, A., Haasnoot, M., Middelkoop, H., Bouma, T., & McEvoy, S. (2021). Ecological Consequences of Sea Level Rise and Flood Protection Strategies in Shallow Coastal Systems: A Quick-Scan Barcoding Approach. *Ocean & Coastal Management.* 210, 105674. DOI: 10.1016/j.ocecoaman.2021.105674.

van de Poel, I. R., & Royakkers, L. (2011). *Ethics, Technology, and Engineering: An Introduction*. West-Sussex: Wiley-Blackwell.

Wang-Erlandsson, L., Tobian, A., van der Ent, R. J., et al. (2022). A Planetary Boundary for Green Water. *Nature Reviews Earth & Environment*. 3(6), 380–92. DOI: 10.1038/s43017-022-00287-8.

WaterAid. (2012). *Water Security Framework*. London: WaterAid.

16

Navigating Pathways to Just Water Futures?

*A Hydrosocial Engagement with the Planetary Boundaries Framework**

NICOLE J. WILSON

16.1 Introduction

Water is essential to all life on earth. Water not only plays a fundamental role in maintaining the global biosphere, climate regulation, and carbon and nutrient cycling (Gleeson et al. 2020a) but also in supporting thriving human communities who maintain diverse and divergent relationships with, and understandings of, what water *is* (Wilson & Inkster 2018; Yates et al. 2017). Changes in the water cycle observed globally are driven by a variety of sources of anthropogenic forcing including greenhouse gases, aerosol, and land use change while future water cycle changes will increasingly be dominated by anthropogenic climate change due to greenhouse gas emissions (Douville et al. 2021). These changes are considered to contribute to the 'intensification' of the global water cycle (Huntington et al. 2018). Such changes have the potential to disrupt processes and relationships across multiple scales – from the local to the global (Vörösmarty et al. 2013). It is argued that in the newly entered era of the Anthropocene,[1] the speed, scale, and magnitude of anthropogenic drivers of risk may trigger transitions that threaten the habitability of the Earth for human society (Barnosky et al. 2012; Gleeson et al. 2020a; Steffen et al. 2018).

* This work was funded by the Canada Research Chair program # 950-232734.

[1] Noel Castree calls PBs a 'collateral concept', closely related to Anthropocene thinking (Castree 2014a, 2014b, 2014c; see also Brown 2017). Crutzen and Stoermer (2000) proposed naming the current geologic epoch the 'Anthropocene' rather than the Holocene to indicate the significant role humans have played in shaping the earth and its atmosphere. While there has been significant debate on the political and ethical implications of not only whether the present period should be considered the Anthropocene, but the question of when this era began has significant implications (Davis & Todd 2017). Proposed dates ranged from the advent of agriculture to the start of the Industrial Revolution in the 1800s, when human activity began having a significant impact on carbon and methane in the Earth's atmosphere. However, in 2016 the Anthropocene Working Group agreed that the Anthropocene began in 1950 when the 'Great Acceleration' or the dramatic increase in human activity affecting the planet was initiated with the invention of the first steam engine (Steffen et al. 2015a). Indigenous scholars critique the Eurocentric framing of present approaches to dating in that it misses an opportunity to acknowledge the beginning of the colonial period in 1610 which initiated trajectories of dispossession and genocide along with the transformation of the environment through colonial occupation (Davis & Todd 2017). Anthropocene thinking is increasingly applied to hydrologic systems (e.g., Falkenmark et al. 2019; Vörösmarty et al. 2013).

The planetary boundaries (PB) framework proposes one approach to characterizing and managing the current anthropogenic pressures on the hydrologic system globally (Rockström et al. 2009b). The PB for human freshwater use (water PB) relies on the global average of annual surface flows in rivers concerning environmental flow needs (Rockström et al. 2009a; Steffen et al. 2015b). Proponents of the PB framework consider the water PB critical; however, this proposed boundary has been critiqued for oversimplifying in ways that fail to reflect the heterogeneity of water's distribution (Gleeson et al. 2020b), the complete range of human impacts on the global water cycle (Gleeson et al. 2020b), and the complexities of developing global water governance to manage the water PB (Zipper et al. 2020). Recent works have sought to revise the water PB to transcend such critiques. For instance, Gleeson et al. (2020b) propose that a water PB would be more useful and scientifically sound if it were rearticulated to highlight the ways climate and ecosystems respond to alterations in water in all its forms ranging from atmospheric water, soil moisture, groundwater, and frozen water, as well as surface water. However, the complexity of human–water relationships and the relationship between socio-cultural, political, *and* hydrological dimensions across scales remain undertheorized.

While PBs are presented as 'green limits' – or in this case 'blue limits' – rooted in science, they are ultimately a normative concept rooted in particular understandings of human–environment relationships (Brown 2017; Leach et al. 2013). From the start, it has been made clear that PBs are not thresholds or tipping points (Rockström et al. 2009a, 2009b). Even where a threshold is likely to exist or is considered measurable (e.g., climate), boundaries are set in a precautionary manner at a 'safe distance' from thresholds to account for risk and uncertainty. In this way, PBs are intended to allow humans time to respond to 'early warning signs that it might be approaching a threshold' (Steffen et al. 2015b). Planetary boundaries therefore involve societal choice regarding how they are set and to what extent the precautionary principle is incorporated.

While the normative nature of PBs has been explicit from the start (Rockström et al. 2009a, 2009b), the politics of PBs including assumptions about global governance and privileging global over local concerns has resulted in considerable debate (Brown 2017; Leach et al. 2013). Melissa Leach et al. (2013) suggest that we need to identify a complementary social boundary to accompany each PB. In doing so they stress the contested and dynamic nature of 'navigating alternative pathways' in the space between social and planetary boundaries, where pathways are to varying extents 'aligned with different cultures, visions, and values, and with different costs, risks, and distributions of power and benefits between social groups' (Leach et al. 2013: 86). The definition and choice between alternate pathways are, therefore, 'deeply political' and the process has significant potential to produce uneven and unjust outcomes across human communities. Unexamined assumptions about the nature of water and human relationships with it shape normative understandings of a 'safe operating space for humanity' and the necessary and possible pathways within it. Building on the PB framework and relevant critiques, Rockström et al. (2023) propose Earth System Boundaries for a number of domains including freshwater, developing a *safe* and *just* boundary that is more conservative than the biophysical water PB. Despite these advancements, many questions remain: How do PBs account for diversity and difference among and between human communities and their hydrosocial relationships? How might a water PB create or reinforce existing water inequities and injustices?

In this chapter, I begin with the normative assumption that the goal of responding to anthropogenic pressures is to advance more just and equitable water futures. I argue that a more nuanced picture of the socio-political dynamics involved in human–water relationships is needed to understand whether the water PB has the potential to advance water justice. Against this backdrop, I apply a hydrosocial approach to illuminate connections between the socio-cultural, political, and material dimensions of water in relation to the PB framework. Based on this analysis, I find that present framings of the water PB through Modern Water and Western techno-scientific understandings of water, privilege the hydrosocial relationships of some human communities over others in ways that can reinforce and even create new water inequities and injustices. More specifically, I engage with three interconnected dimensions of hydrosocial relations including (1) the heterogeneity of both water's materiality and human relationships with it, (2) the inherently political nature of water, and (3) the production of water through particular relations and technologies. I conclude that the water PB is only useful to the extent that it can advance transformative change that addresses present inequities in hydrosocial relationships.

16.2 Hydrosocial Engagement with the Water Planetary Boundary

The rise of the Modern Water paradigm in the early twentieth century promoted the narrow understanding of water as a resource that can be controlled, managed, and exploited by humans (Linton 2010). Modern Water is a normative concept that ignores the heterogeneous and, often unequal, nature of human–water relationships in favour of 'bringing water's social and evolutionary possibilities into the service of liberal forms of life' (Schmidt 2017: 7). Unexamined, these assumptions about water within the PB framework have the potential to influence any suggested actions or solutions that might be taken to address present challenges facing water systems globally.

The hydrosocial framework builds on the hydrologic cycle but moves beyond Modern Water through explicit attention to the social context and dynamics of water and resituates water as fundamentally political (Linton & Budds 2014). In this way, a hydrosocial framework not only positions humans as endogenous to the hydrologic cycle but sees water as a socio-natural process where water and society co-constitute each other across spatial, geographic and temporal scales (Linton & Budds 2014). The hydrosocial approach is a conceptual framework for engaging with the process of 'making and remaking' which offers insights into the production of water, the ways we come to know it and the power relationships that are embedded within processes of change (Linton & Budds 2014).[2] In applying this approach, I engage scholarship within the fields of political ecology (Budds & Sultana 2013; Harris 2015) and critical Indigenous studies (Curley 2021; Whyte 2017b) to carefully trace water ontologies, politics, and inequality, in the context of processes of capital accumulation and settler colonialism.

There are three key aspects of hydrosocial relationships that help us think about the water PB. First, water is heterogeneous in both its materiality and human relationships with it (Linton & Budds 2014; Strang 2004). As discussed, critiques of the water PB have highlighted the oversimplification of water's materiality in the provisional water PB (Gleeson et al. 2020a, 2020b). But the suggested revisions to the water PB stop short of acknowledging the hybrid nature of water or the ways that water's diverse states, forms, and qualities make it act and give meaning in distinct ways (Linton & Budds 2014). For example, relationships with surface water may differ from relationships with groundwater. Furthermore, different human communities may relate to the same forms of water in very different ways. Indeed, Indigenous peoples maintain diverse relationships with water, but these relationships often centre on understandings of water as a living entity that has agency or 'spirit' and carries deep spiritual and cultural meaning both generally and specific to particular bodies of water (Chiblow (Ogamauh annag qwe) 2019; Craft 2018). These relationships with water are not projections of symbolic meaning onto water's materiality but literal expressions of distinct worlds that flow from Indigenous onto-epistemologies (Todd 2014; Watts 2013). While scholars of the water PB refer to water using anthropomorphic metaphors such as the reference to freshwater as the 'bloodstream of the biosphere' (Gleeson et al. 2020a: 2), they stop short of acknowledging diverse water ontologies. Here, Modern Water's assumption of water as a singular material substance or H_2O is also accompanied by the modernist assumption that social relations are contained within the same modernist 'ontological matrix' within which diverse

[2] I draw on scholarship from the field of Science and Technology Studies that assumes that reality is always in the process of being made through assemblages of humans and more-than-human actors (Haraway 2007; Latour 2014; Whatmore 2002).

populations construct symbolic meaning (Blaser 2009: 879), but multiple water ontologies in which distinct water worlds are at stake (Wilson & Inkster 2018). The imposition of Modern water through the water PB framework is an example of onto-epistemological violence as it imposes boundaries around what is 'real'. Anything outside that falls outside the boundaries of water as H_2O is 'unreal' (Hunt 2014).

Second, water is inherently political rather than merely a material substance that is the *object of politics* (Linton & Budds 2014). In other words, water cycles through socio-political processes in ways that internalize and express water politics and conflict. Scholarship on the political ecology of water illustrates the fundamentally political nature of water by examining and articulating the ways that some human communities' access and relationships with water are privileged over others (Boelens et al. 2016; Budds & Sultana 2013), including processes of capital accumulation (Ranganathan 2016) and settler colonialism (Daigle 2018; Leonard et al. 2023; Wilson et al. 2021). This exclusion fails to acknowledge not only Indigenous rights and jurisdiction but also the epistemological and ontological politics of water or the conflicts that develop when different water worlds seek to maintain their own existence as they come into contact with each other (Wilson & Inkster 2018; See also Blaser 2009).

The scalar politics of governing the water PB bring many dimensions of water justice into focus. Indeed, the PB framework has been critiqued for containing within it an implicit assumption of global governance (Brown 2017; Leach et al. 2013). Water scholars have also suggested that water governance at the global level is required to address anthropogenic impacts on water globally (Cooley et al. 2014; Gupta & Pahl-Wostl 2013). However, present water governance structures tend to occur at the local, national, or regional levels of the administrative scale (Dellapenna et al. 2013).[3] Following Gupta (2008), global problems need to be expressed in ways that attend to the expression of the issues at relevant levels of the administrative scale to help 'ensure the legality, legitimacy, and effectiveness of institutional solutions' (Gupta 2008: 231). While water issues may be defined as a global problem, the problem takes varying forms at differing levels on

[3] It should be noted that there is not a total absence of global water governance mechanisms. Today numerous transnational or international institutions play a role in water governance at the global or regional levels, including formal institutions (e.g., those created by the United Nations or other international organizations and/ or by multilateral or bilateral treaties) and informal institutions (e.g., civil society groups, market-based actors, and epistemic communities) (Dellapenna et al. 2013). While governance is often thought of narrowly in relation to regulatory power, institutions without regulatory power can play various roles (Young 2010). Dellapenna et al. (2013) points out that while existing global institutions are weak with respect to regulation they have 'soft governance functions' in that they are 'relatively good at agenda setting, sharing information, mobilizing people, and, to a certain degree, in mobilizing resources' (Schubert & Gupta 2013). For example, Sustainable Development Goal 6 on Clean Water and Sanitation (United Nations Department of Economic and Social Affairs 2022).

administrative scales as it can be unpacked as a distinctive set of problems at each level. Zipper et al. (2020) seek to reconcile the water PB with the complexities of existing governance structures through the idea of a locally 'harmonized' water boundary that would be 'consistent with both global and local water sustainability goals and provide a framework for determining whether local water management is consistent with local socio-environmental processes and Earth System function' (Zipper et al. 2020: 8–9). While this analysis is useful in its attempt to reconcile the global nature of the water PB with locally meaningful limits and thresholds, it fails to acknowledge the significant potential for the governance of the water PB to create and/or replicate existing political inequalities present in governance-structures.

Third, water is produced or co-created by relations and technologies. Such considerations help us think about how water is produced through particular sets of relations and technologies. A hydrosocial framework moves beyond dualistic understandings of relationships between water and society as something that pre-exists to examine how water, society, and social power are co-constituted (Castree 2005). For instance, Linton (2010) shows how drinking water technologies might privilege specific kinds of water relations. They discuss how a public drinking fountain might maintain water as a public good by providing free access, while a commercial bottled water vending machine makes water into a commodity. Each of these technologies produces different types of water and very different forms of access and citizenship.

The water PB develops out of new understandings of 'global water'. Although water challenges have historically been understood and responded to locally, recent advancements in geophysical approaches (e.g., remote sensing, models, data integration systems) are rapidly changing our ability to recognize locally relevant patterns while at the same time producing a 'unified view of the global hydrologic cycle' (Vörösmarty et al. 2013: 542). Remote sensing has played a transformative role in how we see the world. Along with other technologies, such as television and telephones, satellites have been called 'world making' devices as the ability to sense at a distance became central to 'human' faculties in the late twentieth century – rendering humans as 'cyborg entities defined not by where they live but by the telematic flows they receive and emit' (Tuathail 1997: 44). Such advances are useful in considering the ways that global forces are driving local-scale water challenges and at the same time the ways that local stressors scale up to drive changes in water systems globally (Vörösmarty et al. 2013). However, neither these technologies nor the forms of water they produce are politically neutral. In examining the political economy of remote sensing technologies including the use by a range of actors including governments, scientists, and commercial and non-governmental actors, Bennett et al. (2022) argue that '[t]echnology's reflection of societal and cultural biases demonstrates the need to provincialize remote sensing by examining how decisions made across the technical process, from satellite design to image analysis, and even commercialization, affect results' (733). In other words, we must ask ourselves how the production of global water through these technologies creates different forms of access and citizenship. Whom does it privilege?

16.3 Towards Just Water Futures?

While the water PB may provide a metaphor for catalysing responses to anthropogenic impacts on water and other PBs, it is only useful if it results in more equitable and just water futures. In other words, as a 'boundary object' that facilitates communication across spheres while allowing different actors to maintain the diversity of their views (Morseletto et al. 2017), PBs are only useful as far as they advance transformative change – or a 'fundamental' change in natural or human systems (Walker et al. 2004). Such shifts require consideration of the root causes of anthropogenic change, the factors that construct scarcity and vulnerability – exposing some human communities to greater impacts of change – and must ultimately reconfigure human–water systems towards more just and equitable futures (Shah et al. 2018). I, therefore, examine how the theory and practice of water justice can help evaluate the potential for the PB framework to advance more just water futures.

Water justice is a relational and contextual concept that acknowledges the material, economic, cultural, and political dimensions of water while taking seriously the political contestation, negotiation, and struggle related to the inequitable distribution of water access (Sultana 2018; Zwarteveen & Boelens 2014). Water justice helps us think through the critiques of the water PB identified by a hydrosocial perspective. First, in helping us reconsider *how water is known* a hydrosocial critique of the water PB identifies the need for onto-epistemological justice or the idea that we must take seriously the heterogeneity of water and the existence of multiple water worlds including ways of being, knowing, and relating with water that fall outside of the assumptions of Modern water (Wilson & Inkster 2018). For instance, planetary-scale understandings of water are co-produced by Western scientific relations and technologies (e.g., remote sensing). Thus, we must ask ourselves how the water PB might marginalize Indigenous onto-epistemologies rooted in understandings of water as a living entity with which people have reciprocal responsibilities (Chiblow (Ogamauh annag qwe) 2019; Yazzie & Baldy 2017). Diverse ways of knowing and being are needed to imagine more just water futures that might not be conceived of in assessments that rely on Western scientific knowledge alone (Wilson et al. 2024).

Second, a hydrosocial critique asks us to consider the implications of the water PB for the *distribution of waters benefits and harms* also known as 'distributive justice' (Fraser 2000; Schlosberg 2004). According to a 'trivalent' understanding of environmental justice, distributive justice is closely connected to procedural (who is involved or has influence?) and recognitional justice (who or what is valued or respected?) (Fraser 2000; Schlosberg 2004). Gupta et al. (2021) also propose three additional domains of justice relevant to the PB framework that includes responsibilities between humans and the more than human world (interspecies justice), between the generations (intergenerational justice) and between diverse segments within present human populations (intragenerational) (see also Gupta et al. 2023).

While questions of water justice have been accounted for to a greater extent (Rockström et al. 2023), there are many potential water justice questions that should be asked. For instance, the potential of the water PB to advance just water futures is contingent on addressing the politics of scale or rescaling to the global in exacerbating or creating new water injustices. Rescaling water governance to the global level is highly political and requires consideration of whose interests would be privileged through such a hydro-scalar shift (Cohen 2014; Harris & Alatout 2010).

Third, water justice must consider how unjust histories shape the potential for harm in the present context. Shue (2014) notes that these 'compound injustices' occur 'when an initial injustice paves the way for a second' (4). In other words, any action taken as the result of the mobilization of the water PB framework must be considered in light of existing water injustices and the potential for that action to produce more or less just water futures for marginalized groups including consideration for intersecting identities along the lines of gender, race and ethnicity, class, Indigeneity, and more (Harris et al. 2017) or the systemic inequalities created through capitalism, colonialism, and white supremacy. Failing to address systemic inequalities created through capitalism, (settler-)colonialism, and other forms of inequality will mean that any governance system intended to address anthropogenic impacts on the water system has the potential to reproduce these inequalities. For instance, through colonial histories, Indigenous peoples have been excluded from existing colonial water governance frameworks at all administrative levels from the local to global (Wilson 2020).

Assessments of the water potential for the PB to advance just water futures must also advance climate justice, given that future water cycle changes will increasingly be dominated by anthropogenic climate change (Douville et al. 2021). Climate justice is a growing body of theory and civic action that pays close attention to how different populations have contributed to varying extents to the root causes of climate change and how climate impacts people differently and disproportionately (Sultana 2022). Furthermore, while the water PB seems to naturalize anthropogenic contributions to greenhouse gas emissions, we must look beyond this to examine the root causes of climate change. Potawatomi scholar Kyle Whyte (2017a) reminds us that colonialism and capitalism are inextricably tied together and have laid the groundwork for the carbon-intensive economics which is driving climate change. Assessments of water justice and the water PB must also consider climate justice given that (a) there is a positive feedback loop between water injustice and climate vulnerability and (b) any solution to climate change will have implications for water justice (Mills-Novoa et al. 2022).

Where present framings of the water PB lean heavily toward the creation of new water global governance frameworks, we can learn from present injustices in climate governance. Recent research has shown that the inclusion of Indigenous peoples and their knowledge systems in global climate governance without acknowledging their self-determination and governance systems is likely to replicate existing colonial injustices (Cifuentes 2021; Reed et al. 2022). Given the ways that global climate governance has been ineffective at mitigating climate change while also exacerbating existing inequalities, what faith should we have in the potential for a new global water governance framework for the water PB to advance water or climate justice?

16.4 Conclusion

The PB framework proposes one approach to characterizing and managing the current anthropogenic pressures on the hydrologic system globally (Rockström et al. 2009b). The understanding of the water PB has evolved to acknowledge issues of equity and responsibility in responding to present water challenges (Steffen et al. 2015b; Zipper et al. 2020). However, following Katrina Brown (2017: 121), I argue present examinations of the politics and implications of the water PB still fail to address 'the core issues of the social processes driving change, the socially differentiated nature of that change, its impacts, and agency and policy to act in response'. I illustrated some of the limitations of the water PB by examining how present framings rely on Modern Water and Western techno-scientific understandings of water in ways that privilege the hydrosocial relationships of some human communities over others in ways that can reinforce and even create new water inequities and injustices. More specifically, I have applied a hydrosocial framework to reveal a more nuanced understanding of the socio-political dynamics in human–water relationships including water's materiality as well as its histories, epistemologies, ontologies and scalar politics, and corresponding governance frameworks. Drawing on these hydrosocial critiques, I showed what is at stake for the water PB framework to advance just water futures and highlighted areas where further consideration of the multiple dimensions of justice is needed.

References

Barnosky, A. D., Hadly, E. A., Bascompte, J., et al. (2012). Approaching a State Shift in Earth's Biosphere. *Nature*. 486(7401), 52–58. DOI: 10.1038/nature11018.

Bennett, M. M., Chen, J. K., Alvarez León, L. F., & Gleason, C. J. (2022). The Politics of Pixels: A Review and Agenda for Critical Remote Sensing. *Progress in Human Geography*. 46(3), 729–52. DOI: 10.1177/03091325221074691.

Blaser, M. (2009). Political Ontology. *Cultural Studies*. 23(5–6), 873–96. DOI: 10.1080/09502380903208023.

Boelens, R., Hoogesteger, J., Swyngedouw, E., Vos, J., & Wester, P. (2016). Hydrosocial Territories: A Political Ecology Perspective. *Water International*. 41(1), 1–14. DOI: 10.1080/02508060.2016.1134898.

Brown, K. (2017). Global Environmental Change II: Planetary Boundaries – A Safe Operating Space for Human Geographers? *Progress in Human Geography*. 41(1), 118–30. DOI: 10.1177/0309132515604429.

Budds, J., & Sultana, F. (2013). Guest Editorial: Exploring Political Ecologies of Water and Development. *Environment and Planning D: Society and Space*. 31(2), 275–79. DOI: 10.1068/d3102.

Castree, N. (2005). *Nature*. London: Routledge Taylor & Francis Group.

Castree, N. (2014a). Geography and the Anthropocene II: Current Contributions. *Geography Compass*. 8(7), 450–63. DOI: 10.1111/gec3.12140.

Castree, N. (2014b). The Anthropocene and Geography I: The Back Story. *Geography Compass*. 8(7), 436–49. DOI: 10.1111/gec3.12141.

Castree, N. (2014c). The Anthropocene and Geography III: Future Directions. *Geography Compass*. 8(7), 464–76. DOI: 10.1111/gec3.12139.

Chiblow (Ogamauh annag qwe), Susan. (2019). Anishinabek Women's Nibi Giikendaaswin (Water Knowledge). *Water*. 11(2), 209. DOI: 10.3390/w11020209.

Cifuentes, S. (2021). Rethinking Climate Governance: Amazonian Indigenous Climate Politics and Integral Territorial Ontologies. *Journal of Latin American Geography*. 20(2), 131–55.

Cohen, A. (2014). Rescaling Environmental Governance: The Case of Watersheds as Scale-Sensitive Governance? In C. Termeer, F. Padt, N. Polman, & P. Opdam, eds., *Scale-Sensitive Governance of the Environment*. Chichester: Wiley Blackwell, 172–87.

Cooley, H., Ajami, N., Ha, M.-L., et al. (2014). Global Water Governance in the Twenty-First Century. In P. H. Gleick, ed., *The World's Water Volume 8: The Biennial Report on Freshwater Resources*. Washington, DC: Island Press, 1–18.

Craft, A. (2018). Navigating Our Ongoing Sacred Legal Relationship with Nibi (Water). In Centre for International Governance Innovation & Wiyasiwewin Mikiwahp Native Law Centre, eds., *UNDRIP Implementation: More Reflections on the Braiding of International, Domestic and Indigenous Laws*. Waterloo, ON: Centre for International Governance Innovation, 53–62.

Crutzen, P. J., & Stoermer, E. F. (2000). The 'Anthropocene'. *Global Change Newsletter*. 41, 17–18.

Curley, A. (2021). Unsettling Indian Water Settlements: The Little Colorado River, the San Juan River, and Colonial Enclosures. *Antipode*. 53(3), 705–23. DOI: 10.1111/anti.12535.

Daigle, M. (2018). Resurging through Kishiichiwan: The Spatial Politics of Indigenous Water Relations. *Decolonization: Indigeneity, Education & Society*. 7(1), 159–72.

Davis, H., & Todd, Z. (2017). On the Importance of a Date, or Decolonizing the Anthropocene. *ACME: An International Journal for Critical Geographies*. 16(4), 761–80.

Dellapenna, J. W., Gupta, J., Li, W., & Schmidt, F. (2013). Thinking about the Future of Global Water Governance. *Ecology & Society*. 18(3), 28. DOI: 10.5751/ES-05657-180328.

Douville, H., Raghavan, K., Renwick, J., et al. (2021). Chapter 8: Water Cycle Changes. In Intergovernmental Panel on Climate Change (IPCC), ed., *Climate Change 2021: The Physical Science Basis*. Contribution of Working Group I to the Sixth Assessment Report of the Intergovernmental Panel on Climate Change. Cambridge: Cambridge University Press, 1055–210.

Falkenmark, M., Wang-Erlandsson, L., & Rockström, J. (2019). Understanding of Water Resilience in the Anthropocene. *Journal of Hydrology X*. 2, 100009. DOI: 10.1016/j.hydroa.2018.100009.

Fraser, N. (2000). Rethinking Recognition. *New Left Review*. 3, 107–18.

Gleeson, T., Wang-Erlandsson, L., Porkka, M., et al. (2020a). Illuminating Water Cycle Modifications and Earth System Resilience in the Anthropocene. *Water Resources Research*. 56(4), e2019WR024957. DOI: 10.1029/2019WR024957.

Gleeson, T., Wang-Erlandsson, L., Zipper, S. C., et al. (2020b). The Water Planetary Boundary: Interrogation and Revision. *One Earth*. 2(3), 223–34. DOI: 10.1016/j.oneear.2020.02.009.

Gupta, J. (2008). Global Change: Analysing Scale and Scaling in Environmental Governance. In O. R. Young, L. A. King, & H. Schroeder, eds., *Institutions and Environmental Change: Principal Findings, Applications, and Research Frontiers*. Cambridge, MA: MIT Press, 225–58.

Gupta, J., & Pahl-Wostl, C. (2013). Global Water Governance in the Context of Global and Multilevel Governance: Its Need, Form, and Challenges. *Ecology & Society*. 18(4), 53. DOI: 10.5751/ES-05952-180453.

Gupta, J., Liverman, D., Bai, X., et al. (2021). Reconciling Safe Planetary Targets and Planetary Justice: Why Should Social Scientists Engage with Planetary Targets? *Earth System Governance*. 10, 100122. DOI: 10.1016/j.esg.2021.100122.

Haraway, D. J. (2007). *When Species Meet*. Minneapolis: University of Minnesota Press.

Harris, L. M. (2015). Hegemonic Waters and Rethinking Natures Otherwise. In W. Harcourt, & I. L. Nelson, eds., *Practising Feminist Political Ecologies: Moving beyond the Green Economy*. London: Zed Books, 157–81.

Harris, L. M., & Alatout, S. (2010). Negotiating Hydro-Scales, Forging States: Comparison of the Upper Tigris / Euphrates and Jordan River Basins. *Political Geography*. 29(3), 148–56. DOI: 10.1016/j.polgeo.2010.02.012.

Harris, L. M., McKenzie, S., Rodina, L., Shah, S. H., & Wilson, N. J. (2017). Water Justice: Key Concepts, Debates and Research Agendas. In R. Holifield, J. Chakraborty, & G. Walger, eds., *The Routledge Handbook of Environmental Justice*. New York: Routledge Taylor & Francis Group, 338–49.

Hunt, S. (2014). Ontologies of Indigeneity: The Politics of Embodying a Concept. *Cultural Geographies*. 21(1), 27–32. DOI: 10.1177/1474474013500226.

Huntington, T. G., Weiskel, P. K., Wolock, D. M., & McCabe, G. J. (2018). A New Indicator Framework for Quantifying the Intensity of the Terrestrial Water Cycle. *Journal of Hydrology*. 559, 361–72. DOI: 10.1016/j.jhydrol.2018.02.048.

Latour, B. (2014). Agency at the Time of the Anthropocene. *New Literary History*. 45(1), 1–18. DOI: 10.1353/nlh.2014.0003.

Leach, M., Raworth, K., & Rockström, J. (2013). Between Social and Planetary Boundaries: Navigating Pathways in the Safe and Just Space for Humanity. In International Social Science Council & United Nations Educational, Scientific and Cultural Organization, eds., *World Social Science Report 2013: Changing Global Environments*. Paris: OECD Publishing; UNESCO Publishing, 84–89.

Leonard, K., David-Chavez, D., Smiles, D., et al.(2023). Water Back: A Review Centering Rematriation and Indigenous Water Research Sovereignty. *Water Alternatives*. 16(2), 374–428.

Linton, J. (2010). *What Is Water? The History of a Modern Abstraction*. Vancouver: University of British Columbia Press.

Linton, J., & Budds, J. (2014). The Hydrosocial Cycle: Defining and Mobilizing a Relational-Dialectical Approach to Water. *Geoforum*. 57, 170–80. DOI: 10.1016/j.geoforum.2013.10.008.

Mills-Novoa, M., Boelens, R., & Hoogesteger, J. (2022). Climate Change and Water Justice. In T. M. Letcher, ed., *Water and Climate Change: Sustainable Development, Environmental and Policy Issues*. Amsterdam: Elsevier, 399–418.

Morseletto, P., Biermann, F., & Pattberg, P. (2017). Governing by Targets: Reductio ad Unum and Evolution of the Two-Degree Climate Target. *International Environmental Agreements: Politics, Law and Economics*. 17(5), 655–76. DOI: 10.1007/s10784-016-9336-7.

Ranganathan, M. (2016). Thinking with Flint: Racial Liberalism and the Roots of an American Water Tragedy. *Capitalism Nature Socialism*. 27(3), 17–33. DOI: 10.1080/10455752.2016.1206583.

Reed, G., Brunet, N. D., McGregor, D., et al. (2022). Toward Indigenous Visions of Nature-Based Solutions: An Exploration into Canadian Federal Climate Policy. *Climate Policy*. 22(4), 514–33. DOI: 10.1080/14693062.2022.2047585.

Rockström, J., Steffen, W., Noone, K., et al. (2009a). Planetary Boundaries: Exploring the Safe Operating Space for Humanity. *Ecology & Society*. 14(2), 32.

Rockström, J., Steffen, W., Noone, K., et al. (2009b). A Safe Operating Space for Humanity. *Nature*. 461(7263), 472–75. DOI: 10.1038/461472a.

Rockström, J., Gupta, J., Qin, D., et al. (2023). Safe and Just Earth System Boundaries. *Nature*. 619(7968): 102–11. DOI: 10.1038/s41586-023-06083-8.

Schlosberg, D. (2004). Reconceiving Environmental Justice: Global Movements and Political Theories. *Environmental Politics*. 13(3), 517–40. DOI: 10.1080/0964401042000229025.

Schmidt, J. J. (2017). *Water: Abundance, Scarcity, and Security in the Age of Humanity*. New York: New York University Press.

Schubert, S., & Gupta, J. (2013). Comparing Global Coordination Mechanisms on Energy, Environment, and Water. *Ecology & Society*. 18(2), 22. DOI: 10.5751/ES-05440-180222.

Shah, S. H., Rodina, L., Burt, J. M., et al. (2018). Unpacking Social-Ecological Transformations: Conceptual, Ethical and Methodological Insights. *The Anthropocene Review*. 5(3), 250–65. DOI: 10.1177/2053019618817928.

Shue, H. (2014). The Unavoidability of Justice. In H. Shue, ed., *Climate Justice: Vulnerability and Protection*. Oxford: Oxford University Press, 27–46.

Steffen, W., Broadgate, W., Deutsch, L., Gaffney, O., & Ludwig, C. (2015a). The Trajectory of the Anthropocene: The Great Acceleration. *The Anthropocene Review*. 2(1), 81–98. DOI: 10.1177/2053019614564785.

Steffen, W., Richardson, K., Rockström, J., et al. (2015b). Planetary Boundaries: Guiding Human Development on a Changing Planet. *Science*. 347(6223), 1259855. DOI: 10.1126/science.1259855.

Steffen, W., Rockström, J., Richardson, K., et al. (2018). Trajectories of the Earth System in the Anthropocene. *Proceedings of the National Academy of Sciences of the United States of America*. 115(33), 8252–59. DOI: 10.1073/pnas.1810141115.

Strang, V. (2004). *The Meaning of Water*. Oxford: Berg Publishers.

Sultana, F. (2018). Water Justice: Why It Matters and How to Achieve It. *Water International*. 43(4), 483–93. DOI: 10.1080/02508060.2018.1458272.

Sultana, F. (2022). Critical Climate Justice. *The Geographical Journal*. 188(1), 118–24. DOI: 10.1111/geoj.12417.

Todd, Z. (2014). An Indigenous Feminist's Take on the Ontological Turn: 'Ontology' Is Just Another Word for Colonialism. zoeandthecity.wordpress.com/2014/10/24/an-indigenous-feminists-take-on-the-ontological-turn-ontology-is-just-another-word-for-colonialism.

Tuathail, G. Ó. (1997). At the End of Geopolitics? Reflections on a Plural Problematic at the Century's End. *Alternatives: Global, Local, Political*. 22(1), 35–55. DOI: 10.1177/030437549702200102.

United Nations Department of Economic and Social Affairs. (2022). Goal 6: Ensure Availability and Sustainable Management of Water and Sanitation for All. sdgs.un.org/goals/goal6.

Vörösmarty, C. J., Pahl-Wostl, C., Bunn, S. E., & Lawford, R. (2013). Global Water, the Anthropocene and the Transformation of a Science. *Current Opinion in Environmental Sustainability*. 5(6), 539–50. DOI: 10.1016/j.cosust.2013.10.005.

Walker, B., Holling, C. S., Carpenter, S. R., & Kinzig, A. (2004). Resilience, Adaptability and Transformability in Social–Ecological Systems. *Ecology & Society*. 9(1), 3.

Watts, V. (2013). Indigenous Place-Thought and Agency amongst Humans and Non Humans (First Woman and Sky Woman Go on a European World Tour!). *Decolonization: Indigeneity, Education & Society*. 2(1), 20–34.

Whatmore, S. (2002). *Hybrid Geographies: Natures, Cultures, Spaces*. Thousand Oaks, CA: Sage.

Whyte, K. P. (2017a). Indigenous Climate Change Studies: Indigenizing Futures, Decolonizing the Anthropocene. *English Language Notes*. 55(1–2), 153–62.

Whyte, K. P. (2017b). Is It Colonial Déjà Vu? Indigenous Peoples and Climate Injustice. In J. Adamson & M. Davis, eds., *Humanities for the Environment: Integrating Knowledge, Forging New Constellations of Practice*. Routledge Environmental Humanities. London: Routledge Taylor & Francis Group, 88–104.

Wilson, N. J. (2020). Querying Water Co-Governance: Yukon First Nations and Water Governance in the Context of Modern Land Claim Agreements. *Water Alternatives*. 13(1), 93–118.

Wilson, N. J., & Inkster, J. (2018). Respecting Water: Indigenous Water Governance, Ontologies, and the Politics of Kinship on the Ground. *Environment and Planning E: Nature and Space*. 1(4), 516–38. DOI: 10.1177/2514848618789378.

Wilson, N. J., Montoya, T., Arseneault, R., & Curley, A. (2021). Governing Water Insecurity: Navigating Indigenous Water Rights and Regulatory Politics in Settler Colonial States. *Water International*. 46(6), 783–801. DOI: 10.1080/02508060.2021.1928972.

Wilson, N. J., Shah, S. H., Montoya, T., et al. (2024). Climate–Water Crises: Critically Engaging Relational, Spatial, and Temporal Dimensions. *Ecology and Society*. 29(4), article 13. DOI: 10.5751/ES-15469-290413.

Yates, J. S., Harris, L. M., & Wilson, N. J. (2017). Multiple Ontologies of Water: Politics, Conflict and Implications for Governance. *Environment and Planning D: Society and Space*. 35(5), 797–815. DOI: 10.1177/0263775817700395.

Yazzie, M. K., & Baldy, C. R. (2018). Introduction: Indigenous Peoples and the Politics of Water. *Decolonization: Indigeneity, Education & Society*. 7(1), 1–18.

Young, O. R. (2010). *Institutional Dynamics: Emergent Patterns in International Environmental Governance, Earth System Governance*. Cambridge, MA: MIT Press.

Zipper, S. C., Jaramillo, F., Wang-Erlandsson, L., et al. (2020). Integrating the Water Planetary Boundary with Water Management from Local to Global Scales. *Earth's Future*. 8(2), e2019EF001377. DOI: 10.1029/2019EF001377.

Zwarteveen, M. Z., & Boelens, R. (2014). Defining, Researching and Struggling for Water Justice: Some Conceptual Building Blocks for Research and Action. *Water International*. 39(2), 143–58. DOI: 10.1080/02508060.2014.891168.

Part VII

Biogeochemical Flows

17

Introduction to the Biogeochemical Flows Boundary

JEREMY J. SCHMIDT

Earth's elements flow. Carbon, nitrogen, calcium, and phosphate (among many others) cycle through the biosphere, oceans, lakes, rivers, the atmosphere, rocks, and soils at regional and global scales (see Berner & Berner 2012). In the planetary boundaries framework two elements are the focus: phosphorus and nitrogen. Both are key constituents for life, and both have been significantly impacted by anthropogenic forces, especially the production of food through industrial agriculture. The flows of other elements matter, too. The flow of carbon, for instance, is implicit in the climate change boundary and for biosphere integrity. Yet phosphorus and nitrogen hold a special place owing to the ways their limits and excesses affect terrestrial and aquatic ecosystems. In the planetary boundaries framework, they are measured slightly differently. Phosphorus is assessed both globally and regionally, while nitrogen is treated only at the global scale – although there are proposals to measure nitrogen regionally too (see Schulte-Uebbing et al. 2022). It should be noted, too, that initially phosphorus was only treated globally, through focus on the ocean. After interventions by freshwater ecologists, however, phosphorus was subsequently treated regionally and in terms of freshwater impacts (see Carpenter & Bennett 2011). Each of phosphorus and nitrogen are measured in terms of their sustained flows from rivers to the ocean and measured in Teragrams (1 Teragram = 1 billion kilograms). For phosphorus, the boundary is 11Tg, and for nitrogen the boundary is 6.2Tg (Steffen et al. 2015).

Nitrogen and phosphorus flows differ in important ways for understanding how respective human uses affect people, other species, and ecosystems. Perhaps most salient is that while nitrogen is naturally occurring it can also be synthetically produced for application as a dry fertilizer or as anhydrous ammonia (NH_3), a form of liquid nitrogen applied to soils where it bonds to subsurface moisture. The invention of synthetically produced nitrogen in the twentieth century led to large increases in its use, and to significant pollution owing to runoff from agricultural land as the result of excess application. Phosphorus, by contrast, cannot be synthetically produced at industrial scales. It is primarily mined from existing deposits, and this has been a driving feature of both its exploitation in colonial and post-colonial contexts as well as in contemporary agricultural economies that are increasingly globalized. This presents additional challenges regarding how it is sourced, from where, and the politics of controlling an element that is vital to the global food supply (Cordell & White 2014; Elser & Haygarth 2021).

Nitrogen and phosphorus also have shared features. Excessive release of either into freshwater systems can lead to the eutrophication of lakes. It can also lead to 'dead zones' where rivers meet the sea, such as the Mississippi in the Gulf of Mexico, where the cumulative impact of the fertilizer choices of many farmers affect aquatic ecology and the livelihoods that depend on healthy oceans. Histories of nitrogen and phosphorus use also generate significant inequalities when the planetary boundaries framework is used to determine 'fair shares' for using them in the future or for ecosystem recovery from past uses (Sandström et al. 2023).

Outside of the planetary boundaries framework there are also significant ethical issues to consider, as the two chapters in this section also show in their own ways. These require situating the insights of the planetary boundaries framework in larger considerations of history, colonialism, ecosystem science, and environmental governance.

Teaiwa argues that the phosphorus mining critical to food security must be placed amid histories of colonialism and ongoing forms of extraction. Teaiwa seeks to trouble narratives – disciplines, even – that do not grapple with how gendered forms of dispossession affected leadership and governance by Indigenous women while introducing separations of people from land in the name of putting phosphorus in service to a larger humankind. By contrast, Elser examines phosphorus through two perspectives: one that prioritizes technological innovation and one concerned with environmental degradation. Offering an overview of the debate between these positions from the Green Revolution to today, he outlines how elements of both must be combined in a new kind of 'systems innovation' in order for sustainable phosphorus use to be achieved.

References

Berner, E. K., & Berner, R. A. (2012). *Global Environment: Water, Air, and Geochemical Cycles*. 2nd ed. Princeton, NJ: Princeton University Press.

Carpenter, S. R., & Bennett, E. M. (2011). Reconsideration of the Planetary Boundary for Phosphorus. *Environmental Research Letters*. 6(1), 014009. DOI: 10.1088/1748-9326/6/1/014009.

Cordell, D., & White, S. (2014). Life's Bottleneck: Sustaining the World's Phosphorus for a Food Secure Future. *Annual Review of Environment and Resources*. 39(1), 161–188. DOI: 10.1146/annurev-environ-010213-113300.

Elser, J. J., & Haygarth, P. (2021). *Phosphorus: Past and Future*. Oxford: Oxford University Press.

Sandström, V., Kaseva, J., Porkka, M., et al. (2023). Disparate History of Transgressing Planetary Boundaries for Nutrients. *Global Environmental Change*. 78, 102628. DOI: 10.1016/j.gloenvcha.2022.102628.

Schulte-Uebbing, L. F., Beusen, A. H. W., Bouwman, A. F., et al. (2022). From Planetary to Regional Boundaries for Agricultural Nitrogen Pollution. *Nature*. 610(7932), 507–512. DOI: 10.1038/s41586-022-05158-2.

Steffen, W., Richardson, K., Rockström, J., Cornell, S., et al. (2015). Planetary Boundaries: Guiding Human Development on a Changing Planet. *Science*. 347(6223), 1259855. DOI: 10.1126/science.1259855.

18

Food Fight

Moving beyond Wizards vs Prophets to Achieve a Sustainable Phosphorus System

JAMES ELSER

In 1970 Norman Borlaug was awarded the Nobel Peace Prize 'for having given a well-founded hope – the green revolution'. Borlaug's contribution – his development (through selective breeding) and relentless deployment of high-yield varieties of wheat – was one of the key pillars of the Green Revolution, in which global food production tripled with only a 30 per cent increase in farmland area (Pingali 2012). A recent book (*The Wizard and the Prophet* by Charles Mann) posits Borlaug as a central figure of the twentieth century, an exemplar of techno-centric optimism that would use scientific insights and engineering applications to solve humanity's challenges, including its rapidly expanding population in such effective ways that they seem magical even (Mann 2018). Borlaug was a Wizard. Borlaug's legacy in the world of agriculture persists, manifested in the wizardly promise of genetically engineered crops (the 'Gene Revolution'), precision agriculture, and artificial intelligence on the farm.

Interestingly, less appreciated components of Green Revolution wizardry have impacts on water: increased reliance on irrigation that allowed expansion of cropping into dry regions and increased application of chemical inputs, including pesticides and fertilizers that often move from farms into streams, rivers, lakes, and oceans. While the secondary and broadly diffused impacts of these inputs came to widespread awareness via the impacts of DDT illuminated by Rachel Carson, the impacts of agricultural fertilizer use also became collateral damage of Green Revolution wizardry, seemingly inevitable given the low nutrient content of many soils and the fact that continuous removal of crops leads to depletion of soil nutrients. In Mann's book, alarm about the expanding impacts of agricultural intensification was manifested in the persona of William Vogt. The 'Prophet' of Mann's dialectic, Vogt was an ecologist and conservationist and later became the National Director of Planned Parenthood. Throughout his life he spoke of the dangers of exponential population growth, resource depletion, and environmental degradation that would lead to collapse of civilization. In *The Wizard and the*

Prophet, Mann proposes that the tension between Borlaug and Vogt and their respective proponents manifested itself throughout the twentieth century in competing worldviews in which human material progress is confronted by environmental apocalypse.

Indeed, we can now see Vogt's alarm calls echoing in the concept of 'planetary boundaries', a situation that is not surprising given the massive impacts of modern agriculture on all dimensions of the Earth system. Consider the planetary boundaries for phosphorus. In the first application of planetary boundaries for the major biogeochemical nutrient cycles (nitrogen, phosphorus) (Rockström et al. 2009), human perturbation of nitrogen cycle (largely via the Haber–Bosch reaction for fertilizer production) was recognized to be well beyond sustainable levels due to its impacts on climate change (via the greenhouse gas, N_2O) and its role in stimulating algal blooms in N-limited coastal oceans. It was recommended that N use should be brought to levels somewhat arbitrarily set at 25 per cent of then-current (2009) fluxes. However, Rockström et al. (2009) stated that phosphorus cycling was currently within safe operating limits for humanity but that an 11-fold increase in P fluxes beyond pre-industrial background levels would lead to increased risk of large-scale ocean anoxia. This proposition was somewhat surprising to freshwater scientists as it seemed to ignore P's primary role in driving eutrophication of inland waters, central to drinking water supplies, fisheries, and aquatic biodiversity. In response, Carpenter and Bennett (2011) estimated planetary boundaries for P impacts on freshwaters and concluded that fluxes had already exceeded levels that would induce widespread eutrophication of inland waters. Accordingly, updated assessments of planetary boundaries acknowledge that both the N and the P cycles exceed safe operating levels (Steffen et al. 2015; see also Figure 1.1 in this book's introductory chapter). Indeed, nutrient emissions (N, P) constitute two of the three key control variables that currently fall in the 'high risk' category, well beyond their safe operating limits.

Other voices in the mould of Mann's Vogtian Prophet continue to sound the alarm about humanity's unsustainable phosphorus path (Brownlie et al. 2022; Elser & Bennett 2011) and have even raised concerns not only about the impact of phosphorus on water quality but also about possible looming shortages of phosphate rock for fertilizer production (Cordell et al. 2009; Elser 2012). While those concerns have been dampened by upward revision of estimates of phosphate rock availability (Van Kauwenbergh 2010; but see Edixhoven et al. 2014), the price of phosphate rock nevertheless has undergone unprecedented recent volatility (Elser et al. 2014) and has increased at nearly twice the rate of inflation since 2003, including a very large price spike of more than 700 per cent in just a few months in 2007. Furthermore, the 2010 upward revision of phosphate rock reserve estimates by Van Kauwenbergh (2010) was entirely due to recognition of geological deposits found in Morocco, which now controls more than 70 per cent of known phosphate rock reserves (Garside 2022), placing it in an unusually powerful position with respect to current and future supplies of this essential farm input.

So, whither the Wizard and Prophet in the world of phosphorus? In Mann's telling, this conflict is one between techno-optimists in the world of agriculture and pessimistic doomsayers in the world of environmental science, locked in enduring battle in a polarized political arena (Oury 2019). But is this the case? With respect to phosphorus, I would propose that the dichotomy is not so easily crystallized in this framework. First, the players in conflict both have what seem to be universally embraceable values (it is good to protect others from suffering) and unassailable goals (Who is against 'feeding the world'? Who opposes clean water?). Second, techno-optimists also abound on the water protection side of the coin not just on the agricultural production side. Third, Mann's view of this enduring dichotomy does not recognize emerging efforts to move beyond it to achieve more holistic and systems-level approaches to the challenge of planetary P sustainability. Indeed, the situation described by Mann is reminiscent of a classic critique of 'command and control' in natural resource management (Holling & Meffe 1996) in which increasingly intensive and heavily capitalized interventions are imposed in order to reduce or eliminate undesirable environmental variations, reducing ecological complexity and compromising resilience. Next I discuss these difficulties with Mann's dichotomy.

Those promoting increased agricultural production to 'feed the world' may indeed consider themselves on high ethical ground, but such a position becomes more assailable when considering the agricultural system as it currently operates. Most notably, only ~60 per cent of agricultural production goes into growing food that is directly consumed by humans (mostly as grains such as rice and wheat) while 35 per cent goes to growing feed for animals (Elser & Haygarth 2022). While meat and dairy do carry considerable nutritional value, it is hard to argue against the view that 'feeding the world' would be a lot easier if diets contained less meat and relied more on plant-derived fats and proteins. Ethical dimensions connected to animal welfare are also prominent here. Non-food crop commodities make up the final share of agricultural production that does not 'feed the world', and these include biofuel crops that, arguably at least, are being produced in order to bring humanity within safe operating limits for climate change. Indeed, production of ethanol increased threefold and biodiesel increased fourfold between 2000 and 2021 (Elser & Haygarth 2021). A notable example of this is cultivation of corn for ethanol production in the USA, which constituted 32 per cent of corn production and 10 per cent of fertilizer use in the USA in 2009. Expansion of the Renewable Fuel Standard (RFS) in 2007 further increased fertilizer use by 3–8 per cent (Lark et al. 2022). The benefits of this effort for staying within the climate change safe operating space remain unclear. While early analyses indicated that corn-derived ethanol had lower carbon intensity than gasoline thus allowing it to fall within the RFS programme (Lark et al. 2022), a more recent analysis that takes into account the emission effects of expanding corn cultivation and altering land use indicates that the carbon intensity of corn ethanol is no better than gasoline and is likely 24 per cent higher (Lark et al. 2022). While this report has drawn criticism (though not yet in the peer-reviewed literature), given the well-known and intense local impacts of fertilizer runoff on water quality in corn-intensive regions as well as the downstream impacts of exported nutrient pollution on coastal waters (e.g. the Gulf of Mexico 'dead zone'), continued promotion of ethanol production on croplands that are suitable for cultivation of food for human consumption seems ethically dubious given that beneficial impacts on carbon emissions are probably minimal. Current practices are likely a reflection of strong political forces at play in sustaining programmes like the Renewable Fuel Standard.

Are the Prophets who challenge agricultural Wizards in defence of the environment blameless in these contests? It might seem so since, after all, clean water is essential for human well-being, and non-point source pollution from agriculture is a dominant cause of water pollution in many regions, including the USA (Stets et al. 2020), Europe (Werner & O'Doherty 2012), and China (Liu et al. 2016; Zhang et al. 2022). However, once again the ethical scene is murkier than one might imagine. The main complicating factor, of course, is the imperative discussed earlier – to feed the world. While global hunger has been gradually diminishing (but worsened during the COVID-19 pandemic), 10 per cent of the world's population still suffer from hunger while nearly one-third don't have regular access to sufficient food (United Nations Department of Economic and Social Affairs 2022a). In addition, world population just passed 8 billion and is estimated to reach 9.7 billion by 2050 (United Nations Department of Economic and Social Affairs 2022b). It is difficult to imagine how this population can be fed without extensive fertilizer use, even if progress is made in restructuring diets and targeting fertilizer use to food crops. Furthermore, substantial reductions in fertilizer use would, eventually at least, reduce crop yields per unit of farmed land. In this scenario, increased food production can only take place by further conversion of forest, grassland, and desert ecosystems into croplands, accelerating biodiversity loss (a planetary boundary humanity has already exceeded). So, we are faced with an ethical conundrum: thirst cannot trump hunger but neither can hunger trump thirst.

And what of Mann's assignment of techno-optimists exclusively to the Borlaugian agricultural camp? Are agriculturalists the only ones who think we can invent our way out of the phosphorus dilemma? Looking at the range of responses being mustered to meet the challenge of phosphorus pollution, the answer is clearly 'no'. Many new technological approaches are now in early deployment or active development to meet the phosphorus sustainability challenge on the 'downstream' side of the problem. (Note that various pieces of wizardry are being deployed or developed for agricultural practices to address P sustainability, from genetic engineering to improve crop fertilizer use to precision agriculture to complex soil amendments and microbiome adjustments to cellular agriculture.) Technologies to intercept lost phosphorus from various forms of 'waste' and return it for re-use include struvite precipitation equipment that recaptures P in wastewater treatment plants, anaerobic bioreactors that use animal manure to produce biogas and concentrated nutrient solutions, and urine source separation. So, from among the Prophets also arise Wizards, both, it seems, seeking to apply a range of technological fixes to a complex food–water system that encompasses both natural and anthropogenic ecosystems that are globally linked via trade and environmental flows.

How then can the phosphorus sustainability challenge be met in a fair and ethical manner? I would follow in the spirit of Holling and Meffe (1996) to argue that, while both Prophets ('problem finders') and Wizards ('problem fixers') will always be needed, both on the food side and on the water side of the food–water nexus, we need to move rapidly towards new ways of thinking and doing that embrace the full range of system complexity involved. Einstein apocryphally said, 'The definition of insanity is doing the same thing over and over again and expecting different results.' In the same vein, Einstein or Ram Dass or Marc Prensky or maybe someone else said: 'The problems that exist in the world today cannot be solved by the level of thinking that created them.' We need new ways of thinking about food systems and about water systems and about how both are wrapped into a complex web that connects biogeochemistry, hydrology, nutrition, politics, and culture. In our book *Phosphorus: Past and Future* (Elser & Haygarth 2022), Phil Haygarth and I call for the emergence of a new kind of thinker who we call a 'Systems Innovator', someone who has the systems perspective of a Prophet combined with a Wizard's drive to innovate. Systems Innovators possess not only deep technical expertise but also the 'soft' human skills to communicate knowledge and to navigate the complex ethical, political, and cultural terrains within which P sustainability must emerge. Haygarth and I see Systems Innovators primarily as individuals who provide leadership and serve as 'boundary persona' (akin to 'boundary objects,' sensu Star & Griesemer (1989)) to provide trusted connections between disparate sectors and stakeholders (e.g. grower associations and water resource managers) that need to cooperate to transition towards more sustainable practices. For example, Systems Innovators would recognize that current livestock production is ethically problematic and is a major driver of nutrient pollution and of heavy P fertilizer use (a la Vogt) and that radical technological innovations (a la Borlaug) such as cellular agriculture ('stem cell meat') are likely to emerge from food science Wizards. However, they would also recognize, analyse, and work to minimize concomitant environmental impacts of cellular agriculture, they would acknowledge the potential for strong cultural and psychological barriers to this novel technology, and they would provide strategies by which informed community engagement could shape the political environment within which necessary markets and regulations involving cellular agriculture would emerge. Additionally, 'Systems Innovation Alliances' might emerge as 'boundary organizations' that serve the same function, independent of the particular individuals who populate them. In the domain of the phosphorus challenge, examples might be the European Sustainable Phosphorus Platform (ESPP), which was

> formed in March 2013 through a declaration, signed by over 150 organisations after the first European Sustainable Phosphorus Conference. ESPP ensures knowledge sharing, experience transfer and networking for opportunities in the field of phosphorus management, facilitates discussion between the market, stakeholders and regulators, addresses regulatory obstacles, contributes to policy proposals, circulates information by newsletters, website, conferences and publications, promotes Platform Members' activities, and contributes to define a long-term vision for phosphorus sustainability in Europe. (European Sustainable Phosphorus Platform 2022)

Another example is the Sustainable Phosphorus Alliance (SPA), whose website states that it is

> a member organization addressing the complex problem of phosphorus sustainability. We are North America's central forum and advocate for the sustainable use, recovery, and recycling of phosphorus in the food system. We collaborate with our members and supporters to innovate and implement evidence-based solutions to the phosphorus sustainability challenge, to be shared and accessible to the community. (Sustainable Phosphorus Alliance 2022)

SPA's membership consists of mining companies, water suppliers and agencies, waste treatment companies, and academic research programmes. (Full disclosure: I am the director of SPA, but its mission and activities would, I can hope, persist in my absence.)

The Wizard and the Prophet is a vision of prevailing attitudes of the twentieth century. In the twenty-first century we confront the now-obvious consequences of the insufficiency of the actions and inactions that emerged from past tension between these two poles. This is seen in ongoing climate change, large-scale deforestation and other habitat destruction, and widespread diffuse agricultural pollution that are the red flags of the now-emblematic planetary boundaries, the flashing lights on the dashboard of planetary life support systems about to fail. In the example of phosphorus, we see an imperative to create new ways of being, thinking, and acting to reconcile justifiable moral actions of one group ('Feed the world!') with justifiable moral outrage of others ('Save our drinking water!', 'Save our fish!') in order to return to safe operating conditions. In the coming decades we will see if Systems Innovators, or some similar operators or institutions, emerge in sufficient numbers and with sufficient influence to harmonize these conflicting voices to bring about a sustainable food–water system, one in which nutritious food is abundantly available to all and in which lakes, rivers, and oceans thrive once again.

References

Brownlie, W., Sutton, M. A., Heal, K. V., Reay, D. S., & Spears, B. M. (2022). Our Phosphorus Future: Towards Global Phosphorus Sustainability. UK Centre for Ecology & Hydrology. DOI: 10.13140/RG.2.2.17834.08645.

Carpenter, S. R., & Bennett, E. M. (2011). Reconsideration of the Planetary Boundary for Phosphorus. *Environmental Research Letters*. 6(1), 14009. DOI: 10.1088/1748-9326/6/1/014009.

Cordell, D., Drangert, J.-O., & White, S. (2009). The Story of Phosphorus: Global Food Security and Food for Thought. *Global Environmental Change*. 19(2), 292–305. DOI: 10.1016/j.gloenvcha.2008.10.009.

Edixhoven, J. D., Gupta, J., & Savenije, H. H. G. (2014). Recent Revisions of Phosphate Rock Reserves and Resources: A Critique. *Earth System Dynamics*. 5(2), 491–507. DOI: 10.5194/esd-5-491-2014.

Elser, J. J. (2012). Phosphorus: A Limiting Nutrient for Humanity? *Current Opinion in Biotechnology*. 23(6), 833–38. DOI: 10.1016/j.copbio.2012.03.001.

Elser, J. J., & Bennett, E. (2011). Phosphorus Cycle: A Broken Biogeochemical Cycle. *Nature*. 478(7367), 29–31. DOI: 10.1038/478029a.

Elser, J. J., Elser, T. J., Carpenter, S. R., & Brock, W. A. (2014). Regime Shift in Fertilizer Commodities Indicates More Turbulence ahead for Food Security. *PloS One*. 9(5), e93998. DOI: 10.1371/journal.pone.0093998.

Elser, J. J., & Haygarth, P. (2022). *Phosphorus: Past and Future*. Oxford: Oxford University Press.

European Sustainable Phosphorus Platform. (2022). About the European Sustainable Phosphorus Platform (ESPP). www.phosphorusplatform.eu/platform/about-espp

Garside, M. (2022). Phosphate Rock Reserves by Country 2021. www.statista.com/statistics/681747/phosphate-rock-reserves-by-country.

Holling, C. S., & Meffe, G. K. (1996). Command and Control and the Pathology of Natural Resource Management. *Conservation Biology: The Journal of the Society for Conservation Biology*. 10(2), 328–37.

Lark, T. J., Hendricks, N. P., Smith, A., et al. (2022). Environmental Outcomes of the US Renewable Fuel Standard. *Proceedings of the National Academy of Sciences*. 119(9), e2101084119. DOI: 10.1073/pnas.2101084119.

Liu, X., Sheng, H., Jiang, S., et al. (2016). Intensification of Phosphorus Cycling in China since the 1600s. *Proceedings of the National Academy of Sciences*. 113(10), 2609–14. DOI: 10.1073/pnas.1519554113.

Mann, C. C. (2018). *The Wizard and the Prophet: Two Remarkable Scientists and Their Dueling Visions to Shape Tomorrow's World*. New York: Alfred A. Knopf.

Oury, J.-P. (2019). Borlaug vs Vogt: Limiting the Population, or Developing Precision Agriculture? *European Scientist*, 4 June. www.europeanscientist.com/en/editors-corner/borlaug-vs-vogt-limiting-the-population-or-developing-precision-agriculture/.

Pingali, P. L. (2012). Green Revolution: Impacts, Limits, and the Path Ahead. *Proceedings of the National Academy of Sciences*. 109(31), 12302–8. DOI: 10.1073/pnas.0912953109.

Rockström, J., Steffen, W., Noone, K., et al. (2009). A Safe Operating Space for Humanity. *Nature*. 461(7263), 472–75. DOI: 10.1038/461472a.

Star, S. L., & Griesemer, J. R. (1989). Institutional Ecology, 'Translations' and Boundary Objects: Amateurs and Professionals in Berkeley's Museum of Vertebrate Zoology, 1907–39. *Social Studies of Science*. 19(3), 387–420. DOI: 10.1177/030631289019003001.

Steffen, W., Richardson, K., Rockström, J., et al. (2015). Sustainability. Planetary Boundaries: Guiding Human Development on a Changing Planet. *Science*. 347(6223), 1259855. DOI: 10.1126/science.1259855.

Stets, E. G., Sprague, L. A., Oelsner, G. P., et al. (2020). Landscape Drivers of Dynamic Change in Water Quality of U.S. Rivers. *Environmental Science & Technology*. 54(7), 4336–43. DOI: 10.1021/acs.est.9b05344.

Sustainable Phosphorus Alliance. (2026). Allies Addressing a Complex Problem. phosphorusalliance.org.

United Nations Department of Economic and Social Affairs. (2022a). Goal 2: End Hunger, Achieve Food Security and Improved Nutrition and Promote Sustainable Agriculture. sdgs.un.org/goals/goal2.

United Nations Department of Economic and Social Affairs. (2022b). World Population Prospects 2022: Summary of Results. www.un.org/development/desa/pd/content/World-Population-Prospects-2022.

van Kauwenbergh, S. J. (2010). World Phosphate Rock Reserves and Resources. International Fertilizer Development Center. (IFDC Technical Bulletin 75).

Werner, B., & O'Doherty, J. J. (2012). European Waters – Current Status and Future Challenges: Synthesis. European Environment Agency. www.eea.europa.eu/publications/european-waters-synthesis-2012.

Zhang, J., Zhong, Q., Yu, Y., et al. (2022). Supply-Side Drivers of Phosphorus Emissions from Phosphorus Supply Chains in China. *Ecosystem Health and Sustainability*. 8(1), 2107578. DOI: 10.1080/20964129.2022.2107578.

19

Between Justice and Food Security

*Planetary Legacies of Mining Phosphate Islands**

KATERINA TEAIWA

19.1 What Can Mined-Out Pacific Phosphate Islands Teach Us about the Limits of the Planet?

At the 79th session of the United Nations General Assembly, the President of the Pacific island nation of Nauru, David Adeang, said that 'the international community must not let "fear and misinformation" block the exploitation of seabed minerals that could be critical to the global energy transition' (Pearl 2024). Adeang argued that the minerals in the seabed would provide essential ingredients for electric vehicle batteries and other green technologies declaring:

> The greatest risk we face is not the potential environmental impacts of mineral recovery, but the risk of inaction .... There is a risk of failing to seize the opportunity to transform to renewable energy and to decarbonize our planet. We cannot let fear and misinformation hold us back. (Pearl 2024)

One of the region's most respected political analysts, Associate Professor Tarcisius Kabutaulaka of the University of Hawai'i's Center for Pacific Islands Studies, responded publicly on Facebook asking: 'Mining the ocean floor to save the environment? What kind of twisted logic is this? Coming from Nauru, of all places' (Kabutaulaka 2024).

Kabutaulaka's confusion resonates with many Pacific scholars and policy folks who are well aware of Nauru's long history of phosphate mining and the calamitous impacts that German and Australian colonialism and phosphate extraction has had on Nauruan society, culture, politics, economy, and environment. Mined first by a consortium of British, Australian, and New Zealand agricultural interests from the early 1900s, Nauru took over the mining of its own island in 1968, further degrading their environment and reducing its ecological

* With many thanks to Jeremy J. Schmidt and Simon P. Meisch for their support and great patience, and to Rebecca Monson for much needed feedback on the writing. An earlier, and much shorter, version of this essay was published by Indiana University Press on its blog in 2015.

resilience considerably while similarly eroding Nauruan culture, health, traditional knowledge, and sense of place. A hundred and twenty years after mining began, and following the loss of millions of dollars and millions of tons of the island's surface, it is not surprising that certain Nauruan leaders have turned to the seabed as the next frontier for extraction. Their values align with those of far more powerful actors, particularly in the United States, who have resuscitated the bold, extractive ethos of the industrial age, deliberately ignoring scholars, activists, and communities who have worked so hard to warn others of the limits of our ancestral places, and indeed, the earth's natural resources.

Using the logic of decarbonization to argue for the exploitation (and likely destruction) of another corner of the planet continues a practice of self-sabotage that is well entrenched in the age of calamitous, anthropogenic climate change. I have long studied the impacts of phosphate mining on Nauru's nearest neighbour, Banaba, also called Ocean Island, and my family's ancestral home, in what is now the Republic of Kiribati. It is abundantly clear that the colonial introduction of phosphate mining in tandem with Christianity and western economic, legal, and governance systems, on both these islands, dramatically changed cultures that had survived and thrived for millennia on phosphate islands. These island societies were once focused on safeguarding and regenerating their cultures and critical environmental resources. After phosphate rock was identified by an Australian and New Zealand–based British commercial business in 1900, many in these islands were forced to adopt laws, currencies, and industrial lifestyles that were touted as superior to their once sustainable practices and values.

There were two particularly destructive consequences of these changes. The first was the imposition of patriarchy and the removal of women's political and environmental authority and role to safeguard land and critical resources such as freshwater. The second was the severing of epistemological, ontological, and axiological ties between islanders and their lands and waterways, so that they came to see themselves as above and separate from the islands they once experienced as intimately connected to their own bodies, minds, and spirits. These changes are global in terms of the impacts of European and Asian colonialism and imperialism over Indigenous cultures where these ontologies were fundamental. Such values are captured in Pacific languages where the terms describing people, such as te aba, vanua, whenua, fenua, fanua, and aina, encapsulate both land and people. Banaba, for example, was known as te buto, the 'navel of the universe' (Maude & Maude 1932: 265).

The message, from the British and antipodean phosphate stakeholders, was that mining and the conversion of phosphates into fertilisers were for the good of humankind (K. Teaiwa 2014). It was viewed to be critical for mass agriculture, critical to feeding rapidly expanding global populations, and critical to developing farming-based settler colonies such as Australia and New Zealand. In the United States, the Guano Act of 1856 not only signalled the strategic value of phosphate – in this case in the form of more easily accessible bird and bat droppings, such as guano (Cushman 2013) – but also a strategy for acquiring more territory in a critical region between the United States, Russia, and East Asia. Many of these islands are now part of the Pacific Islands Heritage Marine National Monument covering 1.2 million square kilometres of the Pacific Ocean and governed by the US Fish and Wildlife Service. This is 1.2 million square kilometres of potential ocean for deep sea mining and other forms of marine exploitation.

Between the seventeenth and mid twentieth centuries, colonial, capitalist actors and interests in settler colonies, and across agricultural enterprises and commodity chains, appropriated lands, appropriated labour, and removed the ability for slaves, their descendants, and most Indigenous populations to have access to the resources they needed for sovereign capabilities and self-determination. It was no different in the Pacific. Banabans and Nauruans were forced to give up their islands to miners and farmers from Australia, New Zealand, and across the British Empire, resulting in severely ruined landscapes, waterways, and cultures (K. Teaiwa 2014).

19.2 Planetary Boundaries

In 2012, Kate Raworth, a senior research analyst with Oxfam in Britain, proposed a conceptual and visual framework which brought the concept of planetary boundaries outlined by Johan Rockström and his colleagues at the Stockholm Resilience Centre, together with that of social and moral boundaries (Stockholm Resilience Centre n.d.). Imagined as a global-scale compass shaped like a doughnut (Figure 19.1), the framework aimed to chart the course between sustainability, on the one hand, and justice and equity between nations, on the other (Raworth 2012). Raworth's goal was to create a safe and just space between the social and environmental imperatives in which humanity can thrive. This, she argues, requires equity in the use of natural resources and greater efficiency in transforming those resources to meet human needs.

It is worth sharing Raworth's own summary of her integrated vision which is also a compass of sorts, to guide policymakers through decision-making, and the interlinked systems:

> The Doughnut raises four key implications for the pursuit of human wellbeing in the Anthropocene. First, it highlights the dependence of human wellbeing on planetary health.... The best chance of enabling a life of dignity and opportunity for more than 10 billion people over the coming century therefore depends on sustaining Holocene-like conditions, such as a stable climate, clean air, a protective ozone layer, thriving biodiversity, and healthy oceans. Second, the concurrent extent of social shortfall and

> ecological overshoot reflects deep inequalities – of income and wealth, of exposure to risk, of gender and race, and of political power – both within and between countries. The Doughnut helps to focus attention on addressing such inequalities when both theorising and pursuing human wellbeing. Third, the Doughnut implies the need for a deep renewal of economic theory and policymaking so that the continued widespread political prioritisation of gross domestic product growth is replaced by an economic vision that seeks to transform economies, from local to global, so that they become regenerative and distributive by design, and thus help to bring humanity into the Doughnut. Last, the Doughnut might act as a 21st century compass, but the greater task is to create an effective map of the terrain ahead. Thanks to ongoing socioecological systems research, this century is likely to be the first in which humanity begins more fully to understand and appreciate the complex interdependence of human wellbeing and planetary health. (Raworth 2017)

Agricultural scientists have long argued that industrial and mass agriculture is necessary to feed the planet's rapidly expanding population, predicted to increase to almost 10 billion people by 2050. But mass agriculture utilizing limited resources such as phosphate has primarily transformed diets, particularly in the developed world, increasing consumption of meat and dairy products that require large-scale animal and mono-cropping operations. One would have hoped that in the twenty-first century there would be more sophisticated, equitable, and sustainable techniques harnessed to feed the planet than those applied during the Green Revolution of the 1950s that relied on the breeding of high-yield crops, and application of pesticides and fertilizers. But in addition to the pollution generated, and the overconsumption of water and fuel to power these industries, agricultural technologies have actually generated excessive and inessential food and related products that provide an abundance of choice for industrial nations while depriving Indigenous and other smaller societies of the relatively sustainable agriculture, soil- and water-based knowledges, food varieties, and techniques they once employed.

Political choices across many powerful countries today indicate that the manipulation and re-shaping of our natural environment for human consumption based on wants, rather than needs, have not changed much over time. The lack of real commitment to climate change goals and mechanisms that emerged from the Conference of the Parties (COP) meetings since the 1990s is but one indicator of the global appetite, or lack thereof, for real change. While many global consumers express care for sustainably sourced and manufactured products, they are simultaneously as likely to vote for leaders with values that undermine any real gains made on the sustainability and equity fronts. Or, they are unaware or apathetic to the ways in which the wealthiest of industries and players, from the tech conglomerates and retail, to oil, gas, and weapons manufacturers, are deeply uninterested in sacrificing wealth for planetary and species health.

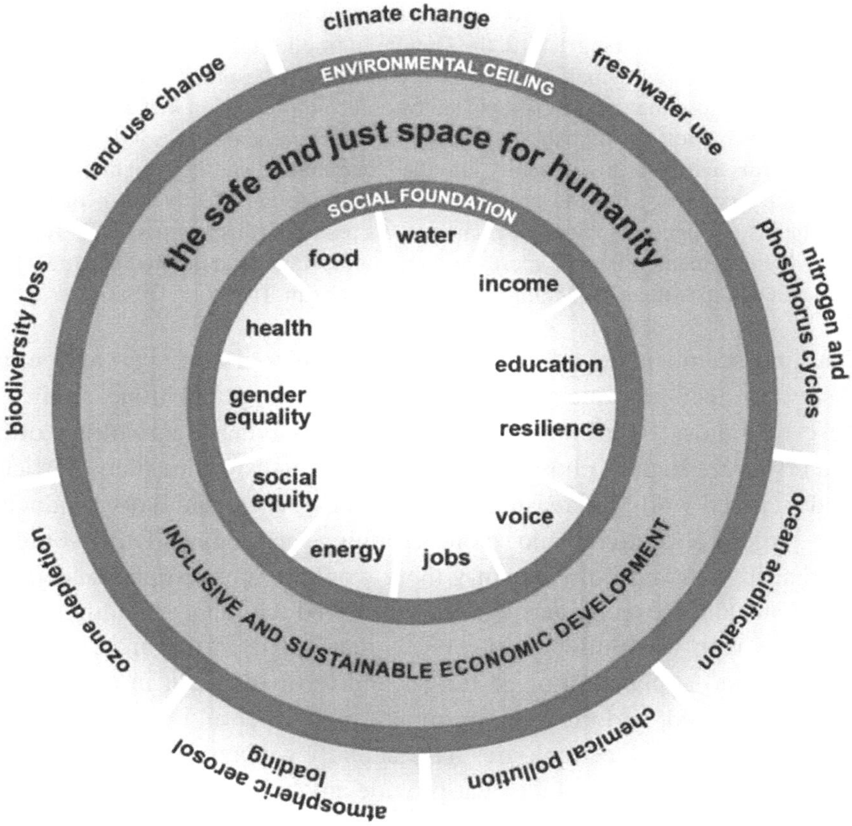

Figure 19.1 The 'Doughnut' for a safe and just planet. The 11 dimensions of the social foundation are illustrative and were based on governments' priorities for Rio +20. The nine dimensions of the environmental ceiling are based on the planetary boundaries set out by Rockström et al. (2009). The material 'A safe and just space for humanity: can we live within the doughnut? Kate Raworth – 2012 – www .oxfam.org/en/researcb/safe-and-just-space-humanity – Figure 4 p. 15' is reproduced with the permission of Oxfam, Oxfam House, John Smith Drive, Cowley, Oxford OX4 2JY, UK, www.oxfam.org.uk. Oxfam does not necessarily endorse any text or activities that accompany the materials.

Breaching the thresholds of the planet's nine essential terrestrial, aquatic, atmospheric, and biogeochemical systems, including the nitrogen and phosphorus cycles, is what Rockström and his colleagues strongly cautioned against in their model. Raworth's framework grounded all this in the social, political, and moral contexts that drive many of the extractive decisions and practices in the first place.

But there is far more that drives people, both everyday folks and the disproportionately wealthy, beyond scientific, environmental, social, political, and moral imperatives or threats.

19.3 Phosphate Stories

Phosphorus is not an infinitely abundant planetary resource (Elser & Haygarth 2021). And yet, as highlighted by Rockström and colleagues, along with nitrogen, phosphorus is absolutely essential for life on Earth. The majority of the world's phosphate reserves are concentrated in a few countries including Morocco, with over 70 per cent of the world's known phosphate rock reserves; China, a major producer of phosphate, though its reserves are relatively smaller than Morocco's; the United States with significant reserves, primarily in Florida; and Russia and Jordan, along with smaller sources in other lands. While the size of these reserves is relevant to global markets, the quality of the phosphate is just as important. In fact, the mined-out Pacific islands yielded some of the world's highest quality, and most easily accessible, phosphoric acid, which is the critical ingredient in agricultural fertilisers (K. Teaiwa 2014).

In the late 1800s when British, Australian, New Zealand, French, and German prospectors were scouring the Pacific looking for guano and phosphate rock supplies to manufacture these important phosphate fertilizers, they frequently used the phrase 'for the good of mankind', to justify the destruction of island landscapes, the use of low-paid labour, and the displacement of communities and cultures. One might consider Raworth's attempts to bring questions of equity and justice into conversation with those of food security, environmental and human sustainability and wonder how the needs of small, Indigenous groups fit into the equation. This is particularly challenging when assumptions of scale and significance are at stake: feeding 8 billion humans today and 10 billion in the near future versus a couple of thousand disposable islanders.

Indigenous rights are now more widely recognized than they were on Banaba and on Nauru in the early twentieth century when massive phosphate mines extracting millions of tons of phosphate were in full industrial swing (Williams & Macdonald 1985). Banabans and Nauruans had lived in their home islands sustainably for over two thousand years, and yet their traditional knowledge of resilience and effective use of natural resources, like that of many Indigenous peoples, was and is still not valued.

Today, Banaba in particular is a decimated landscape with rusted, dilapidated infrastructure, asbestos, and dozens of artefacts of science, progress, and industry lying in a state of waste and ruin. In my earlier work, primarily utilizing family, archival, and cultural sources, I took a close, ethnographic, historical, multi-sited, and multi-scalar look at the lives of Banaban landowners, mining company workers, and the families of mining management and labourers, alongside the transformation of our tropical, ancestral island landscape into a global industrial hub and, eventually, a mining ghost town. I found the format of text, of journal articles and a book limiting in terms of the kinds of visual, archival, and oral resources I was drawing upon, and the forms of embodied and cultural tactics that Banabans utilized to resist the displacement of their lands and their own communities. Text alone did not adequately capture how Banabans were forced to cross many cultural taboos, boundaries, and thresholds that would not have been imagined before prospectors from Australia and New Zealand landed on their shores (but see a celebration of that colonial history in Williams and Macdonald (1985)). Ignoring such, often quite gendered, boundaries opened up spaces for extraction and environmental degradation that had never existed before.

For example, young Banaban women were the only people allowed into underground water caves, or bangabanga, to collect fresh drinking water for their communities. These were the only sources of potable water on an island with no wells, rivers, or streams. The bangabanga were some of the first spaces white men requested to explore while prospecting in order to map the required resources to support the mining enterprise. This quickly undermined women's social and political authority as their roles to hold land, water, and culture were thrust aside by the patriarchal and conservative Christian mining and colonial personnel who quickly took over island governance while cultivating their own Banaban, male allies (K. Teaiwa 2014).

Thinking critically about what was lost in this process, I expanded insights from my book into a multimedia exhibition in 2017 (Figures 19.2 and 19.3a,b). I utilized stories and photographs from the vast Australian and New Zealand phosphate archive and my own embodied research on Banaba, Tarawa, Tabiteuea (both sources of phosphate mining labour), and on Rabi where Banabans were displaced in 1945. Like most Banabans, I-Kiribati, and Fiji Islanders, I had grown up learning cultural and other dances and was able to harness methods and tools from the choreographic context, to better understand how islanders remembered and resisted their challenging histories.

The exhibition followed the history and routes taken by the Pacific Islands Company, The Pacific Phosphate Company, and the British Phosphate Commissioners over almost a century of mining on Nauru and Banaba (Figure 19.2). The show has travelled from Sydney in Australia where Nauruan phosphate was discovered and Banaba identified as a geological phosphate twin, to Napier and the Hawke's Bay in 2019 where millions of tons of Pacific phosphate was processed and distributed to farms across New Zealand. It also travelled to Auckland in 2021 where one of the company headquarters were based, and to Honolulu in

Figure 19.2 A map of Banaba. Credit: Jeremy Schmidt.

Hawai'i in 2023–2024 where shipments of phosphate were received for use on American plantations.

My exhibition attempted to move the stories of phosphate beyond academic institutional, scholarly, and disciplinary boundaries, into public spaces where families, and people of all ages, could reflect on one critical ingredient to grow the food that has nourished many generations of settlers in the antipodes. The science of global agriculture in the twentieth century mandated that phosphate fertilizer was key to fuelling the chain of agricultural commodities, but at what costs, ethically and materially, to Pacific Islanders? As humanity seeks to find new technical solutions to food challenges, whose lives, lands, and waterways will be improved or transformed, and whose will be ruined?

All living organisms require daily phosphorus intake to produce energy and the element of 'P' cannot be replaced; there is no synthetic substitute. Phosphate fertilizers that agribusiness was so hungry for in the nineteenth and twentieth centuries were manufactured, globally distributed, and allowed humans to double

(a)

(b)

Figures 19.3 (a and b) Project Banaba by Katerina Teaiwa curated by Yuki Kihara, Joy Enomoto, and Healoha Johnston, 2023–2024. Bishop Museum, Honolulu. Exhibition stills by Gui Taccetti. In these images, a multi-sited and multi-temporal sense of the impacts of mining is reconstructed in a gallery where you can journey through 80 years of mining in visual, textual, and audio-visual forms. Reproduced by permission of the artist, Katerina Teaiwa.

their food production. Food production was doubled, but many were deprived of food and sustainable resources in the process. Much phosphorus, after undergoing chemical transformation, is wasted and not cycled back into the system but enters others, like waterways, and wreaks havoc on those systems (Brownlie et al. 2024). Through mass agriculture, we have profoundly altered the planet's natural cycles for maintaining and keeping the flow of phosphorus and nitrogen in forms that are long-term and sustainable (Egan 2023).

19.4 Repairing Islands?

What can Pacific Islands teach us about the limits of the planet? In many ways the scale of industrial activities on islands across Oceania provides rich opportunities to reflect on global challenges. Banaba is 6 square kilometres of rock that yielded over 22 million tonnes of phosphate for fertiliser. The size of a landscape tells us nothing about its depth or its value for global food production, which, in the case of Nauru and Banaba, was significant. I became an academic not to enrol in a field or discipline and devote my career to furthering its theories and methods, but to find out what happened to Banaba and why. I stayed in academia because such institutions gave me space to investigate why we were wrenched from places that had sustained our ancestors for millennia.

My research on phosphate histories and their impacts on peoples and environments could be summarized into three questions that I contemplate almost daily and that have truly shaped my transdisciplinary career. The first is: how do we research and teach in ways that are inspired and motivated by issues of justice, transformation, healing, and well-being? Neither the scientific imperative of objectivity nor the false rationality of neoliberal economics provides pathways towards Indigenous or environmental justice. As I have become more involved with Banaban activism on the ground and observed economic initiatives by Banabans and Nauruans who now subscribe to colonial and extractive approaches and values, I've focused more on the need for environmental, social, and spiritual healing, repair, and well-being on both Rabi in Fiji, and Banaba in Kiribati. Such opportunities to critically reflect on the intergenerational and structural consequences of mining have rarely happened for Banaban communities.

The second question is: how do we share such work towards healing and repair in forms and formats inspired by the records, materialities, environments, values, practices, and lives of the communities (and places) we work with and belong to? This is an older challenge based on various attempts to draw, dance, perform, and story my way through tertiary teaching, learning, and research while dodging the institutional pressure to be trained, rigorously, in specific theories, methods, and ways of presenting knowledge. My background, before postgraduate studies, was in the natural sciences, dance, media, and visual arts, and as I began to study Banaba, I found myself in spaces such as history and anthropology that had little interest, or skills, in producing knowledge in other forms – including the visual, embodied, choreographic, or oral, that harmonized with what their informants and sources were teaching them.

I have learned a lot from students who increasingly identify themselves as neurodiverse, recognizing that I probably would've needed to do the same as an undergraduate, but didn't have the language to explain why the standard approaches to teaching, learning, and assessment gave me such high levels of anxiety. While the reality of being in a perpetual state of fight or flight in an unstable world is better understood today, I'm not sure any of us has a handle on the intergenerational trauma that embeds this hyper-awareness of being unsafe in our blood, our organs, our bones, muscles, and across all senses. If you are of the land, and the land is of you, I can imagine the clearing, the blasting, the digging, the removing, the drying, the loading, and the shipping of pieces of the land leaves your body feeling quite unsafe. So, I've tried to dance in 'weird' spaces, to push the boundaries of fields filled with scholars who need explanations for what is going on, and for why anyone is trying to move beyond the boundaries of disciplinary genealogies, legacies, and norms (K. Teaiwa 2024). Twenty-plus years of back pain later, I have been experimenting with telling Banaban stories in ways that require audiences to move, to listen, to watch, to wayfind through 124 years of Banaban attempts to have stable, joyful, and efficacious lives.

The third issue, which is a very live and sometimes daily issue depending on what's happening amongst my family, leaders, and communities on Rabi Island, on Banaba, and across Fiji and Kiribati is: how do we move beyond these disciplinary and institutional genealogies, legacies, and norms to empower Pacific islanders to work towards healthy, non-extractive futures (K. Teaiwa 2024)? The recent and now-persistent interest of Australian and other mining companies to explore and potentially extract remaining phosphate and limestone pinnacles from Banaba has revealed that very little has changed amongst some Banabans who now view their ancestral island as relatively dead, disposable, and a quick road to imagined wealth. They speak of 'my land' and 'our lands' like property in the capitalist sense, but with an assumption that there is a deep and forever wellspring of phosphate for long-term prosperity. These contemporary Banabans do not resemble pre-mining ancestors who, even in the face of devastating droughts at the end of the nineteenth century (Maude & Maude 1994), climbed aboard slave ships travelling to Tahiti and Hawai'i and committed to returning and repopulating their home island because it was te buto and te aba – the body of the land and the body of the people. Many Banabans today are locked in an almost 124-year-and-counting pattern of extraction, dispossession, celebration, desperation, creative survival, and the search for saviours and 'big men', like the miners and colonial officials who mined out their lands, to lead them to a biblical promised land (Silverman 1971). Thankfully, there are also courageous Banabans on Rabi and across Fiji and Kiribati, such as the Banaban Human Rights Defenders. Led by inspired youth following global climate activist Itinterunga Rae Bainteiti, and elders such as Pelenise Alofa and Toanuea Taratai, these folks embody the care, resistance, and activism of their Banaban forebears, including many women, who attempted to safeguard their lands, culture, and environment.

19.5 Islands and Planets

The framework of planetary boundaries, its warning about critical thresholds for nine essential processes for the global system, is a powerful, useful, and great forecasting tool that also completely misses the things that lead to such boundaries being crossed – the politics and the social and cultural values driving capitalism, consumption, colonialism, land grabbing, and other damaging human practices. Kate Raworth's doughnut gestures towards but does not fully grasp the materiality or spirituality of severing peoples from places. We need a moral, social, ideological, and spiritual boundary that indicates clearly where human values and practices have exceeded the well-being of planet and people.

In 2025, particularly after the re-election of Trump in the United States, and an administration in the world's most powerful government filled with conservative Christians, tech billionaires, climate deniers, and leaders tearing down programmes that focus on social, environmental, and planetary equity, sustainability, and well-being, ruination at a much larger scale than tiny Banaba or Nauru is most likely. War, death, and genocide have and continue to rage across our world and such state-sanctioned, state-supported violence certainly crosses critical life-giving thresholds.

The concept of planetary boundaries has certainly helped visualize the intense impact of the Anthropocene on planetary systems, and Banaba with its rich phosphate has been one significant source of phosphorus for eight decades. But if the flows of phosphorus are critical to those humans thinking about and coming up with solutions for living within environmental means, surely what happened to Banaba, to 'Ocean Island', as tiny as the island is, is an important and sobering lesson for the whole planet.

Islanders and most Indigenous peoples understand the power and purpose of both connections and boundaries. I learned much of this from my elder sister, Teresia Kieuea Teaiwa, who passed away in 2017 after a short, but deeply impactful, life of Pacific scholarship, teaching, activism, and poetry that critiqued the colonial, political, and gendered manner in which Pacific peoples, and particularly Pacific women, had been robbed, forced, and gaslit out of their lands, dignity, sovereignty, and self-determination at all levels. She taught me that what happens to the land happens to the people; what happens to the ocean and to the island, happens to the whole planet. Teresia's poem "To Island" (T. Teaiwa 2007) was transformed into a poignant, animated story by Mana Moana, a collective of artists who produced video works for COP27 to amplify and support Pacific initiatives to drive global action to reduce greenhouse gas emissions. She wrote:

> Shall we make 'island' a verb? As a noun, it's so vulnerable to impinging forces. Let us turn the energy of the island inside out. Let us 'island' the world! Let us teach the

inhabitants of planet Earth how to behave as if we were all living on islands! For what is Earth but an island in our solar system? An island of precious ecosystems and finite resources. Finite resources. Limited space. The islanded must understand that to live long and well, they need to take care. Care for other humans, care for plants, animals; care for soil, care for water. Once islanded, humans are awakened from the stupor of continental fantasies. The islanded can choose to understand that there is nothing but more islands to look forward to. Continents do not exist, metaphysically speaking. It is islands all the way up, islands all the way down. Islands to the right of us, islands to the left. Yes, there is a sea of islands. And 'sea' can be a verb, just as 'ocean' becomes a verb of awesome possibility. But let us also make 'island' a verb. It is a way of living that could save our lives. (T. Teaiwa 2022)

References

Brownlie, W. J., Alexander, P., Cordell, D., et al. (2024). National Phosphorus Planning for Food and Environmental Security. *Current Opinion in Biotechnology*. 90, 103226. DOI: 10.1016/j.copbio.2024.103226.

Cushman, G. T. (2013). *Guano and the Opening of the Pacific World: A Global Ecological History*. Cambridge: Cambridge University Press. DOI: 10.1017/CBO9781139047470.

Egan, D. (2023). *The Devil's Element: Phosphorus and a World Out of Balance*. New York: W. W. Norton.

Elser, J. J., & Haygarth, P. M. (2021). *Phosphorus: Past and Future*. New York: Oxford University Press.

Kabataulaka, T. (2024). Facebook Post on 26 September 2024. Last accessed 24 April 2026 at www.facebook.com/1133771441/posts/10232833236789918/?rdid=tMrQAuMYo89VRQ3D.

Maude, H. C., & Maude, H. E. (1932). The Social Organisation of Banaba or Ocean Island, Central Pacific. *The Journal of the Polynesian Society*. 41(4), 262–301.

Maude, H. C., & Maude, H. E., eds. (1994). *The Book of Banaba: From the Maude and Grimble Papers, and Published Books*. Books on the I-Kiribati, Published for the I-Kiribati 7. Suva: Institute of Pacific Studies.

Pearl, H. (2024). Nauru Uses High-Profile UN Address to Press Case for Sea-Bed Mining. *Benar News*, 25 September. www.benarnews.org/english/news/pacific/pac-nauru-un-09252024005706.html.

Raworth, K. (2012). A Safe and Just Space for Humanity: Can We Live within the Doughnut? Oxfam Discussion Paper. Oxford.

Raworth, K. (2017). A Doughnut for the Anthropocene: Humanity's Compass in the 21st Century. *The Lancet Planetary Health*. 1(2), e48–e49. DOI: 10.1016/S2542-5196(17)30028-1.

Rockström, J., Steffen, W., Noone, K., et al. (2009). Planetary Boundaries: Exploring the Safe Operating Space for Humanity. *Ecology and Society*. 14(2), 32. DOI: 10.5751/ES-03180-140232.

Silverman, M. (1971). *Disconcerting Issue: Meaning and Struggle in a Resettled Pacific Community*. Chicago: University of Chicago Press.

Stockholm Resilience Centre. n.d. Planetary Boundaries. www.stockholmresilience.org/research/planetary-boundaries.html.

Teaiwa, K. (2024). Not Here for the Disciplines: Researching with and for the Pacific. In T. C. Davis & P. Rae, eds., *The Cambridge Guide to Mixed Methods for Theatre and Performance Studies*. Cambridge: Cambridge University Press.

Teaiwa, K. M. (2014). *Consuming Ocean Island: Stories of People and Phosphate from Banaba*. Bloomington: Indiana University Press.

Teaiwa, T. (2007). To Island. In G. Baldacchino, ed., *A World of Islands: An Island Studies Reader*. Charletown: Institute of Island Studies, 514.

Teaiwa, T. (2022). *To Island*. Mana Moana; Pasifika Voices and Storybox. www.manamoana.co.nz/artwork/to-island. Last accessed 28 April 2026 at www.pasifikavoices.com/film/to-island/#:~:text=The%20Mana%20Moana%20%E2%80%93%20Pacific,to%20reduce %20greenhouse%20gas%20emissions.

Williams, M., & Macdonald, B. (1985). *The Phosphateers: A History of the British Phosphate Commissioners and the Christmas Island Phosphate Commission*. Carlton: Melbourne University Press.

Part VIII

Ocean Acidification

20

Introduction to the Ocean Acidification Boundary

SIMON P. MEISCH

The oceans mirror many of the environmental problems described in the planetary boundaries framework and reported widely in the media, such as overfishing and deep-sea mining (biosphere integrity), plastic islands or microplastics (novel entities), shipping (atmospheric aerosol loading), or dead zones lacking oxygen (biogeochemical flows). Less present in the media is the problem of ocean acidification addressed by the planetary boundary. Like the atmosphere, the oceans are sinks for CO_2 produced by the burning of fossil fuels.

CO_2 absorption changes the chemistry of the oceans, lowering their pH. The resulting acidification also leads to fewer of the calcium carbonate shells produced by numerous creatures sinking carbon to the ocean floor before they dissolve back into the water. The boundary's control variable relates to this by describing the 'carbonate ion concentration in surface seawater' and, more specifically, the 'average global surface ocean saturation state with respect to aragonite' (Richardson et al. 2023). Aragonite is one of the most abundant forms of calcium carbonate. The boundary is at more than 80 per cent of the pre-industrial aragonite concentration, defined as 3.44. In 2023, the level was 81 per cent, meaning the ocean boundary was, at that time, still one of the few that had not been crossed. Thus, Pelejero et al. (2010: 332) describe ocean acidification as the '"evil twin" of global warming' – it is a moot question, though, as to which is of the two is more evil. This also means that reducing greenhouse gas emissions will reduce the pressure on ocean acidification.

Acidification of the oceans and declining concentrations of calcium carbonate are becoming a problem for marine life. Many creatures depend on the presence of these calcium compounds in the water to build their shells or skeletons. Lower pH not only erodes the existing structures of these creatures, but they also find less calcium to rebuild them. These include plankton, corals, crustaceans, and molluscs, which are vital to marine biodiversity, marine food chains, and, ultimately, human food security (Rockström et al. 2009). So, the control variable of the ocean boundary is the global average saturation state of calcium carbonate in surface water. The framework sets the preindustrial Holocene base value at 3.44 and the boundary value at 2.75. Since 2009, the actual values have fluctuated between 2.8 and 2.9 (Richardson et al. 2023). In 2025, the ocean acidification boundary was first recognized as having been crossed (Findlay et al. 2025).

Both chapters on ocean acidification take the problem structure of the ocean acidification boundary as the starting point of their ethical reflections. Konrad Ott describes it as post-normal, that is, a situation when problem-solving has to cope with uncertain facts, disputed values, high-stakes, and urgent decisions (Funtowicz & Ravetz 1993). Ott explores the issue of ocean acidification from an environmental ethics perspective. He discusses different ways of justifying why ocean acidification is an ethical problem and what follows from this. In doing so, he aims to show how well the most common environmental ethical approaches can address the issue. They mostly differ in their scope of the moral community, that is, to whom we have direct moral obligations: to humans only, or also to individual animals, species, or ecosystems? Depending on the answer one gives, the reasons for protecting oceans will change. Ultimately, Ott intends to specify the normative building blocks of deep anthropocentrism as his preferred approach to the issue.

Rachel Haug Fossbakk, too, addresses the issue of decision-making relating to uncertainty and risk perception. She argues that the human response to ocean acidification is inadequate. This is despite the fact that there is already enough knowledge to act. Her hypothesis is that action requires an adequate perception of risk, which many people do not have when it comes to ocean acidification. She concludes by asking who has a responsibility to act, and on what ethical basis. Rachel Haug Fossbakk then advocates an ecocentric perspective by which she understands a situation 'where humans are part of nature and need to live sustainably as part of it on the same term as other species'.

References

Findlay, H., Feely, R. A., Jiang, L.-Q., Pelletier, G., & Bednaršek, N. (2025). Ocean Acidification: Another Planetary Boundary Crossed. *Global Change Biology*. 31(6), e70238. DOI: 10.1111/gcb.70238.

Funtowicz, S. O., & Ravetz, J. R. (1993). Science for the Post-Normal Age. *Futures*. 25(7), 739–755. DOI: 10.1016/0016-3287(93)90022-L.

Pelejero, C., Calvo, E., & Hoegh-Guldberg, O. (2010). Paleo-Perspectives on Ocean Acidification. *Trends in Ecology & Evolution*. 25(6), 332–344. DOI: 10.1016/j.tree.2010.02.002.

Richardson, K., Steffen, W., Lucht, W., et al. (2023). Earth beyond Six of Nine Planetary Boundaries. *Science Advances*. 9(37), eadh2458. DOI: 10.1126/sciadv.adh2458.

Rockström, J., Steffen, W., Noone, K., et al. (2009). Planetary Boundaries: Exploring the Safe Operating Space for Humanity. *Ecology and Society*. 14(2), 32. DOI: 10.5751/ES-03180-140232.

21

How Can We Curb Ocean Acidification in a Time of Multiple Competing Crises?

RACHEL HAUG FOSSBAKK

21.1 Introduction

Around the end of the eighteenth century, with the design of the steam engine by James Watt, the geological age of the Anthropocene started (Crutzen 2006). This has led to unprecedented changes in the natural environment, the most serious of which is climate change. We have tried to change our paths since then though. In fact, more than 50 years have now passed since the 1972 Stockholm Conference (Rockström et al. 2021) – the United Nations Conference on the Human Environment. This was when global leaders for the first time made the environment and the triple planetary crisis of climate, nature, and pollution a focus issue. At that time, in 1972, they warned – and this still holds true today – that 'Through ignorance or indifference we can do massive and irreversible harm to the earthly environment on which our life and well-being depend' (United Nations 1973).

Since then, state leaders have created the United Nations Environment Programme (UNEP) entered into force both a climate and biodiversity agreement (United Nations Framework Convention on Climate Change (UNFCCC) and the Convention on Biological Diversity (CBD)) with many iterations; signed the UN 17 Sustainable Development Goals (SDGs) in 2015 with corresponding targets, as part of the realization of Agenda 2030; and in 2022 and 2023, state leaders also signed the Kunming-Montreal Global Biodiversity Framework (GBF) and the Agreement under the United Nations Convention on the Law of the Sea on the conservation and sustainable use of marine biological diversity in areas beyond national jurisdiction (BBNJ). Still, we have continued to saturate Earth's capacity to support future generations (Intergovernmental Panel on Climate Change 2018). The planet is more polluted than earlier, there are fewer fish in the ocean, the temperature is increasing, methane levels are at their highest in 800,000 years, polar regions are melting, plastics are ubiquitous, and the ocean is more acidic than ever (Barnes & Kaiser 2009; Hönisch et al. 2012; Worm & Branch 2012; MacLeod et al. 2021; Intergovernmental Panel on Climate Change 2022).

In fact, six of nine planetary boundaries that are needed for humans to be able to develop and thrive for generations to come have been breached already (Richardson et al. 2023). Then, in 2025, the ocean acidification boundary was crossed (Findlay et al. 2025). This is a boundary that is also specifically mentioned in the SDGs, with target 3 of SDG14 Life below water specifying the need to 'Minimize and address the impacts of ocean acidification, including through enhanced scientific cooperation at all levels'. It was in fact the only by-product of climate change and greenhouse gas emission that was singled out to be included under this goal (United Nations Department of Economic and Social Affairs 2022). Still, there is a lack of concrete knowledge of how breaching this boundary will impact either non-human populations or the corresponding human population that depends on these (Buckley et al. 2017).

In light of this, the current chapter considers ocean acidification within the planetary boundaries framework and assesses the lack of responses that led to, and may further exacerbate, the breaching of this boundary. We argue that this in part can be explained by the human-centric perspective of global environmental governance, where effective governance is linked to risk perception within a human context. In terms of ocean acidification, we argue that humans still have a low personal risk perception, coupled with a lack of knowledge (or interest) about it, which directly hinders its agenda setting and implementation of effective governance. We conclude with considering whether the increased saliency of biodiversity challenges, as brought to the surface with the signing of the GBF in 2022, can bring more attention to the role of stressors such as ocean acidification on sustainability and the consequences the breach of this planetary boundary can have on not only humans but also non-human populations.

21.2 Ocean Acidification and Planetary Boundaries

Planetary boundaries are related to critical Earth-system processes that jointly define the operating space within which what some call the Anthropocene can be maintained (or the Holocene as it is officially known as). However, findings demonstrate that the current level of resource use must be dramatically reduced if all people are to lead a good life for not only the current, but also for future generations. Still, since the start of multilateral negotiations towards managing humans' exploitation of nature in Stockholm in 1972, and even within the planetary boundaries framework itself, there is a marked human well-being and growth perspective which effectively hinders its saliency. The SDGs, for example, demonstrate a distinctly human, and not planetary focus, with 12 of its 17 goals being social objectives, and only 4 having an environmental focus and the final one on the process of implementation. They therefore have a focus on pursuit of human development, a good, reserved for humans. However, this may in fact undermine Earth-system processes that frame the planetary boundaries themselves, by causing disturbances to Earth system components to a degree that risks changing global environmental conditions markedly (O'Neill et al. 2018). Ocean acidification is driven by climate change and its corresponding pressure indicator of CO_2 emissions, which has already breached said boundaries. This in turn is further driven by the land system change boundary, which, if properly managed, could bring global forest covers to high enough levels so that they better complement the ocean as substantial sinks of atmospheric CO_2 and bring it back to safe operating space in terms of planetary boundaries.

The interlinkages between land system change, climate change, and ocean acidification demonstrate the interconnectedness of human interventions and resultant impacts on nature. These connections and the corresponding breach of ocean acidification and corresponding gravity and impact on nature and humans alike, however, lack recognition in society. Ocean acidification and its effects *have* been well documented the last decade, though, within the scientific community especially, with studies ranging from effects on plankton and larvae and economic activities, to the microbial food web and bacterial processes to name some (Burrell et al. 2016; Leo et al. 2017; Narita & Rehdanz 2017; Alvarez-Fernandez et al. 2018; Tiller et al. 2019; Galdies et al. 2020a, 2020b). There has also been ample scientific knowledge reported, proving that the ocean pH is declining at an unprecedented rate compared to the last 300 million years of the Earth's history (Pearson & Palmer 2000; Hönisch et al. 2012; Intergovernmental Panel on Climate Change 2022; Richardson et al. 2023). We also know that this corresponds with the post-industrial revolution increase in anthropogenic CO_2 emissions (Gattuso & Hansson 2011; Zeebe & Ridgwell 2011).

This is because ocean acidification is caused by the uptake of about 30–40 per cent of all carbon dioxide from the atmosphere into the ocean (Doney et al. 2009), which, I concur, on its face sounds like a positive. In fact, about one million tonnes of CO_2 will enter the ocean per hour today removing it from the atmosphere (Sabine et al. 2004). However, with a continuous increase of CO_2 in the atmosphere, this also means an increased amount of it in the ocean, altering ocean chemistry and causing potential disruptions to oceanic biodiversity through making the waters more acidic, 30 per cent more acidic in fact since the Industrial Revolution, and the rate of changes is faster than any time on record with no sign of slowing down (Gattuso & Hansson 2011; Zeebe & Ridgwell 2011), after having remained constant for millions of years (Pearson & Palmer 2000).

The impact of this on biological, biogeochemical, and ecological processes of the oceans, and the consequences of these impacts on not only nature itself, but also human needs, are only partially known, and poorly quantified. Though some organisms are more sensitive to lower pH, others are better able to adapt to it. Some again, even similar species, may compensate the effect of acidification through some internal mechanisms (Tiller et al. 2019; Galdies et al. 2020b). The IPCC, in part B.4.2, also stated with high confidence that 'The level of ocean acidification due to increasing CO2 concentrations associated with global warming of 1.5°C is projected to amplify the adverse effects of warming, and even further at 2°C, impacting the growth, development, calcification, survival, and thus abundance of a broad range of species, for example, from algae to fish'. In fact, the term 'Ocean Acidification' was mentioned 103 times in this report, demonstrating that it is a known and important effect of climate change, with several negative impacts on both nature and humans. And to be able to stop it, we need to see significant reductions in the atmospheric CO_2 concentration within a very short time frame, as specified by among others the Monaco Declaration and the latest IPCC report (Monaco Declaration 2009; Intergovernmental Panel on Climate Change 2018).

21.3 Risk Perception

Despite the availability of this knowledge, more than 50 years after Stockholm, we are still to a large extent choosing to continue within the vein of business as usual, and ocean acidification still does not reach the consciousness of policymakers to the extent that other multilateral agreements do; those that may be more tangible in their implementation and goals like the Montreal protocol to tackle the hole in the ozone layer, or the forthcoming plastic treaty (Green 2009; Raubenheimer & McIlgorm 2017; Tiller et al. 2019, 2022; Simon et al. 2021). We, therefore, look to the risk perception literature to assess how the science–policy interface can be exploited to convey the importance of ocean acidification to the public, which in turn may ensure its rise to the agenda setting stage of public policy.

Our assessment starts with the premise that the concept of personal risk in given situations not only refers to the perceptions of a probability for an event happening, such as ocean acidification, affecting marine biodiversity negatively. It also concerns our perceptions of what the consequences of this will have on a personal level, including both objective and perceived risk. The former is based on knowledge about the issue, while the latter refers to the individual's perception thereof; and there is an unbridged gap between these two groups. When there is a substantial amount of divergent scientific literature on an environmental topic and its effect on human health, this can lead to no consensus about whether or what

'Four Problems of Risk'		*Knowledge*	
		Certainty	Uncertainty
Consent	Complete	Problem: Technical Solution: Calculation	Problem: Information Solution: Research
	Contested	Problem: (dis)Agreement Solution: Coercion or discussion	Problem: Knowledge and consent Solution: ?

Figure 21.1 Four Problems of Risk. Risk perception, problems, and solutions as products of knowledge and consent adapted from Wildavsky and Dake (1990) 'Four Problems of Risk'. The term 'consent' is used by the original authors. We understand it as having the meaning of 'consensus' or 'agreement'. However, generally, we understand that 'consent' implies a lack of objection, whereas consensus' is an active agreement to something.

mitigation measures need to be taken, such as in the case of ocean acidification being 'just' an impact of climate change. The latter could result in cultural constructions of risk events that can differ greatly within society and between sectors that depend or not on ocean resources (Adams 1995). For the purposes of ocean acidification, we will assess it within the framework of the knowledge-consent vs. certainty-uncertainty figure (Figure 21.1).

The top-left box is when individuals in question have both complete certainty around the knowledge about a given topic, such as ocean acidification, and complete consent within the group that this is the complete amount of knowledge about the issue area. All the alternatives around the topic are known already, and the objectives as well. In a case like this, for a given environmental problem, the solution is centred on calculating the outcome that is most efficient to solve it. An example is the hole in the ozone layer resulting in the Montreal protocol. The 'hole' in the atmospheric ozone layer, from ozone depletion because of the use ozone-depleting substances such as chlorofluorocarbons (CFCs) (Molina & Rowland 1974), led to global public outrage. The reason for this was perception of personal risk, and a fear of millions of new skin cancer patients and other human health effects (for example, infectious diseases, suppression of the immune system and eye disorders). This risk led to the issue being taken seriously in public and political arenas (Morrisette 1989). In this case, the challenge was technical, with a need for innovation in product development, and the solution was calculating how to solve it. The alternatives were known, and the objectives were agreed upon, and as such, finding a solution was a matter of calculating the most efficient outcome that would ensure that the problem was solved.

In the bottom-left box, there is also certainty about the environmental challenge and its importance in terms of scientific knowledge around it like in the case of the ozone hole. However, there is disagreement with regard to consent. There is also some contestation around the value of existing knowledge with respect to the cost-benefit of the consequences of it. The example here could be the COVID-19 pandemic, and the solution was differentiated responses in different countries, based, among others, on risk relative to number of cases. Individuals knew during this time period that there was a global pandemic and that it resulted in people losing their lives. However, there was disagreement regarding the extent of the problem in various regions or countries. Also, some people questioned whether economic security was more important in terms of how best to respond, for example if lockdowns should happen or not. Some countries therefore implemented economically strict and coerced shutdowns (for example Korea). Other countries, such as Sweden, on the other hand, implemented limited or partial shutdowns that led to higher rates of mortality in the shorter-term (Hale et al. 2020).

In turn, the top-right category is when there is complete consent in the public that there is an environmental challenge that needs to be solved. However, the knowledge about the risks of this environmental challenge is limited, and there is insufficient information for some to assess the risks. A solution in such a case is to default to more research being needed to address knowledge gaps. This is likely where the case of ocean acidification is currently located and why there is not yet an encompassing treaty for either and that the solution is further research to assess to what degree they represent a risk for human populations. The certainty around knowledge pertaining to risk is still not strong enough to drive the public to demand immediate action, as was the case of the Montreal Protocol.

Finally, the bottom-right category is one when there is neither enough knowledge in place nor consent on the cost-benefit of consequences. Different social relations groups fall into different groups in this table when it comes to perceptions of personal risk around environmental issues. An example of this is arguably the UNFCCC and solutions to climate change. Though there is no clear-cut solution to a challenge where there is insufficient consent or knowledge, one of the actions that the global community put in place was that of establishing the Intergovernmental Panel on Climate Change (IPCC). This was to prepare comprehensive scientific, technical, and socio-economic knowledge about climate change and its impacts that were accessible to the public as well. In Tiller et al. (2019), the authors quote cast member of Weekend Update of *Saturday Night Live* TV show Michael Che saying, in response to the IPCC report on climate change that had just been released:

> I keep asking myself 'why don't I care about this?' Don't get me wrong, I 100 percent believe in climate change, yet I'm willing to do absolutely nothing about it. I think it's because they keep telling us we're gonna lose everything, and nobody cares about everything. People only care about some things. (Meyers 2018)

Though said in a comedy setting, the statement still rings true when it comes to public perceptions of climate change policy and global action in general, and risk perception of individuals.

Considering this, and though we have established that the human dimension of ocean acidification, through socio-economic impacts, could be substantial (Turley & Boot 2011), the risk perception is low for the issue area. This is despite the fact that, for fisheries and aquaculture, for example, the phyto- and zoo-plankton communities and resultant transfers to the whole trophic web could affect the potential for sustainable future fisheries (Blaustein 2010; Doubleday et al. 2019). Modification of the sea bottom because of ocean acidification, disappearance of corals, and reduction of rocky substrata for coastline protection against natural hazards could also be affected. It could affect climate regulation negatively as well, by altering the fluxes of sequestration of carbon into the deep ocean through the biological pump (Sarmiento & Gruber 2002). And it could even affect human health, though this is a complex study linked to decline in high seas opportunities for exploiting biodiversity and genetic resources that could be used for new medicinal, pharmaceutical, industrial, and agricultural applications (Blaustein 2010; Doubleday et al. 2019). Combined though, these are dramatic risks – not just for nature but also for human activities and livelihoods. Yet, the feeling of risk on individuals is still arguably low, leaving ocean acidification to not have enough saliency for policymakers to put it on the agenda. This has been reflected in an EU study on the impacts of climate change, where results precisely reflected this, with respondents being least aware of ocean acidification (Buckley et al. 2017).

We can see this clearly when we do a media search on Retriever looking at the term ‘Ocean Acidification’ over time in all international media channels as well (Figure 21.2). We can see that after the SDGs in 2015, there was a sharp increase in media discussions about the issue area, perhaps because of it being explicitly mentioned in target 14.3 as mentioned. This is then followed by a sharp decline until 2020 when it begins to increase in saliency again. McCombs and Shaw (1972) discussed early how news media is able to influence the salience of topics on the public agenda, and thereby help set the agenda for policymaking. Also, increased media attention is believed to result in increased concern for a particular issue (Brown & Deegan 1998).

21.4 Responsibility

Who is responsible to stop ocean acidification, though, if humans don’t feel any personal risk with regard to it and governance actions have a limited human centric approach? And how should this be reflected in governance? Granted, we have tried to curb ocean acidification, directly and indirectly, at national, regional, and global

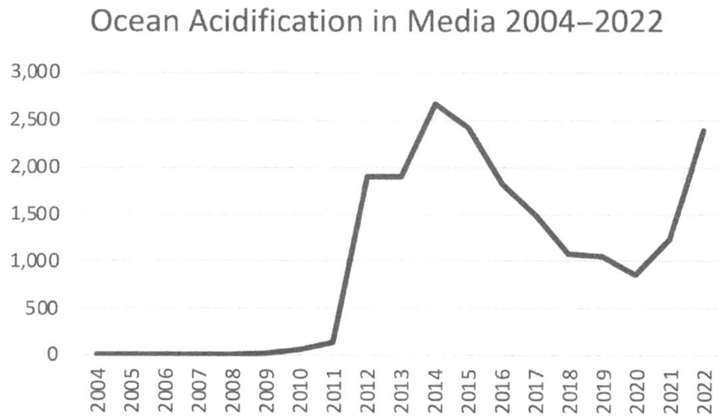

Figure 21.2 Ocean acidification in Media.

levels (Tiller et al. 2019; Galdies et al. 2020a, 2020b). However, we have only created agreements within the framework of planetary boundaries that protect *humans* from anthropogenically induced change that is affecting nature (and us in turn) – and not ensured that we protect nature itself from human use. To manage the expected negative impacts on humans, for example, would focus on biodiversity conservation and habitat restorations, as well as adaptation to changes to the Ocean (Falkenberg et al. 2020). Verma (2019) highlights five guiding principles that identify an anthropocentric world view, namely: (1) humans are at the centre, with the highest status; (2) humans are most capable of managing earth; (3) humans are the most important species on earth, in charge of the rest of nature on the planet; (4) unlimited resources on earth are for humans only, and healthy nature can only be accomplished through a healthy economy; and (5) if we humans are to succeed as a species on earth, we have to be managers of resource extraction and benefits deriving thereof.

In a similar fashion, Verma contradicts this world view with that of ecocentrism, where humans are part of nature and need to live sustainably as part of it on the same terms as other species. Prioritizing ecocentrism is also highlighted by Kotze and French (2018), for example, who emphasized that 'a more ecocentric orientation . . . is urgently required to enable a more ecocentric rule of law to better mediate the human-environment interface in the Anthropocene'. In fact, risk as we have discussed it in this chapter has primarily referred to risk to humans – not nature itself – though that is where most negative effects will materialize first. This, in, turn will lead to, down the chain, humans also being affected, but we are not the primary and immediate recipients of the risks associated with ocean acidification. This is also likely the reason for the lack of focus at the global governance stage, despite its damaging threats. And we know, it's not from a lack of knowledge, nor has it been since the creation of the Intergovernmental Panel on Climate Change (IPCC) in 1988 and the adoption of the UNFCCC in 1992 at least. However, with the signing of the Kunming-Montreal Global Biodiversity Framework, the global community took a sharp turn in that direction by including both sustainable use (human centric) and protection in and of itself. Target 8 specifically emphasizes that it will

> [m]inimize the impact of climate change and ocean acidification on biodiversity and increase its resilience through mitigation, adaptation, and disaster risk reduction actions, including through nature-based solution and/or ecosystem-based approaches, while minimizing negative and fostering positive impacts of climate action on biodiversity. (CBD 2022)

This agreement is scheduled to reach its targets by 2030 and will work in tandem with the BBNJ agreement for areas that are outside national jurisdiction. By 2030, perhaps risk perception will have reached the consciousness of humans about this planetary boundary as well, and we can see reversals of the environmental damages, as we have done with the boundary for atmospheric aerosol loading after the implementation of the Montreal protocol. If we collectively decide that the future high is worth the temporary pain of making the required changes, that is.[1]

References

Adams, J. (1995). *Risk*. London: University College London Press.

Alvarez-Fernandez, S., Bach, L. T., Taucher, J., et al. (2018). Plankton Responses to Ocean Acidification: The Role of Nutrient Limitation. *Progress in Oceanography*. 165, 11–18. DOI: 10.1016/j.pocean.2018.04.006.

Barnes, D. K., & Kaiser, S. (2009). Melting of Polar Icecaps: Impact on Marine Biodiversity. In P. Safran, ed., *Fisheries and Aquaculture. Encyclopedia of Life Support Systems: Volume V 10. of Food and Agricultural Sciences, Engineering and Technology Resources*. Oxford: Eolss Publishers, 345.

Blaustein, R. J. (2010). High-Seas Biodiversity and Genetic Resources: Science and Policy Questions. *BioScience*. 60(6), 408–13. DOI: 10.1525/bio.2010.60.6.3.

Brown, N., & Deegan, C. (1998). The Public Disclosure of Environmental Performance Information: A Dual Test of Media Agenda Setting Theory and Legitimacy Theory. *Accounting and Business Research*. 29(1), 21–41. DOI: 10.1080/00014788.1998.9729564.

Buckley, P. J., Pinnegar, J. K., Painting, S. J., et al. (2017). Ten Thousand Voices on Marine Climate Change in Europe: Different Perceptions among Demographic Groups and Nationalities. *Frontiers in Marine Science*. 4, 206. DOI: 10.3389/fmars.2017.00206.

Burrell, T. J., Maas, E. W., Teesdale-Spittle, P., & Law, C. S. (2016). Assessing Approaches to Determine the Effect of Ocean Acidification on Bacterial Processes. *Biogeosciences*. 13(15), 4379–88. DOI: 10.5194/bg-13-4379-2016.

[1] Paraphrased from the lyrics of "Blank Space" by Taylor Swift.

CBD. (2022). Kunming-Montreal Global Biodiversity Framework: Conference of the Parties to the CBD. Fifteenth Meeting – Part II, Montreal, Canada, 7–19 December 2022. Agenda item 9A. www.cbd.int/doc/c/e6d3/cd1d/daf663719a03902a9b116c34/cop-15-l-25-en.pdf.

Crutzen, P. J. (2006). The 'Anthropocene'. In E. Ehlers & T. Krafft, eds., *Earth System Science in the Anthropocene*. Berlin: Springer, 13–18.

Doney, S. C., Fabry, V. J., Feely, R. A., & Kleypas, J. A. (2009). Ocean Acidification: The Other CO2 Problem. *Annual Review of Marine Science*. 1, 169–92. DOI: 10.1146/annurev.marine.010908.163834.

Doubleday, Z. A., Nagelkerken, I., Coutts, M. D., Goldenberg, S. U., & Connell, S. D. (2019). A Triple Trophic Boost: How Carbon Emissions Indirectly Change a Marine Food Chain. *Global Change Biology*. 25(3), 978–84. DOI: 10.1111/gcb.14536.

Falkenberg, L. J., Bellerby, R. G. J., Connell, S. D., et al. (2020). Ocean Acidification and Human Health. *International Journal of Environmental Research and Public Health*. 17(12), 4563. DOI: 10.3390/ijerph17124563.

Findlay, H., Feely, R. A., Jiang, L.-Q., Pelletier, G., & Bednaršek, N. (2025) Ocean Acidification: Another Planetary Boundary Crossed. *Global Change Biology*. 31(6), e70238. DOI: 10.1111/gcb.70238.

Gattuso, J. P., & Hansson, L., eds. (2011). *Ocean Acidification*. Oxford: Oxford University Press.

Galdies, C., Bellerby, R., Canu, D., et al. (2020a). European Policies and Legislation Targeting Ocean Acidification in European Waters: Current State. *Marine Policy*. 118, 103947. DOI: 10.1016/j.marpol.2020.103947.

Galdies, C., Tiller, R., & Martinez Romera, B. (2020b). Global Ocean Governance and Ocean Acidification. In W. Leal Filho, A. M. Azul, L. Brandli, A. Lange Salvia, & T. Wall, eds., *Life Below Water. Encyclopedia of the UN Sustainable Development Goals*. Cham: Springer International Publishing, 1–12.

Green, B. A. (2009). Lessons from the Montreal Protocol: Guidance for the Next International Climate Change Agreement. *Environmental Law Review*. 39(1), 253–83.

Hale, T., Angrist, N., Kira, B., et al. (2020). Variation in Government Responses to COVID-19. Blavatnik School of Government; University of Oxford. www.bsg.ox.ac.uk/sites/default/files/2020-05/BSG-WP-2020-032-v5.0_0.pdf.

Hönisch, B., Ridgwell, A., Schmidt, D. N., et al. (2012). The Geological Record of Ocean Acidification. *Science*. 335(6072), 1058–63. DOI: 10.1126/science.1208277.

Intergovernmental Panel on Climate Change. (2018). Global Warming of 1.5°C. Special Report. www.ipcc.ch/sr15/.

Intergovernmental Panel on ClimateChange. (2022). Climate Change 2022: Impacts, Adaptation, and Vulnerability: Contribution of Working Group II to the Sixth Assessment Report of the Intergovernmental Panel on Climate Change: Summary for Policymakers. Cambridge: Cambridge University Press. DOI: 10.1017/9781009325844.

Kotzé, L. J., & French, D. (2018). The Anthropocentric Ontology of International Environmental Law and the Sustainable Development Goals: Towards an Ecocentric Rule of Law in the Anthropocene. *Global Journal of Comparative Law*. 7(1), 5–36. DOI: 10.1163/2211906X-00701002.

Leo, E., Kunz, K. L., Schmidt, M., et al. (2017). Mitochondrial Acclimation Potential to Ocean Acidification and Warming of Polar Cod (*Boreogadus saida*) and Atlantic Cod (*Gadus morhua*). *Frontiers in Zoology*. 14, 21. DOI: 10.1186/s12983-017-0205-1.

MacLeod, M., Arp, H. P. H., Tekman, M. B., & Jahnke, A. (2021). The Global Threat from Plastic Pollution. *Science.* 373(6550), 61–65. DOI: 10.1126/science.abg5433.

McCombs, M. E., & Shaw, D. L. (1972). The Agenda-Setting Function of Mass Media. *Public Opinion Quarterly.* 36(2), 176–87.

Meyers, S. (2018). Saturday Night Live. NBC. S44, E3. 13 October.

Molina, M. J., & Rowland, F. S. (1974). Stratospheric Sink for Chlorofluoromethanes: Chlorine Atom-Catalysed Destruction of Ozone. *Nature.* 249(5460), 810–12. DOI: 10.1038/249810a0.

Monaco Declaration. (2009). Second International Symposium on the Ocean in a High-CO2 World, Monaco, 6–9 October 2008. www.iaea.org/sites/default/files/monacodeclaration.pdf.

Morrisette, P. M. (1989). The Evolution of Policy Responses to Stratospheric Ozone Depletion. *Natural Resources Journal.* 29(3), 793–820.

Narita, D., & Rehdanz, K. (2017). Economic Impact of Ocean Acidification on Shellfish Production in Europe. *Journal of Environmental Planning and Management.* 60(3), 500–18. DOI: 10.1080/09640568.2016.1162705.

O'Neill, D. W., Fanning, A. L., Lamb, W. F., & Steinberger, J. K. (2018). A Good Life for All within Planetary Boundaries. *Nature Sustainability.* 1(2), 88–95. DOI: 10.1038/s41893-018-0021-4.

Pearson, P. N., & Palmer, M. R. (2000). Atmospheric Carbon Dioxide Concentrations over the Past 60 Million Years. *Nature.* 406(6797), 695–99. DOI: 10.1038/35021000.

Raubenheimer, K., & McIlgorm, A. (2017). Is the Montreal Protocol a Model that Can Help Solve the Global Marine Plastic Debris Problem? *Marine Policy.* 81, 322–29. DOI: 10.1016/j.marpol.2017.04.014.

Richardson, K., Steffen, W., Lucht, W., et al. (2023). Earth beyond Six of Nine Planetary Boundaries. *Science advances.* 9(37), eadh2458. DOI: 10.1126/sciadv.adh2458.

Rockström, J., Gupta, J., Qin, D., et al. (2021). Stockholm to Stockholm: Achieving a Safe Earth Requires Goals That Incorporate a just Approach. *One Earth.* 4(9), 1209–11. DOI: 10.1016/j.oneear.2021.08.012.

Sabine, C. L., Feely, R. A., Gruber, N., et al. (2004). The Oceanic Sink for Anthropogenic CO2. *Science.* 305(5682), 367–71. DOI: 10.1126/science.1097403.

Sarmiento, J. L., & Gruber, N. (2002). Sinks for Anthropogenic Carbon. *Physics Today.* 55(8), 30–36. DOI: 10.1063/1.1510279.

Simon, N., Raubenheimer, K., Urho, N., et al. (2021). A Binding Global Agreement to Address the Life Cycle of Plastics. *Science.* 373(6550), 43–47. DOI: 10.1126/science.abi9010.

Tiller, R., Arenas, F., Galdies, C., et al. (2019). Who Cares about Ocean Acidification in the Plasticene? *Ocean & Coastal Management.* 174, 170–80. DOI: 10.1016/j.ocecoaman.2019.03.020.

Tiller, R., Booth, A. M., & Cowan, E. (2022). Risk Perception and Risk Realities in Forming Legally Binding Agreements: The Governance of Plastics. *Environmental Science & Policy.* 134, 67–74. DOI: 10.1016/j.envsci.2022.04.002.

United Nations. (1973). Report of the United Nations Conference on Environment and Development: Stockholm, 5–16 June 1972. A/CONF.48/14/Rev.1. digitallibrary.un.org/record/523249.

United Nations Department of Economic and Social Affairs. (2022). Goal 14: Conserve and Sustainably Use the Oceans, Seas and Marine Resources for Sustainable Development. sdgs.un.org/goals/goal14.

Verma, A. K. (2019). Sustainable Development and Environmental Ethics. *International Journal on Environmental Sciences.* 10(1), 1–5.

Wildavsky, A., & Dake, K. (1990). Theories of Risk Perception: Who Fears What and Why? *Daedalus.* 119(4), 41–60.

Worm, B., & Branch, T. A. (2012). The Future of Fish. *Trends in Ecology & Evolution.* 27(11), 594–99. DOI: 10.1016/j.tree.2012.07.005.

Zeebe, R., & Ridgwell, A. (2011). Past Changes in Ocean Carbonate Chemistry. In J. P. Gattuso & L. Hansson, eds., *Ocean Acidification.* Oxford: Oxford University Press, 21–40.

22

Environmental Ethics and the Case of Ocean Acidification

KONRAD OTT

22.1 Introduction

This chapter addresses ocean acidification (OA) from an environmental ethics perspective. Ocean acidification is seen as a paradigm case for post-normal science. At its core, the chapter adopts and substantiates several ethical building blocks from a discourse-pragmatic approach, as (a) common heritage of humankind, (b) strong sustainability, (c) Aldo Leopold's principle, and (d) ecosystem services, with a special eye on cultural services. The idea to attribute inherent moral value to marine organisms and ecological systems is seen critically.

This chapter emerged from the research programme BIOACID (Biological Impacts of Ocean Acidification) being conducted at Kiel University. Within the final period of the BIOACID research programme, a work package on environmental ethics was included to address the underlying normative issues. This project resulted in a monograph: Frederike Böhm and Konrad Ott, *Impacts of Ocean Acidification: An Analysis from an Environmental Ethics Perspective* (Böhm & Ott, 2019).[1] The following chapter presents the essential lines of reasoning of this book and adds new thoughts. It is organized as follows: Section 22.2 gives a brief overview of the scientific dimension. Section 22.3 presents a post-normal-science framing of OA. Section 22.4 deals with interrelated sources of normativity which stem from environmental ethics. Section 22.5 addresses the problem of defining a planetary boundary for OA.

22.2 The Scientific Basis

> Ocean Acidification refers to a reduction in the pH of the ocean over an extended period, typically decades or longer, which is caused primarily

[1] The book has been published in a book series on *Theory and Practice of Strong Sustainability* in the German Metropolis press (Marburg). The series intends to refine, specify, and apply the theoretical conception of strong sustainability as it has been conceived by Konrad Ott and Ralf Döring (2008) in the footsteps of Herman Daly.

> by uptake of carbon dioxide (CO_2) from the atmosphere, but can also be caused by other chemical additions or subtractions from the ocean. Anthropogenic ocean acidification refers to the component of pH reduction that is caused by human activity. (Intergovernmental Panel on Climate Change (IPCC) 2011, p. 37)

Ocean acidification is one of several global stressors to the future ocean. It occurs alongside ocean warming and climate change, a complex of problems mainly resulting from continued carbon dioxide emissions. Closely related are issues of deoxygenation and eutrophication. In a marine ecosystem context, it is impossible to isolate the impacts of OA from those of co-occurring and interacting environmental pressures. Many studies and risk assessments therefore consider combined and synergistic effects (Hales et al. 2015). Thus, one has to distinguish OA and its impacts (a) in isolation and (b) in context with other pressures. A focus on (a) may underrate the problem.

OA, as climate change, is an unintended side effect of CO_2 emissions. The basic chemistry is well known, CO_2 entering the ocean changes seawater chemistry via several chemical reactions. The main net effects are an increase of H^+ ions and consequently a decrease in pH and an increase in bicarbonate (HCO_3) ions along with a decrease in carbonate saturation. In the last two centuries, global average surface seawater pH has decreased by about 0.1 units (pH 8.21 to pH 8.1) (Rhein et al. 2013). This is equivalent to approximately 26 per cent increase in hydrogen ion concentration. An overall reduction of oceanic pH by 0.4 pH units (pH 8.21 to pH 7.8) seems possible in the worst case (IPCC 2014; Secretariat of the Convention on Biological Diversity 2014). Under the most optimistic IPCC scenario ('RCP 2.6'), ocean acidification is still projected to increase globally until the end of the twenty-first century and may slowly recover after the mid twenty-second century (IPCC 2014).

OA is occurring extremely fast on geological time scales and irreversible on a historical scale (Knoll & Fischer 2011). Natural restoration of ocean chemistry to a state similar to pre-industrial levels would take tens of thousands of years (The Royal Society 2005). From a human perspective, OA is a creeping and long-term process with a time lag between cause and effect. While the process has already started, the full effect will occur in the future. While CO_2 emissions represent the main driver of OA, there are also local factors which may cause or intensify it. Coastal drivers of OA can be pollutants, erosion of soils, sulphur dioxide precipitation, hypoxia, eutrophication as well as changes in circulation patterns, or biological processes as metabolism rates (Duarte et al. 2013). Thus, scales, context, and local variations matter.

At the species level, OA can affect physiological functions like formation of shells and skeletons by calcifying species, internal pH regulation, or photosynthesis. Besides such direct effects on organisms, many are also indirect (for example via food web or habitat change). OA lowers tolerance to ocean warming by coral and crustacean species, thereby enhancing the effect of OA. With high likeliness, calcifying organisms will be the most negatively impacted group of species (Kroeker et al. 2013).

Therefore, some authors suggest establishing thresholds for OA based on the aragonite saturation state of seawater (Rockström et al. 2009). This proposal would define an additional 'planetary boundary'. Such a planetary boundary is not a natural one but relies on normative presuppositions. Global thresholds can serve well for political purposes. If it has been prudent to determine such threshold in climate policy, it might be prudent to argue for an analogous threshold in the case of OA (Section 22.5). Without ethical grounding, any fixation of an OA threshold in terms of average pH, however, looks arbitrary because it can't be derived from scientific data alone (Section 22.4).

22.3 The Framing: Wicked Problems and Post-Normal Science

From an epistemological perspective, OA can be conceived as a case of post-normal science. 'Post-normal science' is a term that refers to wicked real-world problems in which stakes are high, uncertainties large, values in dispute, and decisions hard and urgent (Funtowicz & Ravetz 1993, 1995). OA has characteristics of a wicked problem, 'where time is running out, where decision makers lack agency, where incentives are in place that push us into a short-termist downward spiral, and where all actors in the system are ourselves part of the problem' (Thomas 2014). The term 'wicked' refers to critical features that overburden the usual 'modern' ways of problem solving via law, incentives, pricing, policymaking, and technological fixes. Those features are marked in this section.

From a scientific perspective, assessing OA must distinguish between OA alone ('as such') and OA in the context of multiple ocean pressures, as ocean warming, pollution, overfishing, and so on. Even seen in isolation, uncertainties remain high. *Contextualizing* increases uncertainties about long-term interactions, non-linear effects, cascades of effects, positive and negative feedbacks, and so on (Section 22.2). Riebesell and Gattuso (2015) emphasize knowledge gaps in ecosystem level responses, the potential of long-term adaptation, and effects of multiple drivers in context. OA exceeds natural variability of seawater alkalinity and carbonate saturation; that is, the human driver is one that the marine ecosystems are not pre-adapted to. This makes predictions uncertain. OA and ocean warming may have cascading effects on marine life and ecological processes (Secretariat of the Convention on Biological Diversity 2014).

From a policy perspective, however, OA has been addressed by Sustainable Development Goal 14 (SDG 14) 'Life below water'. Sustainable Development

Goal 14 aims to 'conserve and sustainably use the oceans, seas and marine resources'. Sub-target 14.3 specifically addresses OA in the aim to 'Minimize and address the impacts of ocean acidification, including through enhanced scientific cooperation at all levels' (United Nations Department of Economic and Social Affairs 2022). Sustainable Development Goal 14 and its sub-target 3 serve as a reference point in international governance for the objective of combatting OA. The wording of sub-target 14.3 refers to the impacts of OA, not to OA as such. To determine a threshold (Section 22.5) might suppose a different understanding of sub-target 14.3: 'Minimize OA and address and minimize its impacts'.

From a temporal perspective, OA is an *intergenerational* problem. Whereas in climate change, the members of the human species feel many of the impacts more directly, for example through extreme weather events, heat waves, droughts, and sea level rise, OA foremost affects natural systems. Humans remain largely unaffected by OA. From an anthropological perspective, OA is a *creeping* and *invisible* problem that does not represent a direct hazard to human life and health. Since humans have evolved in terrestrial environments with specific threats, they are not 'wired' to perceive long-term risks within marine systems. From the moral point of view, however, OA does not appeal to our common moral sentiments, convictions, and imagination. OA doesn't victimize individual persons, and it does not fall within the scope of problems that most people in general regard as moral ones. Thus, OA touches limits of moral perception and modes of moral attentiveness. OA falls outside the scope of usual moral problems. It is 'wicked' and 'post-normal' for ethics as well.

22.4 Environmental Ethics

22.4.1 Environmental Pragmatism

Ocean acidification falls into the field of environmental ethics. Environmental ethics is a discipline of applied practical philosophy, which investigates the moral and value-laden relationships of human beings (individuals and collectives) to natural environments (for overview see Krebs 1999 and Attfield 2014). Environmental ethics has mostly focussed on terrestrial and limnological systems. It has not been applied to the ocean very often (laudable exceptions are Norton 2003; Wolf 2003). Dallmeyer (2003: p. 278) postulates a 'second great awakening – this time for the ocean'. Roberts (2013) also points in this direction. In recent years, there have been many pleas for ocean literacy, ocean stewardship, ocean sustainability, ocean recovery, ocean ethics, and ocean health (Auster et al. 2009; Mengerink et al. 2014; Visbeck et al. 2014; Blaustein 2016; Franke et al. 2020).

My approach to environmental ethics is close to environmental pragmatism, which studies human–nature interactions and practices and wishes to modify them towards more sustainable and preservationist modes ('wise use') (Norton 2005, 2015). I also wish, however, to reconcile discourse ethics and environmental ethics (Hendlin & Ott 2016). This discourse-pragmatic approach can select specific lines of reasoning out of the entire universe of environmental ethical discourse if such lines look appropriate for specific challenges and problems, as OA. Not all arguments fit all problems equally well. Humans may have different reasons to protect endangered species or wilderness areas, to choose a vegetarian diet, to perform organic gardening or to distribute scarce environmental resources fairly among humans. This section rests on hopes that appropriate reasons have been selected which fit the case of OA.

One may ask why one should bother morally about OA. Here comes an answer: If a moral person has been informed about OA (Section 22.2), she *should* have appropriate emotions and attitudes about it. OA and ocean warming imply the insight that not even the ocean remains unaffected and safe from anthropogenic climate change. Being terrestrial beings, 'we' humans must recognize that 'we' are responsible for the long-term fate of the 'great otherness' on planet Earth: the ocean. In *'The Sea Around Us'*, Rachel Carson wrote about 'man': 'He cannot control or change the ocean as, in his brief tenancy of earth, he has subdued and plundered the continents' (1951: 20). Retrospectively, this echoes a final illusion of planetary infinity und human limitations. Meanwhile, Carson has been proven wrong by the great and excessive acceleration since the 1950s. *Yes, we can!* – and this 'is' (or should be) *awfully shocking*. Such primordial moral awareness constitutes a generic sense of human responsibility (Ott 2017, following Hans Jonas) which requires to address OA from the environmental ethical perspective more closely.

22.4.2 Common Heritage of Humankind

Respect for the well-being of future generations is a primary normative vantage point, which is covered by the *common heritage of [hu]mankind (CHM)* principle as advocated by Arvid Pardo, Elisabeth Mann Borghese, and others following their lead (Taylor 2012; van Doorn 2021). The concept is anchored in both ethics and international law (Taylor 2012). Implications of CHM in UNCLOS are outlined in Böhm and Ott (2019). Via a CHM-approach, the former concept of ocean as *res nullius* is replaced by a *res communis*-approach (van Doorn 2021).

Common heritage of [hu]mankind is not grounded in individual rights and entitlements. It would be strange to claim a right on a specific pH threshold of ocean waters. Obligations to future generations with respect to the ocean are twofold *pro tanto* (Ott et al. 2022). On the one hand, the ocean is a planetary entity, while, on the other hand, the chain of generations has no definite end. In consequence, a collective entity called 'humankind' has a long-term responsibility against future members with respect to a huge global good called 'ocean' and its qualities, resources, and inhabitants. Such responsibilities can't be reduced to obligations against individual members of future instances of humankind. In any case, the ocean will remain the ocean, but its features, amenities, and disadvantages may become different to future humans.

In principle, we can bequeath ('pro tanto') goods and bads to future generations. From OA research (Section 22.2) we assume that OA is not a change by which future generations will be benefitted. Therefore, OA should prima facie be regarded as change for the worse ('bad'). A bad becomes a moral evil, if it is recognized but remains unaddressed.

22.4.3 Strong Sustainability

Common heritage of [hu]mankind refers to the idea, but not yet to a specific concept of sustainability. To some economists, natural resources and natural capital might be perfectly substitutable by other kinds of capital. According to optimistic assumptions about substitution, the depletion and degradation of natural capital would not be a catastrophe (Solow 1974) if prudent investments in man-made capital, social infrastructure, and human capital were undertaken. This approach has been dubbed 'weak sustainability'. OA might not be a big problem for weak sustainability, since a change in marine pH might be compensated by artificial ways to enjoy the ocean (by cruising ships, digital twins, improved diving and surfing equipment, etc.). Humans may well cope with OA and won't miss former pH values. They shall conform to shifting marine species compositions.

The concept of strong sustainability denies belief in almost perfect substitutability (Daly 1996; Ott & Döring 2008: ch. 3, 4; Ott 2014). Thus, natural capital should be held intact over time. The concept of natural capital refers to stocks and non-living and living funds whose different flows contribute to overall human welfare, including so-called eudemonic values (Ott 2016; Ott & Reinmuth 2021). Eudemonic values strongly overlap with so-called 'cultural services' within the 'ecosystem services approach' (Section 22.4.5). If natural capitals have been depleted in the past, society should invest in natural capitals by ways of ecological restoration. Neumann et al. (2017) argue that SDG 14 implicitly presupposes strong sustainability. From a strong sustainability perspective, the ocean water bodies are a huge non-living fund of natural capital. This non-living fund is a living condition to a huge diversity of marine organisms nested into complex food webs. A substantial fraction of such organisms may, actually or optionally, belong to living funds of natural capital. Thus, OA is likely to impair fractions of natural capital. If so, OA *is not in accordance with the Constant Natural Capital Rule (CNCR) of strong sustainability*. If OA violates, impairs, and compromises the CNCR, it should be minimized, as SDG 14 demands (Section 22.3). Following this line of reasoning, we see OA as an imminent encroaching long-term threat to ocean's natural capitals and the supporting, provisioning, regulating, and cultural services (*Ecosystem Services* and *Cultural Ecosystem Services*) that stem from them. Such threat is pro tanto, and there is neither need nor possibility to prove it with respect to every group of marine species. According to the argument, CHM, SDG 14, and strong sustainability are coherent and fit together in a reflective equilibrium.

22.4.4 The Legacy of Aldo Leopold to the Ocean

Two sub-targets of SDG 14 refer to 'ocean health' but it remains unclear what ocean health should mean precisely. Literally, the concept of health, whatever its meaning within medicine, applies to organisms only. Therefore, the parlance of a 'healthy' ocean is to be interpreted as a metaphor. Metaphors are guiding images ('visions'). They can inspire our common moral imagination (Weston 2012) and may motivate people more strongly than sophisticated ethical arguments. Such metaphors are 'boundary objects' since they interconnect epistemic communities, but they can serve as boundary objects best, if they are translated into scientific concepts.

Bohnsack (2003), Neumann et al. (2017), and Franke et al. (2020) propose to interpret this 'ocean health' metaphor in the spirit of Aldo Leopold's ethical principle stating that actions should respect 'stability, integrity, and beauty' of biotic communities within land and sea (Leopold 1977). With respect to 'ocean health', this translates to (1) resilience (for 'stability'), (2) productivity or fertility (for 'integrity'), and (3) cultural richness (for 'beauty') of the biotic community. To Franke et al. (2020: p. 559) the ocean is 'healthy if and only if it is resilient, productive, and diverse'. OA may affect the ocean for worse on all three parameters.

22.4.5 Ecosystem Services

Parameters of ocean health can be specified and, at least in part, measured by the ecosystem service approach (ESA). While there are numerous definitions of what comprises an ecosystem service (see Haines-Young & Potschin 2010 for a detailed review), the general idea behind ESA is to bridge the gap between nature and human welfare. Contributions in Schubert and Müller (2023) show

how the ecosystem service approach can be applied to marine systems (as the Southern Baltic coast). Natural capital entails stocks and funds which enable beneficial flows. Such flows are dubbed 'services', although nature should not be interpreted as service delivering industry. Thus, 'service' is a technical term within a concept. The ecosystem service approach bridges the gap as follows: first, a distinction is drawn between nature as such and a concept of natural capital, defined as a broad source contributing to human welfare in many ways. Second, it is assumed that stocks (or funds) of natural capital provide a variety of flows which benefit human beings in different respects. Third, those respects are categorized as:

- [Supporting services (e.g. ecosystem structures and functions)]
- Provisional services (food, fibre, timber, fish, shellfish, etc.)
- Regulating services (water, climate, soil, flood regulation, etc.)
- Cultural services (aesthetics, recreation, spiritual, etc.).

The category of supporting services is contested because it consists of ecological functions and structures, which underlie the beneficial flows of the other three categories (Haines-Young & Potschin 2010). Their sustenance is necessary for ongoing beneficial flows, but a separate valuation of supporting services implies double counting (Boyd & Banzhaf 2007). Thus, they have been put in brackets. There is conceptual connectivity between the Leopoldinian categories (fertility, resilience, beauty) and types of ecosystem services. Fertility refers to supporting and provisioning services that result in measurable yields (fisheries). Richness, diversity, and beauty refer to cultural ecosystem services. Flows can also have opposite effects, that is, cause harm or disadvantage. These are termed 'disservices'. OA might not only impair services but may foster disservices. For instance, it has been highlighted that there will be winners (including, for example, jellyfish, epiphytic algae, and seagrasses) and losers among marine organisms in response to OA and ocean warming. Ocean acidification in combination with warming and other drivers of ocean change may cause in jellyfish blooms. This could locally interfere with the recreational value of bathing at the coast (Rodrigues et al. 2013). Algal blooms can be problematic for swimmers' health. Disservices might impair our admiration and appreciation of the ocean.

22.4.6 Cultural Ecosystem Services

Let's take a special look upon cultural services. Recreation and (eco)tourism are the most frequently considered cultural ecosystem services in economic studies, because they allow for contingent valuation. However, these services do not fully represent the category of cultural services. Neumann et al. (2017) specify the cultural value category with coastal ecosystems in mind:

- Aesthetics, beautiful sceneries
- Sense of place, feeling at home
- Leisure and recreation, health improvement
- Inspiration for poetry, painting, and music
- Education and knowledge
- Transformative values, virtue enhancement
- Spiritual encounters with nature
- Cultural heritage
- Specific (local, regional) cultural values.

All these cultural services (= eudemonic values) can be realized at sea and in coastal zones (Ott & Berg 2023). Cultural values (inclusive non-western ones) might be encapsulated in narratives, proverbs, songs, festivals, rituals, and so on. Narratives are highly important for the tradition of values, perhaps especially in oral cultures but are often overlooked from a scientific perspective. Cultural values are intrinsic to activities being undertaken for their own sake, such as sailing, swimming, or diving, even if performed by visitors and tourists. Living close to the sea is often a chosen way of life. By intuition, we regret a person who never saw the ocean.

Gains or losses in ecosystem services are often explored from an economic point of view which aims to place monetary values upon benefits derived from ecosystems. It is frequently pointed out that there are limitations to economic approaches and that especially many cultural services are intangible or difficult to quantify (Jax et al. 2013). The meaning and significance of cultural services must be hermeneutically understood as it touches the background of belief systems ('lifeworld'). Habermas (1981) has warned against a presumptive 'colonization of the lifeworld'. It might count as an instance for such colonization if aesthetics, place-making, spiritual, and transformative attitudes to the ocean are to be represented in an economic cost-benefit-scheme. How to represent awe in $?

In any case, future humankind has a 'pro tanto' entitlement to enjoy cultural ocean values even if it is unpredictable how and where such values may actualize. The conjunction of CHM, SDG 14, strong sustainability, and the ESA approach, including the full set of cultural values, constitutes a 'deep anthropocentric' ocean ethics within the paradigm of discursive pragmatism. If the primordial moral awareness (*Environmental Pragmatism*) becomes enriched by these normative building blocks, they may transform such awareness into attitudes and virtues which are reflected in environmental virtue ethics. Sense of responsibility and (biophilic) appreciation of ocean life may converge. Analysis of such transformation by which a terrestrial being can come 'close' to the ocean is beyond the scope of this article. The demarcation problem remains beforehand.

22.4.7 Inherent Moral Value of Nature

To attribute the category of inherent moral value (IMV) to a natural being (NB) means to respect this being for its own sake irrespectively of how valuable and beneficial it may be to humans. If such attribution is justified, direct moral obligations hold with respect to NB. The natural being, then, has become a 'moral patient'. Moral patients are beings, which lack the capability to join moral discourse but are protected by moral rules which imply obligations to moral agents. Such obligations may be obligations not to kill, not to harm, not to violate, not to capture, not to hunt. The problem of how to attribute IMV is called the 'demarcation problem'. Analysis is given in Ott (2008). The non-anthropocentric ('physiocentric') solutions are:

- Sentientism (all sentient beings are moral patients)
- Zoocentrism (all prehensive animals are moral patients)
- Biocentrism (all living beings are moral patients)
- Ecocentrisms (all ecosystems as such are moral patients)
- Holism (all natural beings are moral patients).

I have sympathies with sentientism, yet I see sophisticated problems in applying sentientism to marine life. Sentientism seems appropriate with respect to marine mammals, sea birds, and even turtles and sharks. If one adopts sentientism, the problem occurs that (most or all) fishes are sentient (Rey et al. 2015) Egalitarian sentientism implies a moral ban on fisheries as a rigid conclusion, given the reliance of many coastal communities on artisan fisheries for nutrition. If one wishes to attribute IMV only to *some* sentient creatures, as marine mammals, turtles, sharks, and so on, one faces the accusation of arbitrariness. If environmental pragmatism conceives sentience as gradual, it is possible to circumvent a complete ban on fisheries. *Gradual* sentientism allows treating members of different sentient species differently according to other features and traits, as proliferation patterns. Therefore, according to gradual sentientism, a mackerel is morally less considerable than a dolphin, a sardine less than a whale. The impacts of OA to sentient marine life are hard to predict.

OA affects many non-vertebrates without sentience. Some non-vertebrates, however, seem to display kinds of awareness (Rozzi 2019). It remains doubtful, however, whether minimal awareness is sufficient for membership in the moral community. One would have to adopt a *biocentric* position in order to respect moral entitlements of all organisms to be protected against the impacts of OA. If one adopts biocentrism, however, one faces the problem of how to overview and assess the impacts of all anthropogenic pressures on the myriads of non-sentient, often single-cell organisms living in the ocean. Ocean life in itself puts a question mark upon biocentric considerability. Reverence for life (Schweitzer 1926) is rather a basic attitude toward life than a proper solution of the demarcation problem.

Ecocentrism faces the problem to delineate marine ecosystems. In holism, one has to respect seafloors and ocean waters as such, and it might be mandatory to restrain any kind of human interference with the ocean to a minimum. *Holism* (Gorke 2003) will consider OA directly from the moral point of view. Holism implies a principle of least possible human interference with nature. It gives IMV to all existing entities. Ocean waters exist. If they are modified by anthropogenic impacts, this is morally wrong in holism. To holism, OA is not just bad, but it is a moral evil in itself. Holism, however, rests on a fallacious justification. Holism states that arbitrariness is not permitted on the moral point of view. Any exclusion of any entities from the moral community should be strictly non-arbitrary. To draw a line of moral considerability (distinguish, exclude) counts as arbitrary until proven otherwise. The burden of proof falls on the side of exclusion, and such burden is specified by a criterion of ontological parsimony ('Occam's razor'). Any statement 'entities E-1 deserve IMV because of their morally relevant features F, while entities E-2 don't deserve IMV because they lack any F' is less parsimonious than 'all entities have IMV'. Thus, holism always wins if one accepts the burden of proof.

On reflection, it becomes questionable whether biocentrism, ecocentrism, and holism are adequate ethical concepts with respect to the ocean and its lifeforms. How to follow the moral demands of some egalitarian sentientism to reduce suffering in nature (Horta 2017) if big fish *must* eat small fish? How to consider all plankton and all marine microorganisms, as biocentrism demands? How to distinguish distinctive ecosystems, as ecocentrism requires, if marine life is in constant flux at the high sea? How to deal morally with the water bodies, sediments, and geological items, as holism demands? Facing the ocean, these ethical concepts are puzzling on ontological grounds. Within the arena of UNCLOS, no agreement on the demarcation problem is to be expected. One should not impose a specific solution on all coastal communities. Moreover, to apply different solutions to the demarcation problem to the impacts of OA to marine life will be highly speculative and conjectural. If there is a reason to keep environmental ethics simple, there is a reason to adopt deep anthropocentrism (Section 22.4.6) at least in the case of OA.

22.5 A New Planetary Boundary: Threshold Problem Revisited

In climate policy, article 2 of UNFCCC has been specified to the goal that GMT warming should be limited to 'well below 2 °C'. We are now in a position to specify the normative building blocks of deep anthropocentrism (CHM, SDG 14, strong sustainability, 'ocean health', ESA) which implies aversion against OA-induced disservices to an analogous boundary. Such boundary might be adopted by UNCLOS, as 'well below 2 °C' has been adopted by the parties of UNFCCC. Such a boundary might make sense, even if OA has not yet crossed the planetary boundaries according to the diagrams of the Stockholm Resilience Centre. Such presumptive planetary OA-boundary has to be *grounded* in environmental ethics (Section 22.4) but must be also *based* on scientific knowledge (Section 22.2).

At current levels of atmospheric CO_2, an overall risk of damaging effects to marine species resulting from the combination of OA and warming is very likely; some foraminifera and pteropods are already affected by decreased calcification. Changes in abundance of these organisms would be relevant for marine food webs. Considering OA in isolation, high-risk levels, by which 30–50 per cent of calcifying species across taxa including molluscs, corals, and echinoderms are affected, are predicted for atmospheric CO_2 levels exceeding 500 ppm. Adding effects of ocean warming to the picture, the transition to the high-risk zone already starts above 400 ppm, that is, close to present CO_2 levels. Right now, we are at the breach of the high-risk zone. The zone of very high risks is looming beyond. Two thresholds of very high risk to marine life (BIOACID 2017) are characterized by (1) increasing effects on vulnerable taxa, marginalization of warm-water reefs, and loss of fish habitat; and (2) negative effects on more than 50 per cent of molluscs and corals. If the combined effects of OA and warming are considered, two 'very high risk' thresholds to marine life overlap and are located at levels around 450 ppm. This indicates that pH 7.97 comes close to the very high-risk margin (Böhm & Ott 2019). If so, there is only 0.1 distance to the zone of very high risk.

The Stockholm Resilience Centre proposed a planetary boundary for OA based on average global surface ocean aragonite saturation as an indicator, following which aragonite saturation should be sustained at 80 per cent or more of the pre-industrial state, ± natural diel and seasonal variability (Rockström et al. 2009). Personally, I support this planetary boundary for OA. With the legal framework of UNCLOS, there should be further debate on a precautious pH-threshold.

References

Attfield, R. (2014). *Environmental Ethics: An Overview for the Twenty-First Century*, 2nd ed. Cambridge: Polity Press.

Auster, P. J., Fujita, R., Kellert, S. R., et al. (2009). Developing an Ocean Ethic: Science, Utility, Aesthetics, Self-Interest, and Different Ways of Knowing. *Conservation Biology: The Journal of the Society for Conservation Biology*. 23(1), 233–35. DOI: 10.1111/j.1523-1739.2008.01057.x.

BIOACID. (2017). Biological Impacts of Ocean Acidification. oceanrep.geomar.de/id/eprint/40047/.

Blaustein, R. (2016). United Nations Seeks to Protect High-Seas Biodiversity. *BioScience.* 66(9), 713–19. DOI: 10.1093/biosci/biw097.

Böhm, F., & Ott, K. (2019). *Impacts of Ocean Acidification: An Analysis from an Environmental Ethics Perspective*. Marburg: Metropolis.

Bohnsack, J. A. (2003). Shifting Baselines, Marine Reserves, and Leopold's Biotic Ethic. *Gulf and Caribbean Research.* 14, 1–7. DOI: 10.18785/gcr.1402.01.

Boyd, J., & Banzhaf, S. (2007). What Are Ecosystem Services? The Need for Standardized Environmental Accounting Units. *Ecological Economics.* 63(2–3), 616–26. DOI: 10.1016/j.ecolecon.2007.01.002.

Carson, R. (1951). *The Sea around Us*. Oxford: Oxford University Press.

Dallmeyer, D. G. (2003). Conclusion: Launching Environmental Ethics into the Mainstream. In D. G. Dallmeyer, ed., *Values at Sea: Ethics for the Marine Environment.* Athens: University of Georgia Press, 278–82.

Daly, H. E. (1996). *Beyond Growth: The Economics of Sustainable Development.* Boston: Beacon Press.

Duarte, C. M., Hendriks, I. E., Moore, T. S., et al. (2013). Is Ocean Acidification an Open-Ocean Syndrome? Understanding Anthropogenic Impacts on Seawater pH. *Estuaries and Coasts.* 36(2), 221–36. DOI: 10.1007/s12237-013-9594-3.

Franke, A., Blenckner, T., Duarte, C. M., et al. (2020). Operationalizing Ocean Health: Toward Integrated Research on Ocean Health and Recovery to Achieve Ocean Sustainability. *One Earth.* 2(6), 557–65. DOI: 10.1016/j.oneear.2020.05.013.

Funtowicz, S. O., & Ravetz, J. R. (1993). The Emergence of Post-Normal Science. In R. von Schomberg, ed., *Science, Politics and Morality: Scientific Uncertainty and Decision Making*. Vol. 17 of Philosophy and Methodology of the Social Sciences. Dordrecht: Springer Netherlands, 85–122.

Funtowicz, S. O., & Ravetz, J. R. (1995). Science for the Post Normal Age. In L. Westra & J. Lemons, eds., *Perspectives on Ecological Integrity.* Vol. 5 of Environmental Science and Technology Library. Dordrecht: Springer Netherlands, 146–61.

Gorke, M. (2003). *The Death of Our Planet Species.* Washington, DC: Island Press.

Habermas, J. (1981). *Theorie des kommunikativen Handelns.* Frankfurt (Main): Suhrkamp.

Haines-Young, R., & Potschin, M. B. (2010). The Links between Biodiversity, Ecosystem Services and Human Well-Being. In D. G. Raffaeli & C. L. J. Frid, eds., *Ecosystem Ecology.* Cambridge: Cambridge University Press, 110–39. http://westcoastoah.org/wp-content/uploads/2016/04/OAH-Panel-Key-Findings-Recommendations-and-Actions-4.4.16-FINAL.pdf.

Hales, B., Chan, F., Boehm, A. B., et al. (2015). Multiple Stressor Considerations: Ocean Acidification in a Deoxygenating Ocean and Warming Climate. The West Coast Ocean Acidification and Hypoxia Science Panel; California Ocean Science Trust.

Hendlin, Y. H., & Ott, K. (2016). Habermas on Nature. *Environmental Ethics.* 38(2), 183–208. DOI: 10.5840/enviroethics201638215.

Horta, O. (2017). Animal Suffering in Nature. *Environmental Ethics.* 39(3), 261–79. DOI: 10.5840/enviroethics201739320.

Intergovernmental Panel on Climate Change (IPCC). (2011). Workshop Report of the Intergovernmental Panel on Climate Change Workshop on Impacts of Ocean Acidification on Marine Biology and Ecosystems. https://archive.ipcc.ch/pdf/supporting-material/IPCC_IAOMBE_WorkshopReport_Japan.pdf.

Intergovernmental Panel on Climate Change (IPCC). (2014). Climate Change 2014: Impacts, Adaptation and Vulnerability: Part B: Regional Aspects. Contribution of

Working Group II to the Fifth Assessment Report of the Intergovernmental Panel on Climate Change. www.ipcc.ch/site/assets/uploads/2018/02/WGIIAR5-PartB_FINAL.pdf.

Jax, K., Barton, D. N., Chan, K. M., et al. (2013). Ecosystem Services and Ethics. *Ecological Economics*. 93, 260–68. DOI: 10.1016/j.ecolecon.2013.06.008.

Knoll, A. H., & Fischer, W. W. (2011). Skeletons and Ocean Chemistry: The Long View. In J. P. Gattuso & L. Hansson, eds., *Ocean Acidification*. Oxford: Oxford University Press, 67–82.

Krebs, A. (1999). *Ethics of Nature: A Map*. With a Foreword by Bernard Willimas. Berlin: De Gruyter.

Kroeker, K. J., Kordas, R. L., Crim, R., et al. (2013). Impacts of Ocean Acidification on Marine Organisms: Quantifying Sensitivities and Interaction with Warming. *Global Change Biology*. 19(6), 1884–96. DOI: 10.1111/gcb.12179.

Leopold, A. (1977). *A Sand County Almanac and Sketches Here and There*. New York: Oxford University Press.

Mengerink, K. J., van Dover, C. L., Ardron, J., et al. (2014). A Call for Deep-Ocean Stewardship. *Science*. 344(6185), 696–98. DOI: 10.1126/science.1251458.

Neumann, B., Ott, K., & Kenchington, R. (2017). Strong Sustainability in Coastal Areas: A Conceptual Interpretation of SDG 14. *Sustainability Science*. 12(6), 1019–35. DOI: 10.1007/s11625-017-0472-y.

Norton, B. (2003). Marine Environmental Ethics: Where We Might Start. In D. G. Dallmeyer, ed., *Values at Sea: Ethics for the Marine Environment*. Athens: University of Georgia Press, 33–49.

Norton, B. (2005). *Sustainability: A Philosophy of Adaptive Ecosystem Management*. Chicago: University of Chicago Press.

Norton, B. (2015). *Sustainable Values, Sustainable Change: A Guide to Environmental Decision Making*. Chicago: University of Chicago Press.

Ott, K. (2008). A Modest Proposal about How to Proceed in Order to Solve the Problem of Inherent Moral Value in Nature. In L. Westra, K. Bosselmann, & R. Westra, eds., *Reconciling Human Existence with Ecological Integrity: Science, Ethics, Economics and Law*. London: Earthscan, 39–60.

Ott, K. (2014). Institutionalizing Strong Sustainability: A Rawlsian Perspective. *Sustainability*. 6(2), 894–912. DOI: 10.3390/su6020894.

Ott, K. (2016). On the Meaning of Eudemonic Arguments for a Deep Anthropocentric Environmental Ethics. *New German Critique*. 43(2), 105–26. DOI: 10.1215/0094033X-3511895.

Ott, K. (2017). Verantwortung im Anthropozän und Konzepte der Nachhaltigkeit. In R. Sierra & A. Grisoni, eds., *Nachhaltigkeit und Transition: Konzepte: Sozio-ökologische Transformation aus deutsch-französischer Perspektive*. Frankfurt (Main): Campus, 141–88.

Ott, K., & Berg, M. (2023). Eudaimonic Valuation of Cultural Ecosystem Services. In H. Schubert & F. Müller, eds., *Southern Baltic Coastal Systems Analysis*. Vol. 246 of Ecological Studies. Switzerland: Springer , 235–43.

Ott, K., & Döring, R. (2008). *Theorie und Praxis starker Nachhaltigkeit*, 2nd ed. Marburg: Metropolis.

Ott, K., Negenborn, C. von, & Matz-Lück, N. (2022). Ethics, Justice, and Human Rights: Normative Considerations in Marine Environmental Change. In P. G. Harris, ed., *Routledge Handbook of Marine Governance and Global Environmental Change*. London: Routledge, 299–312.

Ott, K., & Reinmuth, C. (2021). Integrating Environmental Value Systems: A Proposal for Synthesis. In V. Beckmann, ed., *Transitioning to Sustainable Life on Land*. Vol. 17 of Transitioning to Sustainability Series. Basel: MDPI Books, 43–77.

Rey, S., Huntingford, F. A., Boltaña, S., et al. (2015). Fish Can Show Emotional Fever: Stress-Induced Hyperthermia in Zebrafish. *Proceedings: Biological Sciences*. 282(1819), 20152266. DOI: 10.1098/rspb.2015.2266.

Rhein, M., Rintoul, S. R., Aoki, S., et al. (2013). Observations: Ocean. In Intergovernmental Panel on Climate Change (IPCC), ed., *Climate Change 2013: The Physical Science Basis: Working Group I Contribution to the Fifth Assessment Report of the Intergovernmental Panel on Climate Change*. Cambridge: Cambridge University Press, 255–316.

Riebesell, U., & Gattuso, J.-P. (2015). Lessons Learned from Ocean Acidification Research. *Nature Climate Change*. 5(1), 12–14. DOI: 10.1038/nclimate2456.

Roberts, C. (2013). *Ocean of Life: How Our Seas Are Changing*. London: Penguin Books.

Rockström, J., Steffen, W., Noone, K., et al. (2009). Planetary Boundaries: Exploring the Safe Operating Space for Humanity. *Ecology & Society*. 14(2), 32.

Rodrigues, L. C., van den Bergh, J. C., & Ghermandi, A. (2013). Socio-Economic Impacts of Ocean Acidification in the Mediterranean Sea. *Marine Policy*. 38, 447–56. DOI: 10.1016/j.marpol.2012.07.005.

The Royal Society. (2005). *Ocean Acidification Due to Increasing Atmospheric Carbon Dioxide*. London: Royal Society.

Rozzi, R. (2019). Taxonomic Chauvinism, No More! Antidotes from Hume, Darwin, and Biocultural Ethics. *Environmental Ethics*. 41(3), 249–82. DOI: 10.5840/enviroethics201941325.

Schubert, Hendrik, & Müller, Felix, eds. (2023). *Southern Baltic Coastal Systems Analysis*. Vol. 246 of Ecological Studies. Cham: Springer.

Schweitzer, A. (1926). *Kultur und Ethik*. München: Beck'sche Verlagsbuchhandlung.

Secretariat of the Convention on Biological Diversity. (2014). An Updated Synthesis of the Impacts of Ocean Acidification on Marine Biodiversity. 75 of Technical Series.

Solow, R. M. (1974). The Economics of Resources or the Resources of Economics. *The American Economic Review*. 64(2), 1–14.

Taylor, P. E. (2012). The Common Heritage of Mankind: A Bold Doctrine Kept within Strict Boundaries. In D. Bollier & S. Helfrich, eds., *The Wealth of the Commons: A World beyond Market and State*. Amherst: Levellers Press, 353–60.

Thomas, J. B. (2014). Ocean Health as a Wicked Problem. *Huffington Post*, The Blog. www.huffpost.com/entry/ocean-health-as-a-wicked_b_4770052.

United Nations Department of Economic and Social Affairs. (2022). Goal 14: Conserve and Sustainably Use the Oceans, Seas and Marine Resources for Sustainable Development. United Nations Sustainable Development Goals. sdgs.un.org/goals/goal14.

van Doorn, E. (2021). *Legal Implications of the 'Common Heritage' Principle for Atlantic Bluefin Tuna*. Berlin: Duncker & Humblodt.

Visbeck, M., Kronfeld-Goharani, U., Neumann, B., et al. (2014). Securing Blue Wealth: The Need for a Special Sustainable Development Goal for the Ocean and Coasts. *Marine Policy*. 48, 184–91. DOI: 10.1016/j.marpol.2014.03.005.

Weston, Anthony. (2012). *Mobilizing the Green Imagination*. Gabriola Island: New Society.

Wolf, C. (2003). Environmental Ethics and Marine Ecosystems: From a 'Land Ethic' to a 'Sea Ethics'. In D. G. Dallmeyer, ed., *Values at Sea: Ethics for the Marine Environment*. Athens: University of Georgia Press, 19–32.

Part IX

Atmospheric Aerosols

23

Introduction to the Atmospheric Aerosols Boundary

JEREMY J. SCHMIDT

Dust. Volcanic ash. Belching smokestacks. Wildfire plumes that send particles skyward. These are some of the many phenomena captured in the planetary boundary for atmospheric aerosol loading. The boundary includes aerosols produced by both natural and human sources. Aerosols are defined as the tiny solid and liquid particles suspended in the atmosphere and exhibit significant diversity in their composition, distribution, and in their duration in the atmosphere. All of this diversity creates significant uncertainties about their total impact. Calculating these effects is further complicated by whether aerosols are liquid or solid. These and other factors affect how the planetary boundaries framework tallies the impact of atmospheric aerosol loading on the Earth system. The measurement for aerosol loading uses what is known as aerosol optical depth, or AOD, which is a calculation of how aerosols in the atmosphere interact with clouds and ice crystals to variously absorb or scatter radiation from the Sun. By scattering radiation aerosols can cool the climate, an effect that partially offsets global warming driven by greenhouse gas emissions (Bellouin et al. 2020).

The measure of AOD is an attempt to provide an 'integrated measure of the overall reduction in sunlight reaching Earth's surface caused by all absorption and scattering in the vertical air column' (Richardson et al. 2023: 7). In other words, a maximum value for AOD would mean that no radiation reaches the Earth's surface; the radiation, in strict terms, would be extinct. In the 2009 planetary boundaries report the value for aerosol loading remained to be determined. By 2023, it was set in a way that recognized that aerosol loading had an asymmetric impact on the planet, such as by cooling one area but also potentially shifting precipitation patterns elsewhere. Interestingly, accounting for these differences means that the planetary boundary for AOD is a measure of differences between hemispheres and is set at 0.1 and based on annual averages across hemispheres. These interhemispheric impacts, combined with the many kinds of matter that comprise the category of 'aerosols', create significant uncertainties about aerosol interactions with the climate and especially with global and regional water cycles. For instance, aerosols are known to affect the water cycle, and especially the monsoon rains upon which large populations of humans and non-humans rely across South and Southeast Asia (Li et al. 2016). Aerosol loading in the atmosphere can affect the timing and intensity of monsoons, and aerosols in one hemisphere can affect precipitations patterns in the other (Krishnamohan & Bala 2022; Roose et al. 2023).

Currently, atmospheric aerosol loading is one of the few planetary boundaries that remain in a 'safe space' for humanity. This is despite well-known effects on human health of particulates at the bottom of the atmosphere where humans and other species breathe aerosols in and out. Extreme air pollution in urban centres like Delhi, Beijing, London, or Riyadh has significant effects on human health, local economies, and regional ecosystems. These are recognized in the planetary boundaries framework, but they are not the driving concern when it comes to planetary resilience.

The two chapters in this part consider atmospheric aerosol loading from the perspectives of international law and environmental geography, respectively. The first, by Leslie-Anne Duvic-Paoli, troubles the distinction between the air being breathed daily and atmospheric loading of aerosols. In Duvic-Paoli's view, the tools of international law are critical yet underequipped to address the global governance challenge aerosols present. Her concerns then turn to the ethical considerations of fairness and equity in response to what international law can and cannot (yet) do. The next chapter, by Jeremy Schmidt, examines the ethics of purposefully loading the atmosphere with aerosols in attempts to cool the climate – often referred to as solar geoengineering or solar radiation management. After explaining some of the polarizing debates over these proposals, he examines the ethics at work in a growing global agenda that does not necessarily seek to manage solar radiation but instead to experiment with technologies that could do so. Together, the two chapters offer insights into how ethical issues on aerosol loading pivot not only on how calculations are made with respect to the Earth system, but also on the instruments and techniques of governing aerosols, and how the impacts of having more or less of them loaded into the atmosphere are known.

References

Bellouin, N., Quaas, J., Gryspeerdt, E., et al. (2020). Bounding Global Aerosol Radiative Forcing of Climate Change. *Reviews of Geophysics*. 58(1), e2019RG000660. DOI: 10.1029/2019RG000660.

Krishnamohan, K. S., & Bala, G. (2022). Sensitivity of Tropical Monsoon Precipitation to the Latitude of Stratospheric Aerosol Injections. *Climate Dynamics*. 59(1–2), 151–168. DOI: 10.1007/s00382-021-06121-z.

Li, Z., Lau, W. K.-M., Ramanathan, V., et al. (2016). Aerosol and Monsoon Climate Interactions over Asia. *Reviews of Geophysics.* 54(4), 866–929. DOI: 10.1002/2015RG000500.

Richardson, K., Steffen, W., Lucht, W., et al. (2023). Earth beyond Six of Nine Planetary Boundaries. *Science Advances.* 9(37), 2458. DOI: 10.1126/sciadv.adh2458.

Roose, S., Bala, G., Krishnamohan, K. S., et al. (2023). Quantification of Tropical Monsoon Precipitation Changes in Terms of Interhemispheric Differences in Stratospheric Sulfate Aerosol Optical Depth. *Climate Dynamics.* 61(9–10), 4243–4258. DOI: 10.1007/s00382-023-06799-3.

24

The Ethics of Clean Air Governance

A Legal Perspective on the Atmospheric Aerosol Loading Planetary Boundary

LESLIE-ANNE DUVIC-PAOLI*

24.1 Introduction

Almost the entire world's population breathes unsafe air that threatens their health and degrades the environment. Many urban areas are regularly found to have illegally high levels of air pollutants (WHO 2021). Since pre-industrial times, the global concentration of aerosol particles in the atmosphere has more than doubled and air pollution has been identified as the single largest environmental health risk by the World Health Organisation (Tsigaridis et al. 2006: 5143; WHO 2021). Their harmful impacts on human health and the environment have justified the inclusion of 'atmospheric aerosol loading' as one of the nine planetary boundaries, the scientific framework designed by Rockström and his co-authors to identify the processes that regulate the stability of the Earth system and set safe operating conditions for a healthy planet (2009: 17).

The essay identifies selected ethical issues posed by the global governance of air pollution, using the planetary boundaries framework as a conceptual tool. An analysis of legal systems and institutions in the context of the planetary boundaries framework is essential because law, as a regulatory institution, could set limits on human behaviour and thus keep us within planetary boundaries (Chapron 2017; Kotzé & French 2021). In addition, law has an intimate and complex relationship with justice and can offer insights into how to design safe but also just boundaries (see Gupta et al. 2023). Notably, the refined framework, drawn by Rockström et al. and renamed 'Earth system boundaries', features an aerosol loading boundary that considers the safety of the Earth system and also accounts for human well-being, considering variations in the harmful effects of aerosols on human health across countries, communities, and generations (2023).

* This chapter is dedicated to my friend, colleague and co-author, Dr Emily Webster†, with whom I first explored the legal landscape of atmospheric aerosol loading: 'Peril, loneliness, an uncertain future, are not oppressive evils, so long as the frame is healthy and the faculties are employed; so long, especially, as Liberty lends us her wings, and Hope guides us by her star' (Charlotte Brontë, Villette).

A detailed mapping of the legal instruments available to protect the atmospheric aerosol loading boundary has shown that they are inadequate to respond to the scale of the challenge (Duvic-Paoli & Webster 2021), thus unable to mitigate increasing risks of planetary destabilization. On this basis, the essay aims to shed light on the ethical implications of the legal governance of air pollution facing decision-makers in search of fair, equitable, and effective solutions to keep atmospheric aerosol loading within safe limits.

As a result of the failure of environmental law to respond to the challenges of the Anthropocene era, legal scholarship is becoming more vocal about the need for a fundamental paradigm shift in how law operates (see Adelman 2021: 75). A line of criticism has been that environmental law operates in an ethical vacuum that has prevented it from addressing fundamental questions of vulnerability, equity, and fairness. The emphasis of legal positivism on the law as it stands, rather than as it should be, has been criticized for separating law from ethics and discouraging an ethical examination of existing norms (Voigt 2005: 123–124). An evaluation of the law regulating air pollution can therefore offer insightful lessons for the field: it raises multiple ethical questions, ranging from how legal systems can be designed to protect particularly vulnerable communities from the health impacts of air pollution, to whether it is legitimate to rely on global norms, often representative of the interests of powerful countries, to govern localized impacts.

In order to explore the ethical implications of global air governance, the essay uses the planetary boundaries framework as a conceptual tool. The framework proposes quantitative boundaries within which humankind can operate in a stable Earth system, and beyond which the risk of large-scale and irreversible environmental damage is likely. It offers scientifically based thresholds for decision-makers to design sustainable systems and has become a useful concept for legal scholars to imagine new solutions to overcome the ongoing inability of legal systems to mitigate environmental degradation (Kotzé & French 2021: 7–8).

The essay proceeds as follows: it sets the scene with a brief overview of how law regulates the problem of atmospheric aerosol loading (Section 24.2). It then moves to identify three central ethical questions that decision-makers must confront when regulating air pollution – related to why govern, how to govern, and where to govern. It uses the planetary boundaries framework as a conceptual lens to explore the ethical implications behind the motivations driving air quality governance (Section 24.3), the framing of a multifaceted environmental problem (Section 24.4), and the global governance of multi-scale impacts (Section 24.5). In doing so, the essay reflects on how the design of more ethically oriented legal systems can enhance environmental law's response to planetary problems.

24.2 Legal Protection of the Atmospheric Aerosol Loading Planetary Boundary

Before evaluating the ethical questions carried by the planetary boundaries framework, the essay starts with a brief description of legal responses to the problem of atmospheric aerosol loading.

Global cooperation on air pollution has increased in recent years, as evidenced by explicit references to air quality in UN Environment Assembly resolutions (2016, 2018), and in the 2015 Sustainable Development Goals (UNGA 2015: 11.6 and 3.9). However, responses to the problem have not been unified: no global legally binding instrument exists on air pollution, contrary to the international treaties governing climate change (UNFCCC 1992) or biodiversity (CBD 1992).

Instead, global governance on air pollution is fragmented; it is regulated by a multiplicity of legal instruments, each limited materially and/or geographically. This reality is ill-suited to the logic of the planetary boundary that portrays each environmental problem as a coherent and self-contained ensemble. However, a comprehensive mapping of the instruments available to ensure that safe thresholds are not exceeded reveals substantial gaps in the governance of the planetary boundary (Duvic-Paoli & Webster 2021). Indeed, instruments relevant to the planetary boundary have restrictive mandates to regulate specific air pollutants or polluting industries (such as aviation or shipping). They are also spatially fragmented, as air pollution is governed through a patchwork of regional frameworks, each differing widely in terms of their maturity and effectiveness. Most regional air pollution frameworks lack legally binding mandates; instead, they operate either as scientific projects (such as the Acid Deposition Monitoring Network in East Asia) or as political networks (such as the Southern African Development Community Regional Policy Framework on Air Pollution). This is explained by the fact that, in relation to air pollution, states generally prefer instruments that facilitate multilateral cooperation while preserving national discretion and accommodating regional differences. However, the general lack of regulatory ambition means that existing multilateral instruments and institutions for the governance of air quality are insufficient to be transformational.

An important exception is the Convention on Long-Range Transboundary Air Pollution (CLRTAP), a sophisticated regional treaty adopted in 1979 under the auspices of the UN Economic Commission for Europe (UNECE 1979). Through an advanced web of eight protocols, the CLRTAP regime regulates different sources of pollution (fixed, mobile, new, existing), different types of pollutants (sulphur dioxide, nitrogen oxides, volatile organic compounds, heavy metals, persistent organic pollutants), and mitigates various environmental problems (acidification, eutrophication, tropospheric ozone). The Convention has been remarkably successful at reducing emissions of air pollutants by 40–80 per cent since 1990 in Europe (UNECE 2019: 3). The decrease in sulphur emissions by at least 80 per cent is noteworthy, although nitrogen emissions remain high (UNECE 2019: 3). The Convention illustrates the positive role that international law can play in improving air quality; however, the scope of the treaty is geographically restricted to Eurasia and North America, which means it cannot adequately accommodate the global scale of the problem as outlined by the planetary boundaries framework.

Now turning to the potential aggregate effect of domestic legislation on the protection of the planetary boundary, it appears to be very limited. Indeed, domestic legislation on air pollution across the world is largely immature. A UN Environment Programme report explains:

> In at least 34% of countries, ambient air quality is not yet legally protected. Even where legally adopted, standards are difficult to compare: 49% of the world's countries define air pollution exclusively as an outdoor threat, geographic coverage of air quality standards varies, and over half of countries allow deviations from these standards. Furthermore, institutional responsibility for attaining standards is weak globally – only 33% of countries impose obligations to meet legally mandated standards. Monitoring is critical to knowing if standards are being attained but is not legally required in at least 37% of countries. (UNEP 2021b)

Not only is air pollution poorly regulated but even when regulation is in place, it does not guarantee good air quality – in the European Union, for example, member states regularly exceed their legally binding air quality targets (European Environment Agency 2022).

This brief overview of global responses to atmospheric aerosol loading shows that existing legal instruments are inadequate to respond to the scale of the planetary challenge. The question of how to design more effective legal systems is becoming increasingly pressing. In this endeavour, paying attention to the ethical implications of different regulatory interventions is essential to ensure that the measures do not only prevent environmental degradation but do so in a manner that is fair for the communities involved.

24.3 Why Regulate Air Pollution?

The essay now turns to the reasons underpinning the governance of the planetary boundary. Understanding the motivations driving the regulation of an environmental problem is important because they dictate different policy approaches with distinct ethical implications. The planetary boundaries framework justifies the adoption of a boundary on atmospheric aerosol loading on two grounds: the effects of aerosols on (1) health, as a major cause of disease and death globally; and (2) the environment, including on ecosystems, cloud formation, weather patterns, and the climate (Rockström et al. 2009: 17).

The justification aligns with the largely anthropocentric reasons used to substantiate political action on air pollution internationally. The harm done by air pollution to human health, leading the World Health Organisation to declare air pollution a 'silent public health emergency' (Ghebreyesus 2018), is the primary reason for regulatory interventions on air quality. Combined ambient and household air pollution contributes to around eight million premature deaths each year globally from increased mortality from stroke, heart disease, chronic obstructive pulmonary disease, lung cancer, and acute respiratory infections. This burden of disease is deemed to exceed the burdens of malaria, tuberculosis, and AIDS combined (WHO n.d.).

The motivations driving decision-makers for acting upon a policy problem have fundamental consequences on legal design. Notably, when the main objective is to tackle air pollution to protect human health, then ambient air quality standards are used to set limit values below which aerosol concentrations are considered safe and acceptable in terms of their effects on human health. The level of concern regarding the human health impacts of aerosols determines the extent to which regulatory interventions protect communities most sensitive to air pollution, such as children or the elderly. Alternatively, it can be considered acceptable to compromise the health of vulnerable communities to pursue other public policy objectives related to the development of air polluting activities, such as transportation or agriculture (see Lovett 2012). Similarly, taking action in the context of scientific complexity and uncertainty – the atmospheric aerosol loading was one of only two boundaries that could not initially be quantified (Rockström et al. 2009: 18) – raises difficult ethical choices about whether, and if so how, to protect human health from uncertain risks without compromising human development and well-being associated with aerosol-producing activities. In other words, how should agricultural activities be regulated when the health consequences of resulting atmospheric aerosols are difficult to quantify, and restrictions might impact access to healthy and affordable food?

Another difficulty relates to the fact that the raison d'être of the planetary boundary is dual, to protect human well-being while maintaining planetary stability. And yet, the two are not necessarily aligned: indeed, air quality standards aimed at protecting human health, especially when set low, do not necessarily avoid environmental harm and stricter air quality targets can, at times, be needed to prevent the environmental consequences of atmospheric aerosol loading. To do so, one commonly used tool is that of 'critical load' found, for instance, in the CLRTAP Protocol to Abate Acidification, Eutrophication and Ground-level Ozone (UNECE 1999). It assesses the effects of air pollutants on ecosystems, working out their tolerance to air pollution before damage becomes irreversible. However, as mentioned earlier, higher levels of protection can negatively impact human well-being in other ways and environmentally based motivations tend to have lower traction as the benefits of regulation are less directly visible and are more difficult to be quantified. In addition, Rockström et al.'s framework for safe *and* just boundaries suggests that considerations over societal fairness might require higher levels of regulatory interventions, even in cases when we remain in a safe operating space from an ecological, as well as potentially a human health, perspective (2023: 7).

24.4 How to Frame the Air Pollution Problem?

The essay moves to explore the framing of the air pollution problem, showing how varying understandings of the origins and nature of the problem carry different ethical implications in the context of law-making.

The planetary boundaries framework is generally praised for offering a simple, yet powerful conceptualization of environmental thresholds able to reach beyond the scientific community and capture the imagination of decision-makers (Biermann & Kim 2020: 513). The inclusion of a stand-alone boundary on aerosols has the advantage of putting air pollution on the same level as other environmental problems, such as climate change or biodiversity loss, for which political awareness is higher. The clarity of the message is, however, weakened by the reference to the technical term of 'aerosol loading' to describe what is more commonly known as air pollution (that includes, but is not limited to, aerosol loading).

The planetary boundaries framework presents its environmental thresholds as scientifically determined and, hence, objective. The role of decision-makers is framed as being responsible for a transgression of the boundaries, but no guidance is offered on how this can be achieved (Biermann & Kim 2020: 503–504). Lawmakers are faced with the complex task of designing measures regulating the multiple activities and actors responsible for air pollution. Many of the air polluting activities, such as transportation and industrial activities, are routine activities that are seen to be indispensable to modern life, and, thus, would be difficult to prohibit out right. Standard setting tools that set legal limits on the emission of certain substances are not enough to encourage the transformational changes needed to reduce air pollution at source, that is, to regulate fossil fuel combustion and land-use changes. As such, despite what the planetary boundaries framework might imply, the protection of the atmospheric aerosol loading boundary does not merely rest on technical choices but rather requires complex economic and societal transformations.

Moreover, identifying the origins of air pollution is not a neutral exercise and rather carries major ethical implications due to the multiple sources of atmospheric aerosol loading. Jasanoff revealed that 'the finger of blame often points downward – toward the deficient practices of poor, local communities' (2018: 11). She explained that the pollution in Delhi is due to the annual burning of crop stubble by farmers who are too poor to clear their fields for replanting by less polluting means; but instead of helping farmers buy the equipment they need, states prohibit polluting activities that continue nevertheless due to a lack of affordable alternatives. As a result, legal interventions, often adopting a command-and-control approach that sets specific emission reduction targets, generally fail to incentivize and support transitions to less polluting methods. They also tend to be ineffective because compliance with rules perceived to be unjust and unpractical is generally low.

A final comment related to the framing of the air pollution problem relates to its interactions with other environmental problems. The planetary boundaries framework presents aerosol loading as a self-standing problem operating in isolation from other environmental problems, and interactions between the different boundaries are scientifically difficult to quantify (Richardson et al. 2023: 2–3). And yet, air pollution is not a self-contained problem; it is closely related to pesticide use, oceanic pollution, waste management, and climate change. As a result, regulatory interventions on air quality can have a positive impact on other environmental problems. But the opposite is also true, and, in this case, complex trade-offs need to be considered. Notably, a reduction in emissions of sulphur dioxide carries destabilizing risks for the climate because they tend to cool the climate (IPCC 2022: C.2.3). Under these circumstances, a major intergenerational equity dilemma emerges as lawmakers must decide whether to prioritize the health of the current generation from the impacts of air pollution over that of future generations that will suffer the climate consequences.

To summarize, the aerosol loading problem has multiple sources, involves various contributing stakeholders, and interacts in complex ways with other environmental problems. But the apparent neutral and narrow framing offered by the planetary boundary logic gives little guidance on how decision-makers should respond to this multifaceted problem and sidelines the ethical implications of regulatory interventions.

24.5 Where to Govern a Multi-Scalar Problem?

A final set of ethical difficulties in the governance of the planetary boundary concerns scale (Lewis 2012). The planetary boundaries framework views the air pollution problem as global, reflecting the intercontinental and regional impacts of atmospheric aerosol loading, but impacts are primarily local. The justifications proposed for the adoption of the atmospheric aerosol loading boundary are paradoxical. On the one hand, the framework approaches aerosol loading as a global challenge; on the other hand, the choice of the boundary is justified in spatially restricted terms, based on regional tipping points (Steffen et al. 2015: 7; Rockström et al. 2023: 7). The inclusion of atmospheric aerosol loading in the planetary boundaries framework was initially justified by the effects of aerosols on monsoon patterns in Southeast Asia (Rockström et al. 2009: 17), then used as a case study to quantify a regional safe boundary (Steffen et al. 2015: 5). While scientific advancements have resulted in a preliminary proposal to identify a global threshold beyond which humanity would not be operating within a safe zone (Richardson et al. 2023: 7–8), it remains to be refined. Indeed, such quantification is complex due to the 'multiple natural and human-caused sources [of aerosols], differences in chemical composition, seasonality and atmospheric lifetimes, and the consequently very large spatial and temporal heterogeneity in distribution and climatic and ecological impacts of aerosols' (Richardson et al. 2023: 7).

The global framing of the planetary boundary, together with its shortcomings, sheds light on the important question of whether air quality can be adequately, and ethically, protected by international law. A first reason for caution relates to the fact that the planetary boundaries framework implicitly reinforces the commonly held idea that air pollution is an outdoor problem and fails to highlight the importance of indoor air quality. Indeed, the planetary boundary logic operates far away from the legal interventions into the intimacy of the home that would be required to, for instance, reduce pollution from the use of open fires and simple stoves. And yet, indoor air quality impacts the health of individuals, and disproportionately vulnerable communities in low- and middle-income countries but is rarely included in air quality legislation globally – only 7 per cent of countries have indoor air quality standards (UNEP 2021a: 9). By framing air pollution as a global and environmental problem, the planetary boundary rationale renders invisible the local and developmental nature of aerosol loading. In turn, this raises the question of whether international law has a role in regulating domestic activities beyond classic collective action problems. If so, does it have sufficient legitimacy to address issues such as indoor pollution, which lie close to individuals' everyday lives?

Another challenge for global governance is to remain legitimate in the context of a multi-scalar problem. It has been noted that the planetary boundaries framework offers little guidance on 'how to determine a fair share of the safe operating space and concomitant responsibilities among various actors at multiple levels of governance to stay within this space and to avoid breaching the boundaries' (Kim & Kotzé 2021: 7). Beyond the need to ensure that different scales of governance all work together towards an efficient and just protection of the planetary boundary, the sharing of responsibilities between states at the global level is also an ethical issue. Indeed, imposing identical legal obligations on states when major variations occur depending on the geographic level of industrialization and economic and technological capabilities can appear unfair (Okowa 2000: 6). This is particularly important given that, while the global aerosol loading boundary has not yet been crossed, regional transgressions have been identified (Richardson et al. 2023: 9).

Past experience in the development of air pollution law suggests that flexible international obligations, tailored to national circumstances, are perceived as fairer. By way of example, the 1985 Helsinki Protocol on the reduction of sulphur emissions adopted a single target, with a common timetable, for all Parties (UNECE 1985); as a result, countries like the United Kingdom and the United States did not join the protocol because it did not account for the particularities of each country. Rather, participation increased, and environmental outcomes were better when states were given the choice of several options about how to reduce their emissions (such as in the 1991 Geneva Protocol on volatile organic compounds (UNECE 1991)), or by setting national emission standards tailored to the circumstances of each state (as in the 1988 Sofia Protocol on nitrogen oxides (UNECE 1988)). Hence, the global lens of the planetary boundaries framework can help increase collective awareness but should not be understood to encourage a uniform approach to the aerosol loading problem; rather, global instruments that accommodate local specificities tend to offer more flexibility and share the burdens of environmental action more fairly.

24.6 Conclusion

The planetary boundaries framework alerted to the scientific limits of the Earth but remained silent on how this should translate in terms of human behaviour. The revised framework, which proposes both safe and just boundaries, emphasizes that safe limits cannot be considered sustainable where resources are unevenly distributed and inequalities high (Rockström et al. 2023). Law, as a regulatory tool designed for social organization, has a critically important role to play to build just societal systems operating within their environmental limits. While a complex web of international treaties, texts, and institutions seeks to mitigate air pollution, it has, so far, been unable to provide an adequate response to protect the planetary boundary. In our search for more efficient legal systems, significant ethical issues ought to be considered.

This essay explored three sets of ethical questions that decision-makers must confront when regulating air pollution. It questioned the motivations underpinning legal responses to air pollution, showing how legal design varies depending on what is deemed worth protecting. It also emphasized the complexity of governing air pollution problem when its sources are multiple, and it reflected on how to reconcile a global focus with the localized impacts of air pollution.

The planetary boundaries framework used as a conceptual tool for legal analysis makes visible essential dilemmas about how to design more efficient and more just legal systems; even its own shortcomings alert us to the complex choices facing decision-makers. Although the planetary boundaries framework initially offered little in terms of guidance on how to prompt radical change, it opens important avenues to craft a viable role for law and governance and build more ethically oriented legal systems to better protect the environment.

Acknowledgements

I would like to thank Professor Jonathan Gingerich (Rutgers Law School) for helpful comments and suggestions on an earlier version of this chapter.

References

Adelman, S. (2021). Planetary Boundaries, Planetary Ethics and Climate Justice in the Anthropocene. In D. French & L. J. Kotzé, eds., *Research Handbook on Law, Governance and Planetary Boundaries*. Research Handbooks in Human Rights. Cheltenham: Edward Elgar, 65–83.

Biermann, F., & Kim, R. E. (2020). The Boundaries of the Planetary Boundary Framework: A Critical Appraisal of Approaches to Define a 'Safe Operating Space' for Humanity. *Annual Review of Environment and Resources*. 45(1), 497–521. DOI: 10.1146/annurev-environ-012320-080337.

CBD. (1992). Convention on Biological Diversity. www.cbd.int/doc/legal/cbd-en.pdf.

Chapron, G., Epstein, Y., Trouwborst, A., et al. (2017). Bolster Legal Boundaries to Stay within Planetary Boundaries. *Nature Ecology & Evolution*. 1(3), 86. DOI: 10.1038/s41559-017-0086.

Duvic-Paoli, L.-A., & Webster, E. (2021). Atmospheric Aerosol Loading. In D. French & L. J. Kotzé, eds., *Research Handbook on Law, Governance and Planetary Boundaries*. Research Handbooks in Human Rights. Cheltenham: Edward Elgar, 278–293.

European Environment Agency. (2022). Briefing: Europe's Air Quality Status 2022. www.eea.europa.eu/publications/status-of-air-quality-in-Europe-2022/europes-air-quality-status-2022.

Ghebreyesus, T. A. (2018). Air Pollution Is the New Tobacco: Time to Tackle This Epidemic. *The Guardian*. 27 October. www.theguardian.com/commentisfree/2018/oct/27/air-pollution-is-the-new-tobacco-time-to-tackle-this-epidemic.

Gupta, J., Liverman, D. M., Prodani, K., et al. (2023). Earth System Justice Needed to Identify and Live within Earth System Boundaries. *Nature Sustainability*. 6(6), 630–638. DOI: 10.1038/s41893-023-01064-1.

Intergovernmental Panel on Climate Change (IPCC). (2022). Climate Change 2022: Mitigation of Climate Change: Summary for Policymakers. www.ipcc.ch/report/ar6/wg3/downloads/report/IPCC_AR6_WGIII_SPM.pdf.

Jasanoff, S. (2018). Just Transitions: A Humble Approach to Global Energy Futures. *Energy Research & Social Science*. 35(January), 11–14. DOI: 10.1016/j.erss.2017.11.025.

Kim, R. E., & Kotzé, L. J. (2021). Planetary Boundaries at the Intersection of Earth System Law, Science and Governance: A State-of-the-Art Review. *Review of European, Comparative & International Environmental Law*. 30(1), 3–15. DOI: 10.1111/reel.12383.

Kotzé, L. J., & French, D. (2021). Staying within the Planet's 'Safe Operating Space'? Law and the Planetary Boundaries. In D. French & L. J. Kotzé, eds., *Research Handbook on Law, Governance and Planetary Boundaries*. Research Handbooks in Human Rights. Cheltenham: Edward Elgar, 1–19.

Lewis, S. L. (2012). We Must Set Planetary Boundaries Wisely. *Nature*. 485(7399), 417. DOI: 10.1038/485417a.

Lovett, G. M. (2012). Critical Issues for Critical Loads. *Proceedings of the National Academy of Sciences*. 110(3), 808–809. DOI: 10.1073/pnas.1219007110.

Okowa, P. (2000). *State Responsibility for Transboundary Air Pollution in International Law*. Oxford: Oxford University Press.

Richardson, K., Steffen, W., Lucht, W., et al. (2023). Earth beyond Six of Nine Planetary Boundaries. *Science Advances*. 9(37), 2458. DOI: 10.1126/sciadv.adh2458.

Rockström, J., Gupta, J., Qin, D., et al. (2023). Safe and Just Earth System Boundaries. *Nature*. 619(7968), 102–111. DOI: 10.1038/s41586-023-06083-8.

Rockström, J., Steffen, W., Noone, K., et al. (2009). Planetary Boundaries: Exploring the Safe Operating Space for Humanity. *Ecology & Society*. 14(2), 32. DOI: 10.5751/ES-03180-140232.

Steffen, W., Richardson, K., Rockström, J., et al. (2015). Planetary Boundaries: Guiding Human Development on a Changing Planet. *Science*. 347(6223), 736. DOI: 10.1126/science.1259855.

Tsigaridis, K., Krol, M., Dentener, F. J., et al. (2006). Change in Global Aerosol Composition since Preindustrial Times. *Atmospheric Chemistry and Physics*. 6(12), 5143–5162. DOI: 10.5194/acp-6-5143-2006.

UNECE (United Nations Economic Commission for Europe). (1979). Convention on Long-Range Transboundary Air Pollution. treaties.un.org/doc/Publication/MTDSG/Volume%20II/Chapter% 20XXVII/XXVII-1.en.pdf.

UNECE (United Nations Economic Commission for Europe). (1985). Protocol on the Reduction of Sulphur Emissions or Their Transboundary Fluxes by at Least 30 per Cent. unece.org/sites/default/ files/2021-10/1985.Sulphur.e.pdf.

UNECE (United Nations Economic Commission for Europe). (1988). Protocol Concerning the Control of Emissions of Nitrogen Oxides or Their Transboundary Fluxes. unece.org/sites/default/files/2021-10/ 1988.NOX_.e.pdf.

UNECE (United Nations Economic Commission for Europe). (1991). Protocol Concerning the Control of Emissions of Volatile Organic Compounds or Their Transboundary Fluxes. treaties.un.org/doc/Treaties/1998/02/19980202%2005-54%20PM/Ch_XXVII_01_dp.pdf.

UNECE (United Nations Economic Commission for Europe). (1999). Protocol to Abate Acidification, Eutrophication and Ground-level Ozone. aciunece.org/sites/default/files/2025-10/ECE_EB_AIR_72.pdf.

UNECE (United Nations Economic Commission for Europe). (2019). Protecting the Air We Breathe: 40 Years of Cooperation under the Convention on Long-Range

Transboundary Air Pollution. unece.org/sites/default/files/2021-06/1914867_E_ECE_EB_AIR_ NONE_2019_3_200dpi.pdf.

United Nations Environment Assembly. (2016). Res 1/7: Strengthening the Role of the United Nations Environment Programme in Promoting Air Quality. wedocs.unep.org/bitstream/handle/20.500.11822/34316/ AQRes1_7E.pdf?sequence = 1&isAllowed = y.

United Nations Environment Assembly. (2018). Resolution 3/8: Preventing and Reducing Air Pollution to Improve Air Quality Globally. wedocs.unep.org/bitstream/handle/20.500.11822/31023/k1800222. english.pdf?sequence = 3&isAllowed = y

UNEP (United Nations Environment Programme). (2021a). One in Three Countries in the World Lack Any Legally Mandated Standards for Outdoor Air Quality. www.unep.org/news-and-stories/press-release/one-three-countries-world-lack-any-legally-mandated-standards.

UNEP (United Nations Environment Programme). (2021b). Regulating Air Quality: The First Global Assessment of Air Pollution Legislation. www.unep.org/resources/report/regulating-air-quality-first-global-assessment-air-pollution-legislation.

UNFCCC (United Nations Framework Convention on Climate Change). (1992). United Nations Framework Convention on Climate Change. treaties.un.org/pages/ViewDetailsIII.aspx?src = TREATY &mtdsg_no = XXVII-7&chapter = 27&Temp = mtdsg3&clang = _en.

UNGA (United Nations General Assembly). (2015). Resolution 70/1: 'Transforming Our World: The 2030 Agenda for Sustainable Development'. www.un.org/en/development/desa/population/migration/generalassembly/docs/globalcompact/A_RES_70_1_E.pdf.

Voigt, C. (2005). From Climate Change to Sustainability: An Essay on Sustainable Development, Legal and Ethical Choices. *Worldviews: Global Religions, Culture, and Ecology*. 9(1), 112–137. DOI: 10.1163/1568535053628454.

World Health Organization. (2021). *Ambient (Outdoor) Air Pollution*. www.who.int/news-room/fact-sheets/detail/ambient-(outdoor)-air-quality-and-health

World Health Organization. (n.d.). Air Pollution. www.who.int/health-topics/air-pollution#tab = tab_1.

25

The End of Experimentation

Geoengineering and Planetary Boundaries

JEREMY J. SCHMIDT

In May 2021, after a planned experiment in solar radiation management in Sweden was cancelled, the journal *Nature* published an editorial with a plea to 'give research into geoengineering a chance' (Editorial Board 2021). Solar radiation management involves several potential technologies that are designed to reduce the amount of the sun's energy reaching Earth's surface. These include, among other options, injecting aerosols into the stratosphere and brightening clouds over the ocean, both of which are meant to reflect a portion of solar radiation back out to space. The editorial reasoning was straightforward: (1) the best option for addressing climate change is mitigation, but the failure to arrest greenhouse gas emissions makes it increasingly unlikely that sufficient steps will be taken prior to the Earth system crossing critical climate thresholds. So, other options must be considered; (2) geoengineering has risks, including the moral hazard of misplaced confidence in technology, but without conducting experimental tests there is no evidentiary basis upon which to assess these risks against those of unabated climate change. The one-two punch of this line of reasoning supported an increasingly familiar position that does not advocate for geoengineering per se, but instead makes the case for geoengineering research. A few months later, in October 2021, an op-ed in the *New York Times* by David Keith (2021) made the same case from the opposite direction, arguing that geoengineering is the 'least bad' way to cool the planet. The shared premise of these arguments is that it is prudent to pursue experimental research into geoengineering. Indeed, it may even be the ethical thing to do.

Set in the context of the planetary boundaries, solar radiation management complicates concerns about aerosols. In the planetary boundaries framework, concerns over aerosols like dust, sulphates, and other particulates pivot on how they reduce the amount of sunlight reaching Earth's surface and in so doing affect planetary functions. Reducing radiation affects monsoon rains across Asia and affects biosphere integrity. It can affect regional climates and hemispheric patterns of precipitation, while too many aerosols can also have systematic effects on air quality for 'land and marine ecosystems' (Richardson et al. 2023: 8). But in the case of solar radiation management, where injecting aerosols is done intentionally, the concern is not over limiting aerosols but instead reckoning with whether some amount of additional aerosol loading may be needed for a last-ditch effort to stop runaway planetary heating. Here, then, is a case where it is not only a matter of two different planetary boundaries affecting one another, as so many of them do. Rather, at least one boundary is potentially being sacrificed for another: aerosols for climate. Other boundaries will be affected, too; aerosol loading affects the ozone layer, has knock-on effects on the biosphere and water cycle, and could forge unpredictable alliances with novel entities.

Responses to solar radiation management proposals are often anchored in ethical concerns. The ethics of sacrificing some degree of biosphere integrity, regional climate shifts, and hemispheric precipitation patterns all make it a significant moral wager. This is the case regardless of whether one appraises those risks in terms of utilitarian welfare functions or in terms of meeting (or foregoing) different obligations or duties. Additionally, many argue there are moral hazards to deploying solar geoengineering technologies (see Preston 2011, 2013; Tuana et al. 2012). For instance: solar geoengineering could kick-start an uncertain and complex chain of relationships and consequences that create unanticipated empirical and ethical challenges. Further, without a plan for how to stop using the technology there could be a kind of 'lock-in' where the climate would be hit with a warming shock should geoengineering be suspended. Worse, still, for many critics, is that geoengineering could become a license for further pollution since it avoids confronting the main driver of climate change: excessive greenhouse gas emissions. These types of concerns underpinned a response to the 2021 *Nature* editorial, in which Biermann (2021: 30) represented 17 signatories who argued that geoengineering was an unwelcome distraction from decarbonization efforts and that 'the current world order is unfit to devise and implement' the types of agreements required to manage the planet. The riposte from these critics drew on arguments that connect geoengineering to issues of governance, transparency, and democracy across different scales and actors, from states to Indigenous peoples (see also Gannon & Hulme 2018; Whyte 2018; Reynolds 2019). The critical response was followed a short time later with a call for a global 'non-use' agreement on geoengineering, which argues these technologies present intractable governance challenges that prevent their use in a just, globally inclusive manner (Biermann et al. 2022).

Despite the power of these responses, there is an ethical issue that they miss. And if we return to the context of the 2021 *Nature* editorial, this additional issue comes into view. That context includes, especially, a major report from the US National Academies of Sciences, Engineering, and Medicine (2021), which went to special lengths to make an ethical and empirical case for further research into geoengineering, particularly solar radiation management (SRM). The report represents one of the clearest expressions of a central normative assertion in support of geoengineering research: that experimentation is the best route through which to reach ethically justified decisions regarding solar radiation management. When reading the report, it is hard not to notice the pairing of ethics and experimentation. References to ethics occur on 77 pages and experimentation on 104; issues of justice and equity are not far behind, on 36 and 30 pages respectively. But it is not just the co-occurrence of these concerns that warrants attention. Rather, what is significant is that experimentation is deployed as the vehicle through which to reach decisions on ethical issues. As the National Academies of Sciences, Engineering, and Medicine (2021: 18) report states in the context of achieving an effective solar geoengineering (SG) research programme, a programme of experimentation should examine 'not only the atmospheric effects of SG but also the ecosystem, and economic, political, and ethical implications'.

The notion that experimentation is an appropriate venue for assessing empirical and ethical elements of geoengineering is the fundamental concern of this chapter. Major reports, editorials in major science journals, and op-eds in influential newspapers all suggest experimentation is a reasonable path forward – not ideal, perhaps, but prudent under the circumstances. For the rest of the chapter, I am going to argue that experimentation cannot be the venue for determining both the empirical and ethical considerations in geoengineering research. This is because experimentation is itself ethical, not neutral ground for adjudicating moral concerns. The upshot of my argument is that there may be a sound ethical basis for geoengineering research, but the view that experimentation provides that basis cannot simply be asserted. Nor is it sufficient to pepper reports with ethical concerns yet fail to appreciate how experimentation is not a value-free vehicle for addressing them. Rather, an argument is needed; one that addresses at least the two concerns that this chapter examines. The main argument, and where the bulk of the chapter focuses, explains the case for why appeals to experimentation are ethically inadequate. The second and much shorter argument is practical. It responds to arguments that claim those who quickly declared against geoengineering research – such as in the global non-use agreement – haven't yet considered what people in the Global South may have to contribute to debates and decisions regarding solar radiation management. I conclude by returning to considerations of aerosols and planetary boundaries.

25.1 Experimentation as Ethical Disposition

The idea that experimentation is an appropriate arena in which to convene empirical and ethical discussions regarding solar radiation management is not new.

David Keith (2013) deployed this argument in his 'case for geoengineering' as he laid further groundwork for pursuing research on planet shaping technologies (see also Morton 2016). Yet the idea of experimentation has witnessed a resurgence in view of increases to greenhouse gas emissions that continue to climb higher. In 2023, UNESCO's World Commission on Scientific Knowledge and Technology published its assessment of ethics and geoengineering. That report, like the US National Academies assessment, emphasizes that any experiments with SRM should be at small scales (ideally indoors) and constrained by a series of ethical considerations regarding transparency and conflicts of interest, and that they should also account for different cultural and disciplinary concerns (UNESCO 2023). That same year, the United Nations Environment Programme (UNEP) came to a similar conclusion: there was not enough known about SRM technologies to appraise their risks on matters ranging from impacts on the ozone layer to implications for Indigenous peoples; so, its first recommendation was that a more robust scientific review process by a global body of experts in natural and social sciences be put in place (UNEP 2023: 25). The history of solar radiation management and its current momentum, then, both point towards experimentation as needed to answer empirical questions and to thereby provide the evidence through which to settle ethical concerns.

What is the objection to experimentation? It seems like a good antidote to what is currently happening. For instance, in 2022 the start-up company *Make Sunset* lofted balloons filled with sulphur dioxide high into the atmosphere and boasted they could release more aerosols if people wanted to buy the 'cooling credits' they claimed this achieved. Shortly after, another article in *Nature* pounced on the idea, arguing that the scary fact that anybody can pump aerosols skyward is all the more reason to research geoengineering (see Ricke 2023). This wasn't the first time this kind of thing happened. In 2012 the Haida people hired a fly-by-night scientist to dump 120 tons of iron oxides and iron sulphates off the coast of Haida Gwaii (in the Pacific Ocean off Canada's west coast). This geoengineering project by an Indigenous nation was meant to stimulate the ocean life that ultimately draws carbon down to the ocean floor (Gannon & Hulme 2018). There is, of course, an important difference between start-up publicity stunts, ill-defined research, and careful experimentation. Shouldn't we favour systematic programmes rather than just accept ad hoc interventions? Before digging in on this question, it is important to also note that geoengineering research does not always entail real-world experimentation. For instance, atmospheric modelling doesn't necessarily entail actual release of aerosols. Here, however, I focus on experimental arguments involving actual releases because these are among the most salient sources of knowledge once the theories and models recommend a particular technological path for intervening in Earth's solar radiation budget.

A key aspect of my argument against experimentation as a vehicle for ethical decision-making was put succinctly by the late philosopher Bernard Williams. In his work *Philosophy and the Limits of Ethics*, Williams (1985: 172) argued that a society given over to 'experiments in living' does not necessarily increase 'the chances of living in the best way. It is one sort of society rather than another, and there are various forms of living that it rules out; indeed, those ruled out could include those most worth living'. When Williams made this argument he had a specific target in mind – liberal forms of experimental rule proffered by the utilitarian John Stuart Mill. My argument is not beholden to a critique of either liberalism or utilitarianism, but it equally points out the fact that experimentation orients ethical questions about geoengineering in one way and not in others. In other words, experimentation operates as a kind of ethical disposition – a particular way of arranging concerns and the relations among them. In so doing, experimentation rules out alternate, non-experimental modes of ethics by fiat. This ring-fencing is an important part of what makes experimentation inappropriate as a venue for convening ethical decisions.

One objection that a proponent of experimental research into geoengineering might make is to point out the difference between what Williams termed experimental 'ways of living' versus the more constrained aims of a scientific experiment. This sort of objection emphasizes that experimentation in one area is different than saying *everything is experimental*. If that is true, the proponent could then argue, we can hive off the experimental aspect of geoengineering in a way that does not commit us to making ethics experimental too. There are two challenges this objection faces. The first is that scientific experiments are not conducted in social vacuums. Historians and philosophers of science have routinely made this point plainly and powerfully enough over recent decades that the scientific knowledge once declared to be value-free is now increasingly recognized for the many judgments through which theories, technologies, norms, and institutions make science reliable (Daston & Galison 2010; Oreskes 2019; Cartwright et al. 2022). Risks, too, are assessed not with respect to the universe of total possibility but through culturally situated practices and bounded rationalities that make satisfying multiple constituencies not only a matter of presenting experimental results (cf. Rayner & Cantor 1987). As result of this tangle, the idea that experimentation retains some kind of special status amidst value-loaded sciences is not tenable. This is part of why, I would suggest, major reports by scientific and environmental agencies are trying to show that geoengineering research can be ethical in the first place – because it needs normative legitimacy even at the experimental phase.

The second problem with trying to parse geoengineering research from the argument that it commits decisions to one ethical disposition by fiat is that, if it does not, then proponents face a different challenge: exactly how will experimentation inform ethical decisions? Philosophically, there is an important logical difference between what Nancy Cartwright (2007) termed 'hunting causes and using them'. Suppose, for instance, that an experiment suggested with strong evidence that *x* would be the result of applying solar geoengineering at scale. This still would not provide us with the full range of morally relevant information about how the causal explanations of *x*, as determined by the experiment, ought to bear on different uses of the technology – for how long, at what scale, and with what weight assigned to other trade-offs. These kinds of ethical decisions are a species of what Marcus Düwell (2013: 9) terms 'mixed judgments' in which different disciplines and practices offer partial perspectives on a shared moral concern. These broader considerations are among the most salient questions that opponents of solar geoengineering raise. In other words, if proponents of solar radiation management seek to keep experimentation cloistered from its social context (as my first point earlier suggests is hard to do), then they are left no better off in terms of being able to settle ethical questions through answers to empirical questions. The cloistering has the opposite effect: it cuts off the presumptive connection of ethics and inquiry that the case for geoengineering research via experimentation rests on.

The ball is back in the court of those advancing experimentation as the vehicle for making decisions across the empirical and ethical considerations entailed by using aerosols to manage solar radiation. What can be noted in this, the first of my arguments, is that I have not closed the door to experimentation being ethically appropriate. Rather, I have pointed out that this requires a defensible argument that convincingly shows why this is so. To set up experimentation as the site of decision-making about the prospects for planetary geoengineering is to invoke an entire apparatus of inquiry that goes far beyond the boundaries of any single experiment or set of experiments. Additionally, and because experimentation is entangled with the values, theories, and technologies implicated in establishing scientific claims, the argument for experimentation will, in fact, need to confront the central insight provided by Bernard Williams: giving decisions over to experimentation rules out other ways that decisions might be reached. Experimentation entails a particular way of arranging ethical concerns – a disposition – in relation to scientific practices and this is not a morally neutral endeavour.

25.2 Experimentation in Planetary Perspective

In February 2023, it was widely reported that Earth's temperature averaged over 1.5°C above the pre-industrial average for the first calendar year ever. That year, then the hottest on record, saw significant rises in ocean surface temperatures in the north Atlantic and elsewhere. Explaining the rapid rise in temperatures, which are a decade ahead of what was predicted in many climate models, has often turned to aerosols. It has been understood that aerosols have long been having a cooling effect on Earth's temperature, even though the sensitivity of climate to them was not fully understood (see Andreae et al. 2005). One of the first efforts to quantify these effects was published in 2018 and estimated that aerosols were already cooling Earth by 0.5°C (possibly up to 1.1°C) and that reductions in aerosol loading would increase precipitation 2–4.6 per cent and make extreme weather even more likely (Samset et al. 2018). The sixth (and most recent, as of this writing) Intergovernmental Panel on Climate Change (2021) report considered the upper limit of aerosol cooling to be as high 0.8°C. The counterintuitive concern then arose – what are the ethics of reducing polluting aerosols when they are *already* a kind of de facto geoengineering? Would rapid reduction lead to the kind of climate shock that worries those who want a clear and agreed plan for how to phase down any future, purposeful form of solar geoengineering?

In this context, the case for experimental research into solar radiation management takes on a different tenor. There is already unintentional cooling of the earth through historical aerosol emissions. Could this be levered into a need to understand how to avoid rapid temperature increases that could arise from effective reductions in aerosol emissions? If so, that might be a further basis for legitimating geoengineering research. This is because temperature rises that are much faster than anticipated pose the potential for a climate shock for which species, cities, economies, and social stability might not have adequate adaptive equipment or capacity. In this case there is a kind of practical urgency. Reducing emissions is critical to reducing planetary heating, but it is also critical to avoid moral harms caused by rapid climate heating from aerosol emissions reductions. This is especially the case for those most vulnerable to climate change, such as those in the majority world (for example: the Global South). This argument is not simply about 'giving geoengineering research a chance' as the editorial in *Nature* put it. Rather, it is about supporting the best way to learn about interactions of aerosols and climate.

In this context, Táíwò and Talati (2022) recently argued that one of the problems with current debates about geoengineering research is that many critics base their arguments on standards their own arguments don't meet. This is especially the case, they argue, when the argument is made that harms may accrue disproportionately to people and environments in the Global South through decisions made by actors in the Global North. Such arguments, Táíwò and Talati (2022) reason, often succumb to the inconsistency of never having asked people in the Global South their perspective. So, in this way, the argument that geoengineering cannot be done owing to lack of transparent or democratic global governance fails to meet its own standard of confronting undemocratic or non-transparent forms because it simply asserts its decision without consulting those in the Global South. They think it too presumptive to declare no geoengineering research should take place without consulting those most vulnerable to the risks and trade-offs between climate heating versus managing radiation.

My argument is not of the sort Táíwò and Talati (2022) target. I have not claimed no research could ever be justified, only that the argument for experimentation cannot also double for ethical warrant. My argument also holds for practical questions around urgency. Responses to urgent moral questions are not appropriately confined to experimental dispositions either. They require a suite of tactics. If experimentation is to be principal or even prominent among them, it will require an argument for why this is so, not a default response to crisis. Through deliberations of the kind that Táíwò and Talati (2022) don't wish to prematurely exclude, a more transparent and democratic appraisal could include experimentation as part of a governance package. Here, however, experimentation is already situated differently. Experimentation would be one among many aspects of ethical decision-making – transparency, democracy, and the values these rely on – regarding solar radiation management, not its primary disposition.

25.3 Conclusion

The argument for research into solar radiation management needs a justification that does not collapse ethical decision-making to an experimental disposition. This is an issue that is also likely to become increasingly important for the planetary boundaries framework. That framework, at present, recognizes existing aerosol loading is already reducing monsoon rains and that there are large uncertainties and asymmetric effects across hemispheres in the interactions among different types and sources of aerosols, clouds, and atmospheric climate forcing (Richardson et al. 2023; cf. Intergovernmental Panel on Climate Change 2021). It also places aerosol loading among the remaining boundaries that have not yet been crossed – implying, potentially, that adding aerosols could be permissible from the perspective on Earth system stability and resilience that the framework commits itself to. The most recent assessment, however, also states that 'a much better systemic and quantitative understanding of the hydroclimatic, ecological, and biogeochemical effects of asymmetric aerosol forcing is needed to refine the aerosol loading boundary' (Richardson et al. 2023: 8). How will this more systematic, quantitative understanding be achieved? If it is to be through experiments that involve purposeful aerosol loading, a much more robust argument will be needed than just giving over ethical decisions to experimentation.

References

Andreae, M. O., Jones, C. D., & Cox, P. M. (2005). Strong Present-Day Aerosol Cooling Implies a Hot Future. *Nature*. 435(7046), 1187–1190. DOI: 10.1038/nature03671.

Biermann, F. (2021). It Is Dangerous to Normalize Solar Geoengineering Research. *Nature*. 595(7865), 30. DOI: 10.1038/d41586-021-01724-2.

Biermann, F., Oomen, J., Gupta, A., et al. (2022). Solar Geoengineering: The Case for an International Non-Use Agreement. *WIREs Climate Change*. 13(3), e754. DOI: 10.1002/wcc.754.

Cartwright, N. (2007). *Hunting Causes and Using Them: Approaches in Philosophy and Economics*. Cambridge: Cambridge University Press.

Cartwright, N., Hardie, J., Montuschi, E., et al. (2022). *The Tangle of Science: Reliability Beyond Method, Rigour, and Objectivity*. Oxford: Oxford University Press.

Daston, L., & Galison, P. (2010). *Objectivity*. Princeton, NJ: Princeton University Press.

Düwell, M. (2013). *Bioethics: Methods, Theories, Domains*. London: Routledge.

Editorial Board. (2021). Give Research into Solar Geoengineering a Chance. *Nature*. 593(7858), 167. DOI: 10.1038/d41586-021-01243-0.

Gannon, K. E., & Hulme, M. (2018). Geoengineering at the 'Edge of the World': Exploring Perceptions of Ocean Fertilisation through the Haida Salmon Restoration Corporation. *GEO: Geography and Environment*. 5(1), e00054. DOI: 10.1002/geo2.54.

Intergovernmental Panel on Climate Change, ed. (2021). *Climate Change 2021: The Physical Science Basis*. Contribution of Working Group I to the Sixth Assessment Report of the Intergovernmental Panel on Climate Change. Cambridge: Cambridge University Press.

Keith, D. (2013). *A Case for Climate Engineering*. Cambridge, MA: MIT Press.

Keith, D. (2021). What's the Least Bad Way to Cool the Planet? *New York Times*. 1 October. www.nytimes.com/2021/10/01/opinion/climate-change-geoengineering.html.

Morton, O. (2016). *The Planet Remade: How Geoengineering Could Change the World*. Princeton, NJ: Princeton University Press.

National Academies of Sciences, Engineering, and Medicine. (2021). *Reflecting Sunlight: Recommendations for Solar Geoengineering Research and Research Governance*. Washington, DC: The National Academies Press.

Oreskes, N. (2019). *Why Trust Science?* Princeton, NJ: Princeton University Press.

Preston, C. J. (2011). Re-Thinking the Unthinkable: Environmental Ethics and the Presumptive Argument against Geoengineering. *Environmental Values*. 20(4), 457–479. DOI: 10.3197/096327111X13150367351212.

Preston, C. J. (2013). Ethics and Geoengineering: Reviewing the Moral Issues Raised by Solar Radiation Management and Carbon Dioxide Removal. *WIREs Climate Change*. 4(1), 23–37. DOI: 10.1002/wcc.198.

Rayner, S., & Cantor, R. (1987). How Fair Is Safe Enough? The Cultural Approach to Societal Technology Choice. *Risk Analysis*. 7(1), 3–9. DOI: 10.1111/j.1539-6924.1987.tb00963.x.

Reynolds, J. L. (2019). Solar Geoengineering to Reduce Climate Change: A Review of Governance Proposals. *Proceedings of the Royal Society A: Mathematical, Physical and Engineering Sciences*. 475(2229), 20190255. DOI: 10.1098/rspa.2019.0255.

Richardson, K., Steffen, W., Lucht, W., et al. (2023). Earth beyond Six of Nine Planetary Boundaries. *Science Advances*. 9(37), eadh2458. DOI: 10.1126/sciadv.adh2458.

Ricke, K. (2023). Solar Geoengineering Is Scary: That's Why We Should Research It. *Nature*. 614(7948), 391. DOI: 10.1038/d41586-023-00413-6.

Samset, B. H., Sand, M., Smith, C. J., et al. (2018). Climate Impacts from a Removal of Anthropogenic Aerosol Emissions. *Geophysical Research Letters*. 45(2), 1020–1029. DOI: 10.1002/2017GL076079.

Táíwò, O. O., & Talati, S. (2022). Who Are the Engineers? Solar Geoengineering Research and Justice. *Global Environmental Politics*. 22(1), 12–18. DOI: 10.1162/glep_a_00620.

Tuana, N., Sriver, R. L., Svoboda, T., et al. (2012). Towards Integrated Ethical and Scientific Analysis of Geoengineering: A Research Agenda. *Ethics, Policy & Environment*. 15(2), 136–157. DOI: 10.1080/21550085.2012.685557.

UNEP (United Nations Environment Programme). (2023). One Atmosphere: An Independent Expert Review on Solar Radiation Modification Research and Deployment. www.unep.org/resources/report/Solar-Radiation-Modification-research-deployment.

UNESCO. (2023). Report of the World Commission on the Ethics of Scientific Knowledge and Technology (COMEST) on the Ethics of Climate Engineering. unesdoc.unesco.org/ark:/48223/ pf0000386677.

Whyte, K. P. (2018). Indigeneity in Geoengineering Discourses: Some Considerations. *Ethics, Policy & Environment*. 21(3), 289–307. DOI: 10.1080/21550085.2018.1562529.

Williams, B. (1985). *Ethics and the Limits of Philosophy*. Cambridge, MA: Harvard University Press.

Part X

Novel Entities

26

Introduction to the Novel Entities Boundary

JEREMY J. SCHMIDT

Radiation from nuclear detonations. Genetically modified organisms. Synthetic biological compounds. Microplastics in clouds, rain, and bodies. Endocrine-disrupting chemicals that mess with how amphibians and other animals – including humans – develop and reproduce. These are some of what the planetary boundaries framework characterizes as novel entities, a category 'restricted to include only entities that, in the absence of the anthroposphere, are not present in the Earth system' (Richardson et al. 2023: 3). This planetary boundary, perhaps more than any other, has expanded considerably with respect to definition and measurement. Its concern with different kinds of entities also raises deeper ethical questions, from those regarding what is truly new when it comes to how anthropogenic forces affect planetary evolution, to the variety of risks, uncertainties, and moral challenges that arise when different materials and technologies are grouped together as novel entities.

Novel entities weren't in the 2009 planetary boundaries framework. Instead, 'chemical pollution' held this space. At that time, however, there was not yet sufficient information to suggest a boundary (Rockström et al. 2009). In the 2015 update, 'novel entities' were introduced, with ozone depletion acting as an example that was closely related to what the new category sought to capture (Steffen et al. 2015). Critically, the 2015 update established criteria for how a growing multitude of chemicals might be understood for their effects on the Earth system. These criteria drew on Persson et al. (2013: 12620, original emphasis), who explored approaches to '*unknown* planetary boundary issues [that] likely exist'. The tens of thousands of chemicals introduced into the environment with very little known about their aggregate individual effects, or the ways they might combine, presented exactly the kind of 'known unknowns' that could not be ignored. Their proposals, adopted in the 2015 framework, had three main points of consideration for determining a planetary boundary: the chemical must have a disruptive effect on vital Earth-system processes, not be discovered until it 'is a

problem at the global scale', and, finally, have an effect that is not readily reversible (Steffen et al. 2015: 7).

As chemical pollution morphed into 'novel entities', researchers examined how this planetary boundary turned on the capacity of Earth processes to assimilate chemical pollutants (Diamond et al. 2015). Reducing pollution was an obvious goal, and when Persson et al. (2022a) published a new global assessment they found Earth was exceeding the planetary boundary for novel entities even when that boundary was restricted only to chemicals. Still, as they acknowledged, their assessment had not really shown '*where* the boundary for novel entities is' (Persson et al. 2022b: 6788, original emphasis). And this is for good reason: novel entities are multidimensional owing to how they might impact biological, biogeochemical, or geophysical aspects of the Earth system. Measuring thousands of interactions individually, let alone in concert, presents a near-intractable challenge. In this way, Persson and colleagues were very much faced with the kinds of issues identified at the outset of this book: a boundary is a normative idea that does not necessarily rely on full knowledge of where a quantifiable threshold might be. Boundaries are normative. So, they can countenance some ignorance while still retaining a commitment to gathering more information and revising as necessary in future. When the planetary boundaries framework was updated in 2023, novel entities were expanded beyond chemicals with a 'control variable' determined by 'the percentage of chemicals released to the environment without adequate safety testing' (Richardson et al. 2023: 6). Unlike other boundaries that have quantified values, this is as far as this boundary's measurement goes – to a ratio of chemicals released to the environment to their testing. Because in the Holocene no such entities existed, and since hundreds of thousands of such chemicals now circulate globally, the only safe boundary was determined to be zero. Zero would remain the boundary unless those chemicals are fully tested and determined to be harmless – a standard derived by taking the Montreal Protocol for ozone as an example for thinking about novel entities.

Not everybody agrees on zero for a safe planetary boundary on chemicals, and neither is there always agreement on what counts as novel or why. Several feminist scholars, for instance, hesitate over normative claims that chemicals are wholly bad since many can provide socially valued goods too, such as pharmaceuticals that have health benefits (Murphy 2017; Balayannis & Garnett 2020). Likewise, the chapters in this part offer different assessments of how claims about novelty do not operate free from broader considerations about human interactions with the planet. These two chapters do this in very different ways. The first, by Andreas Bachmann and Ariane Willemsen, looks at the ethics of novel entities from the perspective of ethics within public administration. In these contexts, decisions on

definitions and descriptors matter critically for assessing moral matters. Fuzziness will not do. In their chapter, they offer a carefully considered critical engagement of how the planetary boundaries framework treats 'novel entities' and what is at stake if different interpretations of that boundary are followed through to their ethical conclusions. The second chapter also takes a critical approach to planetary boundaries but is also a different format to previous chapters. In this case, the chapter is an interview transcript with Erle Ellis, a scientist who examines human-Earth interactions at a range of spatial and temporal scales. For Ellis, questions of novelty must be set within broader considerations regarding *for whom* planetary crises are really a crisis. They must also be set within a broader appreciation of how human and non-human ecologies interact at a range of scales and over time. For Ellis, framing problems at the 'planetary' scale often do not sufficiently identify those responsible for forcing the Earth system into a novel state, such as through anthropogenic climate change, and so thinking through the ethics of the planetary boundaries framework requires situating it amid a broader set of concerns crisscrossing science and society.

References

Balayannis, A., & Garnett, E. (2020). Chemical Kinship: Interdisciplinary Experiments with Pollution. *Catalyst: Feminism, Theory, Technoscience*. 6(1), e33524. DOI: 10.28968/cftt.v6i1.33524.

Diamond, M. L., De Wit, C. A., Molander, S., et al. (2015). Exploring the Planetary Boundary for Chemical Pollution. *Environment International*. 78, 8–15. DOI: 10.1016/j.envint.2015.02.001.

Murphy, M. (2017). Alterlife and Decolonial Chemical Relations. *Cultural Anthropology*. 32(4), 494–503. DOI: 10.14506/ca32.4.02.

Persson, L. M., Breitholtz, M., Cousins, I. T., et al. (2013). Confronting Unknown Planetary Boundary Threats from Chemical Pollution. *Environmental Science & Technology*. 47(22), 12619–12622. DOI: 10.1021/es402501c.

Persson, L. M., Carney Almroth, B. M., Collins, C. D., et al. (2022a). Outside the Safe Operating Space of the Planetary Boundary for Novel Entities. *Environmental Science & Technology*. 56(3), 1510–1521. DOI: 10.1021/acs.est.1c04158.

Persson, L. M., Carney Almroth, B. M., Collins, C. D., et al. (2022b). Response to Comment on 'Outside the Safe Operating Space of the Planetary Boundary for Novel Entities'. *Environmental Science & Technology*. 56(10), 6788–6789. DOI: 10.1021/acs.est.2c02265.

Richardson, K., Steffen, W., Lucht, W., et al. (2023). Earth beyond Six of Nine Planetary Boundaries. *Science Advances*. 9(37), 2458. DOI: 10.1126/sciadv.adh2458.

Rockström, J., Steffen, W., Noone, K., et al. (2009). Planetary Boundaries: Exploring the Safe Operating Space for Humanity. *Ecology & Society*. 14(2), 32. DOI: 10.5751/ES-03180-140232.

Steffen, W., Richardson, K., Rockström, J., et al. (2015). Planetary Boundaries: Guiding Human Development on a Changing Planet. *Science*. 347(6223), 1259855. DOI: 10.1126/science.1259855.

27

Novel Entities as a Planetary Boundary

Ethical Considerations

ANDREAS BACHMANN AND ARIANE WILLEMSEN

27.1 Introduction

The perspective of this chapter is that of institutionalized ethics consulting within public administration. The task of this kind of ethics consulting is to draw the attention of the legislator and the implementing authorities to the ethical aspects of the pertinent legal regulations that need to be taken into account. As far as ethical expertise is concerned, ethics is understood and applied as a science.[1] In this respect, ethics consulting within the administration is no different from academic ethics. Where it differs is in the specific perspective: advice is given under the conditions of the specific needs of the administration.

The following considerations are based on the experience of working for the Swiss Federal Ethics Committee on Non-Human Biotechnology (ECNH) and the ethics unit of the Swiss Federal Office for the Environment (FOEN). Both the ECNH and the FOEN's internal ethics unit were established in the 1990s in the same context, namely in view of the emerging moral issues in dealing with the then new biotechnology and gene technology and the concretization of the Swiss constitutional mandate to take account of the dignity of living beings. While the ECNH addresses ethical questions of non-human biotechnology based on its legal mandate and advises all Federal offices affected by the mandate, the FOEN's internal ethics unit deals with moral aspects of the office's broad range of tasks.

More specifically, the ECNH is particularly concerned with the ethical aspects of various technological approaches and products in the field of non-human biotechnology. These include, among others, nanotechnology, synthetic biology,

[1] In contrast to the empirical sciences, that is, the natural and social sciences, ethics is a normative science. Its aim is not to explain the world, but to justify how we as humans should live and act in this world. It examines everyday moral beliefs, what are commonly referred to as 'moral intuitions', in a systematic way with the aim of determining which of them are justified and for what reasons, and which of them are not. It uses specific methods and theories to do so. Its claim is to be able to determine in this way those moral rules and norms that are universally valid.

genetically modified organisms (GMOs), and gene drives. For example, the Committee assesses the ethical acceptability of the idea of combining living organisms or biologically active material with non-living material (chemicals) or of creating life purely from chemical components and DNA blueprints (ECNH 2010). It is working intensively on GMOs, in particular the new genetic engineering methods, with a view to crops and livestock (ECNH 2016, 2018, 2022). Recently, genetically modified microbes have also become an issue, which are to be used to improve soil quality and plant growth, as well as to produce pesticides as a substitute for chemicals. Another example are genetically modified or synthetic gene drives that can be used to introduce additional genetic information for new traits into a population or to remove essential DNA sequences, thereby creating a lethal factor that causes the offspring to die during development, for example, in the fight against malaria. Genetically modified organisms are also relevant for the FOEN's ethics unit, in particular the question of how new genomic techniques can be regulated appropriately based on the Swiss Constitution for the area of plant breeding and the commercial handling of GMOs. The question of the responsible use of chemicals and plastics also plays a role.

The planetary boundary 'Novel Entities' (NE) forms one of the nine planetary boundaries within the PB approach. In order to be able to assess this planetary boundary, the first step is to outline the approach in general. This includes a brief explanation of why some aspects of this approach are problematic from an ethical perspective. We will then take a look at the NE, in particular chemicals, plastics, and GMOs. While in our view the PB approach is generally to be welcomed, even if it currently plays a marginal role in Switzerland's national environmental regulation, we consider the specific planetary boundary 'Novel Entities' to be problematic from an ethical perspective. Our impression is that it does not fit into the overall approach and runs the risk of initiating substitute discussions that distract from the actual problems.[2]

27.2 Ethical Challenges for the Planetary Boundaries Framework

The planetary boundary approach (PB) attempts to determine the negative effects of human activities on the stability and resilience of planetary biophysical and biochemical processes and to show where boundaries are transgressed creating an unacceptable risk to this stability and resilience. While environmental protection is still primarily a national task, it is evident – as climate change, biodiversity loss, or plastic pollution show – that many processes have taken on a global scale that also requires an appropriate global response. With regard to the development of global

[2] The text reflects the personal considerations of the authors.

management approaches, a lot has happened in recent years.[3] But they are not enough to overcome the challenges we face. In this context, any approach that helps to provide science-based arguments that can increase the pressure to act and to initiate the necessary measures at planetary level is to be welcomed.

Regarding the scientific and methodological aspects of the PB approach, there are challenges particularly due to a lack of data and because the complex interactions between the nine planetary boundaries, and the cumulative effects that may accompany them, are not yet sufficiently understood. This aspect will not be discussed next, though. Rather, it is assumed that the PB framework is a scientifically sound approach that can provide policymakers with important points of reference on the basis of which international regulations can be developed, but which can also be downscaled to a sub-global (national or regional) level.

However, the PB approach is not a purely natural scientific concept. It also contains evaluative and normative aspects, but these are either not disclosed or have no plausible justification, at least not from the point of view of ethics or political philosophy. The advocates of the PB framework should give more weight to these normative aspects and work together with the relevant specialists for their detailed elaboration. Two examples serve as illustration: on the one hand, the concept of 'justice' and, on the other, the relationship between the precautionary principle, the harm principle, and risk assessment. The second example is also relevant for the assessment of Novel Entities in particular.

27.2.1 PB, 'Safe and Just Earth System Boundaries'[4] and the 'Planetary Commons'[5]

In two recently published articles, Rockström et al. have attempted to integrate normative concepts such as harm or justice into the PB approach and to expand it to include international law and regulatory aspects. However, the proposals made are not convincing. For instance, harm is an evaluative or normative notion. They define it as

> negative impacts on humans, communities and countries from Earth system change in addition to background rates. (Rockström et al. 2023: 103)

This definition raises two questions:

First, is 'negative impact' to be understood as 'is actually negative' or 'is considered to be negative'? This distinction is normatively relevant. The same

[3] For example with regard to the management of chemicals and waste in sustainable development, see United Nations Environment Programme (2019).
[4] Rockström et al. (2023).
[5] Rockström et al. (2024).

applies regarding the concept of justice, where a distinction must be made between 'considering something to be just' and 'actually being just'. The following applies to both concepts: if they are used as genuinely normative – and not as socio-political – concepts, the question must be: 'what is harm or what is just'? And not: 'what is considered to be harmful or just'? What is considered just can be determined empirically, for example, through surveys. The normative question, however, is whether what is considered just is actually justified. It is not permissible to draw direct conclusions about what is actually just from what is considered just.

Second, how does harm to people as individuals, on the one hand, and to countries and communities as collectives, on the other, relate to each other? Here, too, the same question must be asked with regard to the concept of justice.

Reading the definition of 'significant harm', one might think that, at least with regard to the first question, it is clear that Rockström et al. understand harm as what is negative, not what is considered to be negative.

> [W]e define significant harm as widespread severe existential or irreversible negative impacts on countries, communities and individuals from Earth system change, such as loss of lives, livelihoods or incomes; displacement; loss of food, water or nutritional security; and chronic disease, injury or malnutrition. (Rockström et al. 2023: 103)

However, one would then have to at least begin to sketch out a theory of objective harm that goes beyond intuitive examples. Instead, there is a back and forth between normative and non-normative (that is, socio-political) statements. This is not a satisfactory normative argument. For that kind of argument, one would have to clearly separate these aspects and show why what is considered just is actually just.

The second question about the relation of harm to individuals, on the one hand, versus countries and communities as collectives, on the other, remains unanswered. This is also a problem regarding a plausible normative argument. For what can be a significant harm for a person from the point of view of individual rights – in the sense of justified moral claims – is not necessarily so from the collective perspective of countries or communities. But what 'just' Earth System Boundaries (ESB) are depends crucially on this normative perspective.

The definition of 'just' proposed by Rockström et al. does not help in this respect. For them 'just' refers to a state in which 'human exposure to significant harm (no significant harm (NSH)) from Earth system change' is minimized (Rockström et al. 2023: 103). Again, it remains indistinct whether 'minimization' is to be understood individually or collectively with regard to significant harm (even if many passages in their text point in the direction of a collective reading). The further explanations do not clarify this central point either:

> The just (NSH) boundaries described here are necessary but not sufficient conditions for Earth system justice, which must also enable access to resources for all and distributional

and procedural fairness. A foundation that enables minimum access to water, food, energy and infrastructure for all humans alongside a safe and just (NSH) ESB ceiling of maximum allowed human pressure on biophysical domains could constitute a safe and just 'corridor' over time. (Rockström et al. 2023: 103)

At first sight this seems plausible, but the devil is in the detail. For example, what is meant by 'minimum access to water, food, energy and infrastructure for all humans'? If this applies to all people as individuals, then an individualistic perspective is being adopted here in normative terms. That would mean: every human individual has a corresponding right. If it is withheld from only one individual, this would be a 'significant harm'. As long as these normative questions have not been answered, it seems hopeless to quantify the ESB with a view to 'justice' – which is an explicit goal of Rockström et al. What is not tenable, at least from an ethical perspective, is an additive definition of justice (or harm for that matter) in which it is simply stated that justice encompasses both aspects: individual and collective. But this is exactly what they do. Intragenerational justice, for instance, is understood as 'justice between countries, communities and individuals' (Rockström et al. 2023: 103). This is explained as follows:

Intragenerational justice includes relationships between present individuals, between states (international), among people of different states (global) and between community members or citizens (communitarian or nationalist .... Achieving intragenerational justice means minimizing significant harm caused by one country to another, one community to another and one individual to another. (Rockström et al. 2023: 103)

Apart from the fact that it is unclear why minimizing significant harm should be a criterion of justice, this 'definition' simply lines up the collective and individual aspects without saying how they relate to each other.[6]

The arbitrary results this leads to can be seen at the 'just' ESB for climate change. In this respect, they are of the opinion that the boundary is an average global temperature rise of 1.0°C. The reasoning is as follows:

Assessment of significant harm from climate change suggests the need for a stricter just (NSH) boundary [than 1.5°C]. At 1.0°C global warming, tens of millions of people were exposed to wet bulb temperature extremes ..., raising concerns of inter- and intragenerational justice. At 1.5°C warming, more than 200 million people,

[6] In the same essay there are some remarks on the topic of justice (Rockström et al. 2023: 113). But these only confirm that different conceptions of justice are simply somehow strung together. Rockström et al. refer to an article by Gupta et al. that is dedicated exclusively to the topic of earth system justice (Gupta et al. 2023). This article also has the aforementioned deficits and therefore does not meet the standards of ethics as a science. Of course, the concept of justice also has non-ethical aspects and can be analysed from a social, economic, political, or legal point of view. But when it comes to the question of what is 'actually just', it is a genuinely moral question to which only ethics can provide an appropriate answer. Our position is that this ethical aspect should play the central normative role in the PB approach. (The considerations of Kalfagianni and Meisch (2020) point in a similar direction.)

> disproportionately those already vulnerable, poor and marginalized (intragenerational injustice), could be exposed to unprecedented mean annual temperatures, and more than 500 million could be exposed to long-term sea-level rise .... These numbers of people harmed vastly exceed the widely accepted 'leave no one behind' principle and undermine most of the Sustainable Development Goals. (Rockström et al. 2023: 104–105)

The problem is that, regarding the tens of millions of people who were exposed to wet bulb temperature extremes, it is not explained what justice concerns this raises. That would be important in order to assess the relevance of the number of people affected. But even then, the question remains: Why is this number relevant at all? If only one or 100,000 or 1 million were affected, why should this matter? If the relevant arguments are not explained in detail, these judgmental statements are arbitrary.

This critique also applies to the explanations as to why 1.5°C of warming is too high from the point of view of justice. The thesis would have to be: the harm to the aforementioned affected parties could be so great that its actual occurrence must be avoided at all costs. But, again, why should the number of people affected be relevant? What if only 100 or 1,000 or a million were affected? Would the justice consideration then no longer apply? Would the harm, if it occurred, not be 'significant harm'? Of course, such a threshold can be defined politically. But that does not change the fact that it is arbitrary, even if one has a collective understanding of justice.

The critique remains the same if another widely accepted (political) principle is invoked as justification, namely the principle of leaving no one behind, and it is argued that the number of those affected far exceeds the number of those permitted. The same questions arise with regard to this principle, at least from an ethical perspective. It shows how problematic it is to transfer ethical concepts such as justice into the political or legal sphere. These concepts retain their core moral content, and the questions associated with them cannot be neutralized by adapting them to what is politically and legally feasible.

Questions of global justice are complex. The answers are controversial within political philosophy. If one wants to discuss questions of justice plausibly in this context, one has to engage with them. One has to work out and justify one's point of view and then show what follows from it. To be able to do this and to avoid arbitrary conclusions, expert knowledge is needed. What should not be done is to adopt common patterns of argumentation and simply string them together, such as 'Intragenerational justice combines justice between countries, communities and individuals through an intersectional lens' (Rockström et al. 2023: 113). One should also avoid mixing psychological and normative arguments: one cannot draw conclusions about what is actually just or unjust from the way people perceive justice and injustice, because here, too, what is right is one thing; what is considered right may be something else. Furthermore, a mixture of political, legal, and ethical aspects should be avoided. These methodological rules and distinctions

are not sufficiently taken into account in the arguments of Rockström et al., and, as a result, the statements regarding the 'just' Earth system boundaries cannot serve as the foundation of a tenable argument.

In Rockström et al. (2024) a proposal is made as to how the Earth's biophysical systems, on which humanity collectively depends, can be 'governed' in such a way that their stability and resilience can be guaranteed in the long term. The aim is to secure the critical functions of the Earth system regardless of national borders. The idea is to treat these systems as 'global commons' and regulate them accordingly.

> The planetary commons include, but significantly expand on, the current global commons by embracing all critical biophysical Earth-regulating systems and their functions, irrespective of where they are located, because they are essential to sustain all life across the planet, including the stability of our societies. (Rockström et al. 2024: 2)

The idea is to extend the territories already protected as 'global commons', such as the Antarctic or the deep seabed, to systems that are critical for sustaining life and the habitability of the planet, such as boreal permafrost, tropical rainforests, or coral reefs. From an ethical point of view, if there is a duty to achieve the goal, this duty extends to the instruments and measures required to achieve the goal as effectively and efficiently as possible while making the transition phase as fair as possible. There are good reasons to believe that it is indeed a moral duty to protect the stability and resilience of the Earth system and its associated systems and subsystems in the long term. The question is not whether the extended global commons should be protected, but whether this approach is the best possible or at least a promising approach to achieve the goal; or whether there are better alternatives. The main challenge is that this global system would continue to be supported primarily by nation-states. This raises the question of political feasibility, especially because the central motive required is not national self-interest, but the moral idea of 'planetary stewardship obligations on states and civil society to collectively safeguard Earth system regulatory functions' (Rockström et al. 2024: 6). Yet this inevitably involves restrictions on state sovereignty. This challenge seems enormous, despite the fact that there are (more or less) successful examples of global environmental regulations. A step-by-step approach as before, in which the nation-states continue to play the central role, may seem to be a more realistic alternative, but the question is whether the goal can be achieved in the time available in this way. This depends largely on how urgently the goal needs to be achieved and how much time is left.

27.2.2 *Precautionary Principle*, Risk Assessment, and Harm Principle

Harm scenarios are of central importance for the ethical assessment of the planetary boundaries since we do not have sufficient causal knowledge of many

complex processes affecting the Earth system. This applies to all planetary boundaries, including the NE. For example, we cannot predict with certainty what will happen if the average global temperature rises by more than 1.5 degrees. We can make even less causal predictions about the NE. If we could make reliable predictions, we could determine the extent of the harm and then define the measures required to prevent it from occurring.

The following options remain: either we have enough knowledge to be able to determine and assess the risks, where risk is a function – usually the product – of the likelihood of harm occurring and the extent of harm. In this case, we can decide whether the risks are acceptable or not, and if not, what needs to be done to reduce them sufficiently. Or we know that significant harm can occur, and that the probability is greater than 0, without being able to specify it any further. In this case, we are in a precautionary situation in which the precautionary principle applies. Or we are in a situation in which the potential harm based on scientifically plausible scenarios reaches such an extent that the probability of its occurrence becomes irrelevant: It is simply a matter of preventing it from occurring.

These are the central aspects from which the planetary boundaries, including NE, must be ethically assessed. The ethical questions to be answered are, for example: What criteria should be used to distinguish acceptable from unacceptable risks? When is a potential harm of such an extent that the probability of its occurrence no longer plays a role because, regardless of this, it is only a matter of preventing its occurrence, and the precautionary principle no longer applies either? (This is because this principle assumes that the object to be analysed could be a permissible risk. For this reason, the precautionary principle states that the occurrence of harm must be avoided as far as possible, but at the same time everything necessary must be done to obtain the data needed to determine the probability of its occurrence, allowing the risk to be determined.) And if the harm is so great, what means may be used to prevent it from occurring? What role do normative criteria such as proportionality or respect for fundamental rights play in such situations?

One of the deficits of the PB approach, including the NE, is that its advocates do not adequately apply the precautionary principle and inadmissibly mix it with risk assessments. For example, Richardson et al. write:

> A large body of recent research . . . provides strong evidence supporting the conclusion . . . that the climate change and biosphere integrity boundaries are in a zone of rapidly increasing and systemically linked risks. This strengthens the rationale for using the precautionary principle to set the planetary boundaries at the lower end of the zone of increasing risk. For example, for the climate change planetary boundary, we retain the boundary of 350 parts per million (ppm) CO_2 with the zone of increasing risk ranging from 350 to 450 ppm before reaching high risk. (2023: 2)

The distinction between two risk zones only makes sense if we know the risks. In this case, however, the precautionary principle is not required. Rather, it would then be a matter of deciding, whether and if so, why the risks become unacceptable over 350 ppm. In this case, the 350 ppm is a risk threshold. The precautionary principle, on the other hand, is needed if these risks, especially the probabilities, cannot be determined based on the data available, but at the same time we must assume significant potential harm. Since Richardson et al. understand precaution in colloquial terms, they believe that the aim here is to build in a safety margin. This is correct insofar as in precautionary situations measures should be taken to prevent the occurrence of harm as far as possible and at the same time allow the possibility of collecting the data required to determine the risk. This presupposes the assumption that the risk is not unacceptable from the outset. At this point, it becomes evident how decisive the harm potential is. If it is catastrophic, it can be argued on the basis of the harm principle that everything must be done to prevent its occurrence and that, unlike in precautionary and risk situations, the probability is not relevant unless it is zero (because that would mean that the damage cannot possibly occur) or one (because that would mean that the occurrence of the damage cannot possibly be prevented by human means).

27.3 Problems of a Planetary Boundary for New Entities from an Ethical Perspective

The inadequate distinction between the precautionary principle, risk assessment, and harm principle also has negative impacts on the planetary boundary of NE. The criterion for this boundary is determined with reference to the precautionary principle, because 'quantifying a single boundary for the multitude and diversity of novel entities is an intractable problem' (Persson et al. 2022a: 6788). However, a closer look shows that the relevant principle, if any, is the harm principle and not the precautionary principle. This is because the boundary is introduced based on the idea that NE could pose a threat to the stability and resilience of the Earth system and could therefore cause harm of immense proportions. Nowhere are probabilities of even a qualitative nature mentioned. With regard to chemical pollution, for example, Persson et al. state:

> Chemical pollution has the potential to cause severe ecosystem and human health problems at different scales, but also to alter vital Earth system processes on which human life depends. (2022b: 1510)

Even if this thesis is based on a plausible scientific foundation, it is not comprehensible how one can simultaneously claim that 'the safe operating space of the

planetary boundary of novel entities is exceeded' (Persson et al. 2022b: 1510). The justification given for this is not convincing. It is claimed that,

> the safe operating space of the planetary boundary of novel entities is exceeded since annual production and releases are increasing at a pace that outstrips the global capacity for assessment and monitoring. (Persson et al. 2022b: 1510)

Assuming it is empirically correct, the argument nevertheless is not plausible. In view of their claim that 'quantifying a single boundary for the multitude and diversity of novel entities is an intractable problem', and in view of the potential extent of harm, their conclusion should be to demand a complete ban on the registration of new NE. Even more so as:

> Humanity has repeatedly been surprised by unintended consequences of this release, e.g., with respect to the release of insecticides such as DDT and the effect of chlorofluorocarbons (CFCs) on the ozone layer. (Richardson et al. 2023: 6)

With regard to the planetary boundary of NE the question is rather: Even if certain chemicals do have this potential, are there plausible scientific hypotheses regarding the other substances assigned to the NE category that they have a harm potential large enough to directly affect the stability and resilience of the Earth system? This question would have to be answered separately for each substance. The harm could also result from a cumulative effect of several of these substances. Only for those substances for which there are plausible scientific hypotheses that they have a potential for severe harm could it possibly be justified to ban their marketing in accordance with the harm principle. For nuclear waste and for GMOs it seems even more questionable to assume such a direct effect on the Earth system. As long as this direct link between NE and the stability and resilience of the Earth system has not been satisfactorily shown, it is not possible to see how a separate planetary boundary with regard to NE can be justified.

What is not in dispute is that some of the NE can and do have a significant impact on core planetary boundaries such as biosphere integrity (biodiversity loss) or climate change and in this sense pose a global threat. However, this does not require or justify a specific planetary boundary 'Novel Entities'. This is true all the more for an appropriate approach to GMOs. And it is also true in the case of Gene Drives, where a planetary boundary may seem plausible at first sight. Gene drives are designed to achieve a specific benefit by reducing or eliminating entire populations, for example, to protect the affected population from malaria by controlling malaria-transmitting mosquitoes. In this respect, they can have a considerable influence on entire ecosystems. It is conceivable that this could set in motion dynamics that have a negative impact beyond the boundaries of the ecosystem and are difficult or even impossible to control. However, a direct impact

on the stability and resilience of the Earth system seems unlikely. More likely are negative effects on other planetary boundaries such as biodiversity in particular. GMOs must be evaluated from the point of view of the risks associated with them. And globally, they must additionally be assessed from the point of view of climate- and biodiversity-friendly agriculture that produces enough to feed an increasing world population in a way that ensures everyone's right to adequate food in the long run. What the concept of NE can contribute in this context remains in the dark. Advocates of the PB approach would have to clarify this specifically with regard to GMOs. Without such clarification the NE concept does not seem relevant in normative terms. Rather, it runs the risk of leading to a substitute debate that misses the indisputably existing enormous problems caused by some NE.

In our view, the ethical conclusion would be to dispense with a planetary boundary for NE and instead to consistently apply the precautionary principle to each individual substance and demand appropriate risk assessments when dealing with the very different substances that are summarized under the term NE.

References

Federal Ethics Committee on Non-Human Biotechnology (ECNH). (2010). Synthetic Biology: Ethical Considerations. Bern. www.ekah.admin.ch/inhalte/ekah-dateien/dokumentation/publikationen/e-Synthetische_Bio_Broschuere.pdf.

Federal Ethics Committee on Non-Human Biotechnology. (2016). New Plant Breeding Techniques: Ethical Considerations. Bern. www.ekah.admin.ch/inhalte/ekah-dateien/dokumentation/publikationen/EKAH_New_Plant_Breeding_Techniques_2016.pdf.

Federal Ethics Committee on Non-Human Biotechnology. (2018). Precaution in the Environmental Field: Ethical Requirements for the Regulation of New Biotechnologies. Bern. www.ekah.admin.ch/inhalte/ekah-dateien/dokumentation/veranstaltungen/Veranstaltung_7._Mai_2018/EKAH_Broschu__re_Vorsorge_Umweltbereich_e__18_Web_V2.pdf.

Federal Ethics Committee on Non-Human Biotechnology. (2022). Climate Change, Agriculture and the Role of Biotechnology. Bern. www.ekah.admin.ch/inhalte/dateien/EKAH-Bericht_Klimawandel__Landwirtschaft__Biotechnologie_2022_EN.pdf.

Gupta, J., Liverman, D. M., Prodani, K., et al. (2023). Earth System Justice Needed to Identify and Live within Earth System Boundaries. *Nature Sustainability*. 6(6), 630–638. DOI: 10.1038/s41893-023-01064-1.

Kalfagianni, A., & Meisch, S. P. (2020). Epistemological and Ethical Understandings of Access and Allocation in Earth System Governance: A 10-Year Review of the Literature. *International Environmental Agreements: Politics, Law and Economics*. 20(2), 203–221. DOI: 10.1007/s10784-020-09469-5.

Persson, L., Carney Almroth, B. M., Collins, C. D., et al. (2022a). Response to Comment on 'Outside the Safe Operating Space of the Planetary Boundary for Novel Entities'. *Environmental Science & Technology*. 56(10), 6788–6789. DOI: 10.1021/acs.est.2c02265.

Persson, L. M., Carney Almroth, B. M., Collins, C. D., et al. (2022b). Outside the Safe Operating Space of the Planetary Boundary for Novel Entities. *Environmental Science & Technology*. 56(3), 1510–1521. DOI: 10.1021/acs.est.1c04158.

Richardson, K., Steffen, W., Lucht, W., et al. (2023). Earth beyond Six of Nine Planetary Boundaries. *Science Advances.* 9(37), 2458. DOI: 10.1126/sciadv.adh2458.

Rockström, J., Gupta, J., Qin, D., et al. (2023). Safe and Just Earth System Boundaries. *Nature.* 619(7968), 102–111. DOI: 10.1038/s41586-023-06083-8.

Rockström, J., Kotzé, L., Milutinović, S., et al. (2024). The Planetary Commons: A New Paradigm for Safeguarding Earth-Regulating Systems in the Anthropocene. *Proceedings of the National Academy of Sciences.* 121(5), e2301531121. DOI: 10.1073/pnas.2301531121.

UNEP (United Nations Environment Programme). (2019). Global Chemical Outlook II: From Legacies to Innovative Solutions. Implementing the 2030 Agenda for Sustainable Development. wedocs.unep.org/bitstream/handle/20.500.11822/28113/GCOII.pdf

28

The Crisis of Planetary Narratives

A Conversation with Erle Ellis

ERLE ELLIS

JEREMY J. SCHMIDT: I'm thankful to have you here, Erle, and to discuss your work and its relationship to planetary boundaries.

We were hoping to start with one of the ideas that you're well known for, which is the idea of anthromes which help to capture the extent to which humans are reshaping patterns of the global biosphere. It's a novel idea.

I remember when it came out. I had just recently left McGill, where Navin Ramankutty was at the time, your co-author, and so I was really interested in the idea.

I would like to hear more about why you and Navin thought the idea was needed, how you came up with it, and how you developed it.

ERLE ELLIS: It literally filled a need that I had to communicate with the ecological science community (see Ellis & Ramankutty 2008; Ellis et al. 2010). As an ecologist, I was an oddball. In 1990, I started working in China, studying the long-term ecological changes that had occurred in village landscapes during the transition from traditional agriculture to more intensive industrial agriculture.

My interest was in the idea that we could understand sustainable agriculture by studying a place where you had this very long, deep history of agriculture. I focused on the nitrogen cycle in villages. I was doing village ecology.

It didn't really fit the same model as was used to incorporate people into ecological science at the time.

I would go to an Ecology of Society of America conference, and there was urban ecology and there was agroecology, and urban ecology was at the scale of urban areas, and agroecology was at the scale of fields. There was also landscape ecology,

but it was mostly focused on 'natural patterns' or disturbance of the natural patterns – all those models. When I would give a talk, it just didn't fit anywhere, and it was always this kind of marginal thing. It was so obvious to me from my work in China and in other places, that most of the planet is in fact transformed into different types of ecology shaped by people. Why wasn't that more broadly recognized in ecology? It was something that bothered me. On my first sabbatical, I had a chance to think about how I could express this. How could I show the ecological community just how important the transformed ecologies were? Not just as a kind of degradation or disturbance of ecology, but as what the ecology is like everywhere.

I basically tried to bring in all the evidence I could get at global scale about human populations and different kinds of land use to just see what patterns I could illustrate from the data. Navin was one of the first people to produce a global crops and pastures data set. We started working together on this. I mostly did the analysis and came up with the results, but then when we were discussing how you present this, I was thinking about it in terms of anthropogenic landscapes – a kind of landscape ecology. Navin had also been working a lot on biomes. In a discussion, it kind of emerged, 'Maybe these are really biomes.' They're anthropogenic biomes. They're the global ecological patterns shaped by people. Anthropogenic biomes and then anthromes – as a contraction – emerged from that discussion. And from the beginning, it was remarkable how good the uptake was.

It was very interesting. I've never had an uptake like that before, or even after, in that it was very grassroots. It wasn't just a lot of news outlets that were interested in it. It was every blogger. There were just hundreds and hundreds of bloggers who made blogs about it. People liked it. It fitted a lot of people's thinking.

JEREMY: I think it's fascinating, and it's helpful to pick up on that history of the concept, especially given your work in Anthropocene Working Group that you were part of for a time but have now resigned from. Obviously, there are disagreements, significant ones around how to interpret human relationships to the planet and the impacts that people are having on how the planet functions and how to mark those functional changes. I wonder if you could speak about your approach to the Anthropocene more broadly: are humans doing something new to the planet or is this something that's quantitatively different but qualitatively not something totally novel?

ERLE: I think you're asking that partly because I've written thoroughly on the deep roots of the Anthropocene. That's my view. And I think it's in many ways very healthy that there are many different viewpoints about

> this. I don't think there's ever going to be one viewpoint about human relations with the planet, or about human relations with nature. That's just the way people are. There are a lot of diverse ways of thinking about things, and that's not a problem. That's kind of a solution in a way.

In terms of this relationship, when did people cross the line between just interacting with the ecology of the planet in the way that every other species does? No species can exist without altering the planet. It's not possible.

What is different about our relationship with the planet is that it can just about continually scale up. That kind of scaling relationship, where people go from being a lot like a lot of other mega omnivores (which is our biological niche) to having a social niche. Which is also not so unusual. There are a lot of social insects that have a social niche – but becoming an open-ended social niche. So, how big can a human society be? There's no answer to that. If you look at the other social species, they all have a limit to how big their societies get, and that limit is genetically determined. The behaviors of sociality in most species are genetically determined, like how big a wasp's nest is going to be. Know what species it is, and you can know approximately how big that nest is going to be. That's true of all social species except people. People have evolved the capacity through cultural evolution to have relationships with each other that are cultural – learned socially – and that means they can evolve very quickly.

Culture evolves much quicker than genetics, and it's much more open-ended. Again, how big can a human society be? We don't know. We're discovering this all the time, but the scaling of human societies began a long time ago. You could argue it began around the time of agriculture. But that's very debatable because there are hunter-gatherer societies that also scaled up pretty large – that resemble much the early agricultural large-scale societies. It's not useful to put a simple date on it, but it's that process of being able to scale that really matters. Of course, we do a lot of things to enable that scaling that are not unique, just unique in their scale. The most dramatic one is the use of energy from the nonbiological world. That's exceptional.

That's what's happening right now in the transition to going beyond using the biosphere for energy, which all other species do except for plants. Remember, plants changed the entire planet – or rather 'photosynthesis', at first by cyanobacteria, the first photosynthetic organisms that changed the whole planet forever. By harvesting energy from the sun – a kind of abiotic energy – that process scaled up the entire biosphere. The biosphere could be much bigger when it is powered by solar energy, and that's potentially true of human societies. There are also nuclear and other forms of abiotic energy. The scale of what's possible

energetically keeps increasing. From burning wood to cook food to coal to solar, I see this as part of a long process of becoming a planetary force. But arguably the most transformative process that really changes the relations between people and the planet is this kind of scalable sociality, this ability to scale up operations in a sense, including our metabolism, and all these other capabilities.

JEREMY: I think that's fascinating: the focus on the scalability of the social. In the context of this book, the scalability of the social and how people cooperate, and the means to which they cooperate, become really important. In the past you have introduced these ideas using the idea of the electrical night – Earth's night sky lit by human activity. The analogy was wonderful because this lit-up planet is an unintended consequence, as you put it, of one way that we've organized ourselves.

One of the things that you have also done is to contrast competing narratives of crisis versus a couple of different phrases of cooperation, innovation and so on. I wonder if we could have a bit of a conversation around your ideas about why crisis has limits on it or has certain kinds of connotations to it versus thinking about planetary boundaries themselves in other ways as well. The planetary boundaries concept kind of builds in a bit of crisis, but it puts risk at the heart of the framework.

ERLE: It's intended to form a crisis narrative. There are different ways to look at a crisis narrative. Crisis narratives obviously have utility. It's not like you should never use a crisis narrative. The question is, does it fit the solution space? Is the declaration of a crisis and the use of a crisis narrative going to further the solution to the challenges that are raised in the crisis? If you have a crisis like a dam that is leaking, the idea of declaring a crisis is that everyone needs to respond. I think that's fundamental to every crisis idea. People need to respond urgently to a crisis. That's the whole point of it. To the extent that it encourages people to do this, to cooperate with each other to solve the problem, then that crisis narrative is serving a useful purpose.

If a crisis is declared over and over again, and even worse, if the crisis is not equally everybody's crisis, it can become a very divisive thing: people who agree that there's a crisis and people who have other things to deal with. Simply put, climate change isn't the very biggest problem for people who don't have clean water and people who have incomes below the global poverty line. They have a lot of problems to deal with. Their number one concern isn't that the richest people in the world are emitting so much carbon into the atmosphere that the climate is changing. They're not causing that problem. They can't solve that problem because they're not

creating it and the idea of declaring that as a crisis of the planet – a crisis of humanity – is not actually a correct narrative. This is just a crisis for the people in power.

I wish it was more of a crisis for the people who are most responsible for causing this problem. In fact, the main aspect of the narrative is this idea that it's a planetary crisis that humanity needs to deal with, when in fact it's really just the failures of the elites of the world to solve their own problems. Instead, they're making it out as if it's this kind of broader planetary problem. The problem is they're not taking responsibility.

It's not a surprise that there's a lot of division in discussions about things like climate change. The 1.5° narrative is intentionally developed as a planetary crisis narrative. At the time it was agreed upon by the different countries of the world, it was already known by just about every expert in climate change that it would be nearly impossible, if not actually impossible, to achieve a 1.5° planet. We were going to go past that. That was already known. So, you've got this set out as the boundary between safety and disaster and then you're going to cross it. That's a way of creating this kind of narrative of crisis.

Is it now leading to global cooperation to deal with it? Actually, what you're seeing more and more of at these conferences is a clear dividing line between the developing world that needs more energy to just maintain very low levels of development versus the wealthiest countries in the world that are continuing to burn fossil fuels at pretty much the same rates, even as at the same time, they're declaring a mandatory end to this. It just creates division not cooperation. The narrative of cooperation is not really there anymore. Of course, when we're past 1.5° for the next 25 years, which will almost certainly be the case, it's not going to seem like a climate crisis anymore. It's going to just seem like a giant failure of society because that's what it will represent.

JEREMY: One of the other normative ideas that has come out of discussions around the Anthropocene and around planetary boundaries has contrasted with the crisis narratives, the idea that we should be pursuing the recognition that we're kind of entangled with this situation. There's no escape route from the fact that we're with the planet and we live with the planet. The goal is not to either leave in the physical sense unless you're a billionaire or to restore it to some pre-human state. There are opportunities to make things better and to have a future that is a good Anthropocene, as many, including you and Elena Bennett, have called it (see Bennett et al. 2016).

ERLE: That's been true for a long time. The only question is, what are we going to make of this? That can involve things like ambitions to have a

large conservation area, a large amount of the planet that is managed just to restore or allow ecologies to exist that are not human related, that are dedicated to the other species there. That's a model for a better Anthropocene for the rest of life. There's not going to be a planet that's not shaped by people. The question is just how we shape it and what kind of agency are we going to have in doing this. Most of the changes that we've made at planetary scale have occurred as unintended consequences. They have not been intentional. We've changed a lot of things intentionally, but the unintended consequences are most of the things that we discuss when we talk about the Anthropocene. Climate change is nobody's project.

JEREMY: I wonder if we could push on this further. One thing that crisis narratives try to do is to connect these unintended consequences back to those who are responsible in different ways.

ERLE: Let's be clear here. You're correct that crisis narratives can target the causes. Some of them can, but environmental limits-based narratives do not do this. They say that there's something wrong, in a sense, with the relationship between people and the planet in this very basic way, without assigning any responsibility beyond humans in general. That's the wrong model. Climate change is not caused by humans. Humans are involved in causing climate change, but not all people are creating it equally. The real problem here is not the biology of humans. Humans don't emit carbon dioxide biologically at levels that are a problem. The problem is the burning of fossil fuels.

The problem is not that there are not enough trees taking up the carbon. That's not the problem either. The problem is that fossil fuel combustion needs to end and the countries that are most responsible for emitting most of the carbon from fossil fuels need to invest in energy that does not produce this carbon pollution. That's not part of the planetary boundaries narrative at all. It doesn't assign responsibility. It only sets this kind of planetary level narrative. Let's be clear, most people don't even think about the planet as relevant to them at all. The planet is only seen as relevant by certain people, mostly people like me and you, who study global change, maybe some policymakers, but for the most part, the planet is not really what is actionable. You need narratives that are actionable.

You could talk about ozone depletion as a planetary problem. But that's an example of where the problem – and its resolution – was not actually ever about the ozone layer. The problem was not even the ozone-destroying chlorine from CFCs in the atmosphere. The problem was the production of CFCs. The way that problem was solved is that the producing nations – not all the nations – the

producing nations stepped up, took responsibility, and phased them out. Now, there are a lot of reasons why that happened and it's not happening with fossil fuels. It's true that fossil fuels themselves are inherently much more difficult as a challenge – because their use is much more widespread. CFCs were a patented product, and to some degree, it was much easier to assign responsibility. The bottom line is about assigning responsibility and going after those involved – and those involved actually taking responsibility and dealing with it. That's what got the problem solved, not the planetary narrative.

JEREMY: I like the way that you're framing how to assign or determine responsibility in terms of what is the relationship of producing or burning fossil fuels to the climate or where the production of CFCs is coming from. I wondered about your take on thinking about entanglements and a good Anthropocene. When it comes to this, there's a certain group of people who have responsibilities or the capacity to address those root causes, and those people in addressing those causes, they'll have lots of potential ways to do it. There are ways of addressing things that can also have effects on others. One way of addressing climate change, for instance, is not to stop producing fossil fuels, it's to put a bunch of sulfates up in the atmosphere or something like that, to do some kind of geoengineering. For certain forms of solutions, there are also ethical questions that need to be asked. I just wondered if in your approach to addressing these kinds of root causes you see the considerations of responsibility working in those contexts around the solution spaces.

ERLE: Again, that's a good example of a planetary narrative: the problem is that the planet is too hot. The problem is not that some people are not dealing with their carbon pollution problem. It's a different definition of the problem. The solution, when you start talking about 'It's a planet problem' means the conditions are somehow not right on the planet – it's just too hot. We therefore have to change the planet's conditions. In this narrative, solar radiation management becomes a perfectly legitimate solution because it will solve that particular problem.
It doesn't solve the acidification of the oceans. It doesn't solve a lot of the other really serious problems that are generated by fossil fuel combustion. This 'planet is too hot' narrative only demands a solution to the temperature problem, which is a classic example of dealing only with symptoms but not causes.

I would say that this is exactly the kind of thing that got us here. It's an example of solving one problem by creating a bigger problem, which is something that

human societies are really good at doing. We have solved so many problems by creating bigger problems and kind of pushing out the challenges to some other generation or this sort of thing. Think of it this way, people never had to manage the temperature of the planet before. That just never was something that you had to do. Just like you never had to plant seeds to get the plants that you wanted if you were in an early hunter-gatherer society. When some hunter-gatherer societies did do that, to propagate species that they favored, things began to change. To the extent that you become the ecology that supports you, it becomes more and more your responsibility. In the end, there's no other way to support yourself than to transform the planet into agriculture, for example. You're trying to solve this problem of food shortage by manipulating the plants, and ultimately the planet, to produce more and more.

That's a kind of solution, and that kind of solution has been rolling out indefinitely. With solar radiation management, you're basically going to do the same kind of thing. Instead of solving the problem of temperature increases by getting at the actual cause, which is the people burning carbon-emitting fossil fuels, you say: 'Well, we're just going to manage the temperature of the planet.' There's only one problem with that – because it could be successful in the sense, it can be done. There's almost no question that sulfate aerosols could be manipulated to bring down the temperatures of the planet, and that it might even be okay climate-wise. It might not end up producing huge side effects. But also, it might. We just don't know. That's the big wildcard.

Even if it doesn't produce terrible unwanted side effects like the end of monsoon systems in Asia or some of the other dramatic things that could happen, it does do this: humans will henceforth always be responsible for managing the temperature. If people stop doing it, it's going to cause this really dramatic leap in temperature if you stop putting up the sulfate aerosols. It commits you to a path where you now have to manage the planet's climate or face the consequences in the future. It just puts you on this kind of treadmill of endlessly having to solve the planet problem instead of solving the social problem of transitioning to clean energy and ends up getting us into a whole other ball game. That's a dangerous thing. Plus, you're not solving these other problems that are going to keep getting worse, like ocean acidification.

JEREMY: I think the clarity that you're bringing to this distinction between the framing of a planetary problem versus the understanding of a social problem is key and something that I think is not often articulated.

ERLE: I've heard a similar kind of thing with the Anthropocene. The problem becomes the Anthropocene itself. I've heard people say, 'The Anthropocene is doing this. The Anthropocene is doing that.' That is

just so incorrect. There's no Anthropocene doing anything. The Anthropocene is a description of a syndrome. It's a description of a relationship between societies and the planet. When societies change, the planet is going to change as a result. It describes this entanglement of societies with the planet. We can't operate without changing the planet. The only question is: how are we going to change the planet and in which way? It's really about our societies. It's not about the condition. We've been in this entangled condition for a very long time (see Ellis 2024). That is the challenge of the Anthropocene. It is how you manage that. It's very difficult. Entanglement does not imply simple solutions, optimization, and these kinds of strategic options that you just do, and then the problem is solved. It is true that there's pretty much no solution that is not going to generate some new problems. Just like fossil fuels actually solved a lot of problems. Fossil fuels have really been great for human societies in a lot of ways. They enabled a lot of possibilities that people wanted. Phasing them out is going to cause problems. It's not going to be painless to do that. That's one of the reasons why it's not happening as fast as it needs to deal with climate change. It's just like that and that's not going to change as things go on. There are always going to be hard decisions to make between having one consequence and having another.

Then, of course, the unintended things, some of which are not yet known, just pop up. There's no way to have such a big metabolic force on the planet without changing it. Every change you make to that force is going to change the planet, sometimes in foreseen and intended ways, and for the better as planned. But predictable changes like this are very difficult and rare – and almost always come with unintended things. We kind of have to accept that. That's not a crisis. That is a condition. It's a situation. And realizing you're in that situation and not just treating it as some kind of unforeseen and unforeseeable crisis is pretty important. It makes for much better decisions when you understand the entangled causes and unpredictable effects of this condition – the challenges here – rather than simplifying the problem to just 'the planet is too hot' or something like that.

SIMON P. MEISCH: Could I come back to the question of novelty and anthromes? Would it be fair to say that anthromes have also emerged because humans have always introduced novel substances, such as fertilizers, into ecosystems? When we look at the novel substances that the planetary boundaries framework talks about, is that just a continuation of humans introducing

novel entities into ecosystems? Or would you say there's a difference between novel entities like nuclear waste or antibiotics and plastics?

ERLE: First of all, I don't really think of anthromes as being formed by humans generating and dumping these novel substances. That's not really included in the definition. It's obviously happening in those systems to some degree, but it's not definitional. Think of it this way. The classic biomes – the original concept – describes the global patterns of life. Believe it or not, people learn about this in kindergarten. Even if they don't call it biomes, you learn about tropical rainforests and deserts at that stage. These are the different patterns of life around the world, and these are shaped and defined mostly by adaptations to climate. It's this kind of global ecological pattern shaped by climate.

That's one way we can understand there are different patterns of life around the world. Yet, clearly if you look out of a satellite or aircraft window or whatever, you can see that for most of the planet, the patterns of life are also shaped by human societies in different ways. Agriculture is one of the very biggest ones in terms of extent but there are all sorts of others, including urban infrastructure and mining and all these other things that transform the ecology. If you want to understand the biology of the planet and why it's not the same everywhere, climate patterns are not enough. You have to understand the patterns of the human world. That's where it really comes from. Thinking of people not just as a kind of temporary disturbance, but as the shapers of the ecology of this planet, just like climate shapes the ecology of the planet, it's just a different form. That's where the idea comes from.

It's not really from novel substances. The truth is that the novel term was introduced largely out of work on novel ecosystems, which was something that came from Richard Hobbs and his group out of restoration ecology. It was dealing with these kinds of recalcitrant ecosystems that people had altered in this way or that way and could not feasibly be changed back. Species invasions are often a big part of why ecosystems are getting defined as novel ecosystems, the idea that you couldn't restore them to what came before. There was no real way to do it. That was the novel ecosystem concept.

One thing that happened out of that discussion was that we wrote an article and a book (see Hobbs et al. 2013). We had a workshop in Pender Island. This was 2010 or something like that. It was a long time ago. That resulted in a book. In one of the chapters, a few of us, Emma Marris was one of the authors, and Joe Mascaro

(see Marris et al. 2013). We just didn't really think the concept of novel ecosystems needed to be made a permanent concept. It was a bridging concept. It was a way of helping restoration ecologists who had always used a kind of 'natural baseline' – the 'natural ecosystem' that we can restore in a place. It is essential to have a broader view of the different kinds of ecologies of the current time, with so many you really can't call a 'natural ecosystem.' They're shaped and sustained by societies. That's the number one reason.

I don't use the term 'novel' very much anymore. I've changed a lot of my narrative to now use the term 'cultured' to directly incorporate the idea that it's the culture of the societies interacting with ecology that shapes different ecologies, not just this kind of 'novelty,' whatever that is. There's nothing new about societies doing that. There are definitely new forms. Nuclear waste is a new thing, no question about it. Chlorofluorocarbons (CFCs): new thing to the planet, never existed before. So, novelty is definitely something that's been introduced. A lot of the most novel things are new, but even some old things, like ceramics, are somewhat old but still unique – and therefore novel. That's just one example.

SIMON: So, does it make much sense for you to frame the novel entities boundary in the way that the planetary boundaries framework does: under the heading of novel entities, it subsumes the products of social practices and social problems that are actually unrelated?

ERLE: A planetary boundary neither defines the problems correctly, nor does it say anything about the solutions at all. Actually, it seems like that issue has become an even more important part of the planetary boundaries narratives. If you read the more recent papers, they are even more careful not to say what needs to be done, but they focus more on measurements relating to these arbitrarily defined numbers – these numbers decided by a small group of scientists. You can't decide boundaries like this without having some form of social decision-making. This is just how these things are done. For example, when scientists and society deal with the problem of lead pollution, and need to define allowable limits, it's not just a group of scientists that decide how much lead is allowable in water or food or whatever. This is done as part of society.

Society must decide what it can do and what's optimal, and this gets decided by a social group. Science is informing that, and it's extremely important to inform that – not an optional part. But the idea that politicians, NGOs, industries, and other representatives of society are optional parts of deciding what a boundary is, is also not correct.

The 1.5° planetary climate boundary, the only good thing about it, is that it was decided that way. It was a collective decision. Those are very difficult to make.

Very hard and involve a lot of negotiation. But that process is the only process we have for building larger scales of agency to deal with these challenges. So, the good thing about 1.5° is that it was produced in that way. It's not just a small group of scientists making this declaration in a scientific paper. It's a collective process of important actors at the global scale. I don't have a problem with deciding that. But I think that choosing to put our focus on the temperature of the planet and a number like 1.5° was probably a mistake. It should have been focused on the amount of carbon emitted per person, or the percentage of energy in a nation that is clean energy. If it had focused on something like that, this would have more directly assigned responsibility to those who are actually emitting the carbon and causing the problem, rather than making it this planetary discussion.

Yes, it probably would have been harder to negotiate a target like that. That might have been one of the reasons why they chose to talk about 1.5°. But it doesn't really matter. In the end, you can see the product. Yes, they negotiated something, but it's not leading to the solutions that we need.

SIMON: Erle, many thanks for taking the time to discuss your work with us and your views on the planetary boundaries framework.

References

Bennett, E. M., Solan, M., Biggs, R., et al. (2016). Bright Spots: Seeds of a Good Anthropocene. *Frontiers in Ecology and the Environment.* 14(8), 441–448. DOI: 10.1002/fee.1309.

Ellis, E. C. (2024). The Anthropocene Condition: Evolving through Social–Ecological Transformations. *Philosophical Transactions of the Royal Society B: Biological Sciences.* 379(1893), 20220255. DOI: 10.1098/rstb.2022.0255.

Ellis, E. C., Klein Goldewijk, K., Siebert, S., et al. (2010). Anthropogenic Transformation of the Biomes, 1700 to 2000. *Global Ecology and Biogeography.* 19(5), 589–606. DOI: 10.1111/j.1466-8238.2010.00540.x.

Ellis, E. C., & Ramankutty, N. (2008). Putting People in the Map: Anthropogenic Biomes of the World. *Frontiers in Ecology and the Environment.* 6(8), 439–447. DOI: 10.1890/070062.

Hobbs, R. J., Higgs, E. S., & Hall, C. M., eds. (2013). *Novel Ecosystems: Intervening in the New Ecological World Order*. Hoboken, NJ: Wiley-Blackwell.

Marris, E., Mascaro, J., & Ellis, E. C. (2013). Perspective: Is Everything a Novel Ecosystem? If So, Do We Need the Concept? In R. J. Hobbs, E. S. Higgs, & C. M. Hall, eds., *Novel Ecosystems: Intervening in the New Ecological World Order.* Hoboken, NJ: Wiley-Blackwell, 345–349.

Part XI

Conclusion

29

The Planetary Boundaries

Knowing What the Question Is

SIMON P. MEISCH

29.1 Taking a Pinch of Salt, Yielding the Volume's Insights

Where do we stand now, at the end of this book on ethics and planetary boundaries? To approach this question, we will, as it were, take a proverbial pinch of salt and turn to Douglas Adams' *The Hitchhiker's Guide to the Galaxy*. It may seem surprising, but the work offers fitting inspiration for gathering the diverse insights of this volume's contributions on the complex relationships between ethics, science, and human environments.

In one of the novel's most famous scenes, the supercomputer *Deep Thought* announces 'forty-two' as the answer to the 'Ultimate Question of Life, the Universe, and Everything'. Its creators are left utterly perplexed. *Deep Thought*, unperturbed, gives them food for thought:

> 'I checked it very thoroughly,' said the computer, 'and that quite definitely is the answer. I think the problem, to be quite honest with you, is that you've never actually known what the question is.' (Adams 1992: 128)

It has long been recognized that how we frame a problem is both ethically and epistemologically sensitive. Not knowing what the real question is can easily result in enigmatic or even misleading answers – just like with *Deep Thought*. Moreover, the very framing of questions can (re)produce epistemic injustices and become morally problematic. This prompts us to ask: Does the planetary boundaries framework know 'what the question is' when it answers its own question of 'a safe operating space for humanity'? Or, as those who question the framework might put it, are boundary values simply other versions of 'forty-two'?

Depending on their stance, the authors in this volume assign different roles to ethical reflection. Those who assume that the planetary boundaries framework has understood its question correctly and offered fitting answers tend to focus on the ethical follow-up questions – the ethical implications of accepting (at least some of) the planetary boundaries framework's findings. Others challenge the

framework's knowledge production and explore possible alternatives. This leads us to a broader issue: How should ethics, as an academic discipline, position itself in relation to this framework, to the science it draws upon, and to the problems it raises? And with that, we find ourselves once again back in that scene from *The Hitchhiker's Guide to the Galaxy*.

It is worth noting that the *Deep Thought* scene is not simply a parody of science and technology; it also offers an unflattering portrait of philosophy (Pawlak & Joll 2012). In it, two philosophers protest against outsourcing existential human questions to computational processes. Their objection is not without merit, but their performance quickly descends into posturing and professional vanity. *Deep Thought*, ever cunning, offers them a more lucrative path: public debates – mock battles over the answers it might one day produce. The philosophers agree, motivated less by epistemic virtuosity than by the promise of visibility and reward. While philosophers are not portrayed favourably in this episode, it nonetheless raises further questions that are highly relevant for this volume.

First: What kind of ethics do we need to grapple with the serious questions raised by the planetary boundaries framework? Second: Is it sufficient to accept technical or scientific constructs such as this framework at face value and merely reflect on their implications – as mentioned earlier? Might ethics not also need to contribute to the search for alternatives, helping to move reflections on 'safe operating spaces for humanity' beyond entrenched epistemic paths and towards new horizons of thinking and imagining?

And so, we come full circle – back to the introduction of this volume, where Jeremy J. Schmidt warned against uncritically accepting the planetary boundaries framework as the perimeter of justice. The ethics of planetary boundaries must begin earlier – by critically engaging with the assumptions built into the framework itself. It must also extend further – into epistemic territories left uncovered or ignored by the framework.

29.2 Keeping It Complex

With this volume, we set out to open the discussion on the many different ethical perspectives that emerge from the planetary boundaries framework. Our aim was to move beyond the ethical reflections suggested by the framework's creators and to bring into view a wider set of perspectives (e.g., Rockström et al. 2023). To this end, we invited scholars from a range of disciplines – each with their own methodological approaches and writing styles – and engaged them in ethical dialogue on the planetary boundaries framework.

Reflecting our conversations, one way to think across the many different engagements with the planetary boundaries is to treat the framework as akin to a

boundary object. As the US-American sociologist Susan Leigh Star explains, such objects

> 'sat in the middle' between different groups, very ill structured or sketchy in the common usage. (Star 2010: 608)

Precisely because of this openness, such objects remain malleable enough to enable communication across disciplinary or conceptual divides. To illustrate this point, Star refers to David Ferrier's map of the human brain, which he created based on his studies of great ape brains. Yet, Star (2010: 608) wonders:

> Anatomically, this is the functional equivalent of taking a map of the Paris subway and superimposing it on Cleveland, and using it to talk about traveling around Cleveland, reasoning that all large cities essentially have the same sort of transportation infrastructure, just as all mammals, or primates, have the same localization of function in the brain.

How, Star asks, could such a scientific artefact have achieved the medical success it did? Her answer is:

> [T]he map did not need to be accurate to be useful. It could serve as the basis for conversation, for sharing data, for pointing to things without actually demarcating any real territory. (2010: 608)

In one sense, the planetary boundaries framework performs a similar function – even for those who remain highly critical of it and believe it to be inaccurate. As Ulrich Brand et al. (2021: 268) observe:

> [T]he notion of planetary boundaries went beyond the mere presentation of scientific results to change the frame of popular debates and to inform subsequent research on sustainability issues. However, the diagram is a simplification – while easy to communicate, it suppresses the complexity of different planetary processes as well as their interlinkages.

Treating the planetary boundaries framework as something akin to a boundary object proved to be a productive choice for initiating interdisciplinary dialogue. It did not require the contributors to this book to take any particular stance on the framework, thereby allowing them to enter into the ethical debate on specific boundaries from within their own academic, and non-academic, lifeworlds.

The result is a book that raises many ethical questions and offers possible answers to some of them, while never claiming to have charted the full terrain of an ethics of planetary boundaries. Such an approach may disappoint some readers – who might, in their desperation, recall the two philosophers from *The Hitchhiker's Guide to the Galaxy*. Yet in this volume, we follow the advice of Andy Stirling: when knowledge is uncertain, 'experts should avoid pressures to simplify their advice' and instead 'keep it complex' (Stirling 2010: 129).

29.3 Knowing What the Question Is

As might be expected given the methodological orientation outlined earlier, the contributors to this volume engage with the epistemic resources of the planetary boundaries framework in ethically diverse ways. Some agree that the framework knows what its question is – to borrow the phrase from *The Hitchhiker's Guide to the Galaxy* – and proceed to explore its implications through a range of ethically follow-up inquiries. Others do not share this conviction, and instead foreground ethical concerns regarding the framework's knowledge production processes or unravel the implicit moral judgements articulated within the framework's corpus. Finally, the volume includes contributions that address the phenomena the planetary boundaries describe, without placing the framework at the centre of their ethical reflections.

29.3.1 Ethical Implications of Accepting the Framework

Among the authors who broadly align with the planetary boundaries framework, Konrad Ott, for example, outlines foundational building blocks for the development of ethical guidelines aimed at supporting responsible action in response to ocean acidification. James Elser, by contrast, turns his attention to the type of actor necessary to enable the sustainable use of phosphorus. He introduces the figure of the *system innovator* – an actor who, beyond possessing deep technical expertise, must also be capable of effectively communicating knowledge and navigating complex ethical, political, and cultural terrains. Neelke Doorn and Rachel Haug Fossbakk both engage with risk-ethical dimensions in decision-making processes concerning the management of freshwater systems and ocean acidification, respectively.

Even among those contributions that recognize a given planetary boundary as scientifically appropriate, there is a shared emphasis on the fact that the framework, on its own, cannot offer moral guidance. For scientific knowledge to inform action, it must be complemented by practical knowledge and ethical judgement – a point Jeremy J. Schmidt makes in Chapter 23. Moral orientation cannot be derived from science alone. Rather, sound guidance for action requires what has been termed *mixed moral judgements* – a consistent integration of various forms of descriptive knowledge with normative principles (Düwell 2013; Potthast 2015).

Doorn explicitly addresses the critique that the planetary boundaries framework fails to provide normative direction. She does not regard this as a flaw. On the contrary, she sees the framework as offering a valuable structure – one that is deliberately open, thereby allowing space for normative reflection and contestation within concrete contexts of application.

It is at this point that one fault line within this volume becomes visible. Some contributors challenge the notion that the planetary boundaries framework is a purely scientific analysis of Earth system processes, and the idea that it simply sets the stage for subsequent ethical deliberation. It is to these contributions that we now turn.

29.3.2 Looking for 'Epistemic-Moral Hybrids'

Some contributions critical of the planetary boundaries framework are acutely aware of the extent to which it has become a focal point for the co-production of knowledge (Bremer & Meisch 2017). Critical perspectives therefore draw attention to the epistemic components constituting the framework's foundations. They invite a closer examination of the explicit and implicit normative and evaluative assumptions embedded within the framework. These assumptions require critical scrutiny to prevent the unreflective reproduction of potentially morally problematic notions. In this context, we are particularly concerned with what Thomas Potthast terms *epistemic-moral hybrids* (which Eser's chapter on biodiversity also touches on), that

> are historically contingent specific blends of scientific practices and normative agendas. They comprise a merging of practical actions, concepts, implicit and explicit agendas of protagonists and the epistemological and moral substructure which emerges. (Potthast 2015: 143)

We will now turn to some key points of criticism regarding constitutive elements of the planetary boundaries framework such as its notions of humanity, boundary, science, and temporality.

29.3.3 Implicit Ideas of Humanity

Many of the contributions, in one form or another, return to the central placement of 'humanity' within the planetary boundary framework's conceptual core. Two primary lines of critique emerge. First, given that many of the environmental challenges addressed by the framework are anthropogenic in nature, it seems questionable to focus exclusively on humans while neglecting the concerns of non-human beings. Second, the framework's homogenising view of humanity effaces the social, cultural, and political diversity that characterises human societies, and it also flattens the varied responsibilities, vulnerabilities, and relations of care that exist within and across them.

Several authors criticize the anthropocentric orientation of the planetary boundaries framework, noting that it implicitly presumes that only human interests

count morally. Given that it is human activity that has brought about the critical state the framework identifies, many contributions to this volume support a stronger ethical consideration of non-human life forms and ecological systems. For example, Rachel Haug Fossbakk advocates for an ecocentric approach to ocean governance. Similarly, in their contribution on land use, Joana Castro Pereira and Carlota Houart call for a multispecies ethics and promote the idea of *forestzenship* – a concept grounded in the idea that humans and forests constitute a more-than-human political community. Such approaches assign moral value to the lived, situated relationships between humans and the natural environments they inhabit (see also Chapters 7 and 13). Christine Winter offers a more radical critique of anthropocentric framings by placing relational ethics at the heart of her account. She contends that moral value lies not only in individual organisms – whether human or non-human – but also to the relations that constitute and sustain them including those involving non-living entities such as ozone. What unites these perspectives is their shared concern with the questions of who belongs to the moral community 'in a safe operation space', and on what grounds (which Konrad Ott addresses in a different way in Chapter 22).

Many contributors challenge the reductive dichotomy between 'human' and 'nature' that underpins much of modern environmental discourse. Such a binary not only misrepresents the complexity of actual human–nature relationships, but also reasserts a particular Western ontological framework. By universalizing this view within the planetary boundaries framework, alternative ways of knowing and being in the world – especially those of Indigenous and marginalized communities – are sidelined or rendered invisible. This concern is taken up in different ways by Christine Winter (on ozone), Nicole Wilson (on water), and both Uta Eser and Makere Stewart-Harawira, respectively (on biodiversity). These authors call for acknowledging the moral significance of these relationships as essential to meaningful human lives. Rather than objectifying elements of non-human nature, they emphasize the importance of understanding them within the web of relationships they share with humans.

In addition, many contributors to this volume find another aspect of how the framework views humanity to be problematic. Like the metaphor of the Anthropocene (Lövbrand et al. 2015), the planetary boundaries framework risks presenting humanity as a singular agent, thereby obscuring global inequalities in exposure to environmental harms, as well as in responsibility for their creation. It is far from clear that the 'safe operating space' envisioned by the framework would, in fact, be safe – or just – for all. Nor does the framework adequately account for the fact that some groups and actors disproportionately benefit from the systems that drive planetary degradation, while others bear the brunt of the consequences. Several contributions draw attention to the socio-structural

underpinnings of these dynamics – structures that are implicated in the environmental phenomena the framework seeks to describe, yet which are concealed by its generalising language. These include Uta Eser on biodiversity, Christine Winter on ozone; and Katerina Teaiwa on phosphorus). Erle Ellis (in Chapter 28) similarly argues that it is not humanity in the abstract – the Anthropos – that is responsible for planetary disruption, but specific forms of social organization. While all organisms inevitably modify their environments through metabolic processes, Ellis emphasizes how human cultures have scaled these effects up through historically contingent systems of production and power. If the planetary boundaries framework is to serve as a genuine 'wake-up call', Ellis contends, it must confront these social structures and the asymmetries of responsibility they entail.

Thus, for two key reasons, the contributors to this volume critique how the framework coproduces the idea of 'humanity'. First, it risks masking real suffering and injustice by presenting environmental degradation as a universal, undifferentiated human problem. Second, it deflects attention from the concrete actors, institutions, and systems that drive ecological harm – and from the unequal burdens borne by those with the least responsibility. The focus on 'humanity' effectively naturalizes the causes of ecological crisis, attributing them to the species rather than to the logics of capital accumulation, cost externalization, and global inequality. Several contributions identify the growth-oriented, capitalist mode of production as a primary driver of ecological overshoot – one that must be named and addressed if we are to move beyond mere symptom management.

29.3.4 Implicit Ideas of Science

A further explicit ethical issue arises in relation to the coproduction of boundaries since defining boundaries is never morally neutral. Even retreating to a putative position of scientific objectivity does not help, because, as Richard Rudner argued, scientists *as scientists* make value judgements when it come to the criteria for accepting a hypothesis:

> For, since no scientific hypothesis is ever completely verified, in accepting a hypothesis the scientist must make the decision that the evidence is *sufficiently* strong or that the probability is *sufficiently* high to warrant the acceptance of the hypothesis .... *How sure we need to be before we accept a hypothesis will depend on how serious a mistake would be.* (Rudner 1953: 2; italics in original)

In relation to social risk management, Steve Rayner and Robin Cantor (1987: 3) make a similar case, posing the question, 'How fair is safe enough?' They conclude that the answer to this question cannot be left to the technical experts alone. When

looking for policy options in matters of risk, Rayner and Cantor argue, social debate needs to be grounded 'on the explicit basis of social conflicts over trust and equity, rather than on rival estimates of the probability and magnitude of unlikely events' (1987: 5). It is thus a point of criticism that the development of boundary values for a 'safe operating space' is not transparent and largely determined by technical experts, and that the values thus obtained should be considered the perimeter for ethical considerations (see Chapters 1, 6, and 27).

Related to the aforementioned, there are also criticisms about understandings of science that follow the linear model of scientific expertise and promise certain control over natural and social processes through objective scientific knowledge. Contributions in this volume draw attention to several morally problematic aspects of this model of science. For example, it tends to engage in forms of reductionist knowledge production that removes its objects of study from their contexts to examine them in isolation. Such isolation has moral consequences if it narrows the object of ethical reflection and creates or reproduces forms of epistemic injustice. This means that epistemic agents and their epistemic resources are wrongfully excluded from the production of knowledge and the ethical reflection that goes with it. Contributors such as Christine Winter (on ozone) and Nicole Wilson (on water) therefore argue for understanding these substances in their social relationships and not just as simple molecules. Similarly, Makere Stewart-Harawira discusses the different ontological conceptions of human–nature relations in international biodiversity documents as compared to those of Indigenous peoples, which do not rely on notions of nature–culture divide, and which are precisely what matters when it comes to protecting biodiversity.

29.3.5 Implicit Ideas of Time

On closer examination, we can also see that many contributions in this volume engage with the ethical implications of time, temporality, and the future. This is hardly surprising, given that the planetary boundaries framework is fundamentally concerned with shaping the future in such a way that 'a safe operating space for humanity' can be maintained. Statements about 'the future' are, to some extent, reflections of how individuals and communities conceptualise time itself, as well as the scope of human agency within spatio-temporal structures (Adam 2005; Hanusch & Meisch 2022).

One way to explore the different ways in which time is understood in epistemic resources like planetary boundaries framework is to pay attention to how scholars think about the many possible relationships between past, present, and future – and the values and assumptions that always come with them (Adam 2023: 285). In this respect, time scholar Barbara Adam (2023, 2024) distinguishes four perspectives

Table 29.1 *Epistemic perspectives of researchers on the future (Adam 2023, 2024)*

Temporal orientations	Perspectives on the future	Normative implications
Present futures	Perspective of the present on an immaterial not-yet	The future is understood as open terrain to be shaped by present actions.
Future presents	Perspective of a future person on their present	The future is understood as colonized by the present decisions (or 'time prints of action').
Past futures	Perspective of a past person on their future	The present is understood as the future of the past.
Future pasts	Perspective of a future person on their past	The present is understood as the past of the future and as a time to take responsibility for the future.

(see Table 29.1), which offer a useful framework for organizing the insights emerging from this volume at the intersection of ethics, temporality, and the planetary boundaries discourse. As Adam (2023: 285) emphasizes, 'what we can see and understand by taking these different standpoints depends fundamentally on assumptions about the reality status of the future, that is, whether we accord the future processual reality status or bracket it as an immaterial realm of ideas only'.

From the perspective of time studies, we might say that the core idea of the planetary boundaries framework does not assume the future to be an 'immaterial not-yet'. On the contrary, the model clearly acknowledges that past human decisions have significantly constrained the scope for future action. The research team behind the planetary boundaries framework is concerned with safeguarding the remaining room for manoeuvre – placing it within 'a safe operating space'. Thus, although the future is no longer presented as entirely open, the scientists involved in the framework maintain that it remains, at least to some extent, shapeable. This impulse is taken up by authors such as Neelke Doorn, James Elser, or Rachel Haug Fossbakk, who argue for ethically responsible pathways towards *present futures*.

Christine Winter and Jeremy Schmidt, each in their own way, adopt a *future present* perspective, exploring distinct *time prints* through which the future present is colonized by its past – namely, our current present. Christine Winter elaborates how the planetary boundaries framework contributes to legitimise an atmospheric enclosure, rendering the atmosphere available for private pollution. In doing so, she argues, the scope for action available to future generations will be severely constrained. Jeremy Schmidt, in turn, offers a critical analysis of how present-day decisions regarding geoengineering may profoundly shape the lives of future

people, who could find themselves locked into maintaining inherited technologies they did not choose but must sustain.

Katerina Teaiwa and Alice Benessia are concerned with tensions within the present. They look upon the present as *past future* that, under the banner of human progress, has repeatedly destroyed the lifeworlds of specific communities and imposed heavy burdens upon them. The current devastation on the island of Kiribati or in a Northern Italian landscape is the result of a once-glorious future imagined by the past. Both texts also reveal a struggle for epistemic forms through which responsibility for the future might be assumed in the present – forms that do not rely on the techno-scientific mastery of human lifeworlds. What emerges in their contributions is a present that is at once unsettling and wounded, a moment in which *future pasts* have already begun.

An interesting perspective on the complex interplay between ethics, time, and the governance of planetary boundaries emerges in two contributions focusing on international law and treaties. Louise du Toit (on ozone) and Leslie-Anne Duvic-Paoli (on aerosols) examine how *present futures* might be shaped more equitably through global governance. They also highlight that debates on international regulation must reckon with the *time prints* of the past – the enduring consequences of earlier decisions – now (i.e. in the *past future*) manifest in ozone-depleting substances, CO_2, and air pollution that have steadily accumulated in the atmosphere, profoundly affecting the lives of many who bear no responsibility for their presence.

29.3.6 Implications for the Governance of Planetary Boundaries

The authors of the planetary boundaries framework have, from the outset, been guided by an intention to have a societal impact (Sörlin 2025), a motivation that has become increasingly visible in recent publications that the planetary boundaries are a guide for the future (e.g. Rockström et al. 2024a, b). As noted earlier, however, a tension emerges: while the natural science strand of the corpus continues – against all plausibility – to retreat to a position of scientific objectivity, the growing normative dimension makes increasingly visible the ambition to evolve into a form of *scientific cosmology* (see Allan 2018), one that seeks to influence political action at all levels (on the issue of the planetary boundaries framework as a scientific cosmology, see Meisch & Schmidt, n.d.).

Many contributors to this volume engage critically with knowledge-making processes at the science–society interface of the planetary boundaries framework. A central question is what forms of agency and concepts are required to respond effectively to the challenges posed by each planetary boundary. James Elser introduces the notion of the *system innovator* who acts as a *boundary persona* – a

trusted intermediary who connects actors from disparate domains and thereby facilitates cooperation and collective transition towards more sustainable practices. Taking a more radical position, Christine Winter advocates for what she terms 'an anticolonial planetary (non)boundary'. For her, addressing planetary challenges ethically and politically requires a fundamental reorientation of knowledge production. This involves engaging with relational epistemologies and ontologies that decentre the human and actively challenge Western patterns of thought. An anticolonial approach to planetary studies thus entails the inclusion not only of all human perspectives, but also of the elemental, the geophysical, and the more-than-human world. Building on such relational approaches, Miriam Gay-Antaki calls for more inclusive methodologies of knowledge production and proposes the idea of *border crossing* as an epistemological method. This practice explicitly rejects the conventional Western ideal of an objective, detached observer. Instead, it values knowledge produced by situated knowers – individuals whose insights are shaped by their sociopolitical positions, experiences, and histories. Border crossing invites us to expand our epistemic repertoire by confronting colonial categories and recognizing that all knowledge is situated. The idea of border crossing is informed by the concept of *world-travelling*, developed by Maria Lugones (1987) and Mariana Ortega (2006). Such world-travelling

> has to do with actual experience; it requires a tremendous commitment to practice: to actually engage in activities where one will experience what others experience; to deal with flesh and blood people not just their theoretical constructions; to learn people's language in order to understand them better not to use it against them; to really listen to people's interpretations however different they are from one's own; and to see people as worthy of respect rather than helpless beings that require help. (Ortega 2006: 69)

Elsewhere, Kristie Dotson (2012) draws on this tradition to address more systemic forms of epistemic injustice. She highlights how dominant epistemic systems may appear seamless and authoritative, yet fail to provide the epistemic resources needed for just and adequate responses to complex problems. In such contexts, the task is not to merely refine these existing epistemic systems, but to acknowledge their limits – and to actively seek out alternative forms of knowing that have long been excluded or marginalized. This theme runs through several contributions in this volume, including Alice Benessia (on land use), Tanguy Sandré and Jeanne Gherardi and Miriam Gay-Antaki (both on climate), Makere Stewart-Harawira (on biodiversity), or Katerina Teaiwa (on phosphorus).

Several contributors critically examine the way specific planetary boundaries are framed as problems, drawing attention to related governance challenges. Louise du Toit addresses the issue of *problem shifting*, which occurs when regulatory problems fall outside the scope of international agreements – or are deliberately pushed beyond their remit. Similarly, Leslie-Anne Duvic-Paoli highlights how the

planetary boundaries framework fails to capture the regulatory complexities associated with aerosols. Regarding novel entities, Andreas Bachmann and Ariane Willemsen argue that the way in which the boundary aggregates a wide range of disparate substances under a single planetary dimension risks obscuring the fact that individual substances may pose specific challenges in relation to the domains of other boundaries – for instance, gene drives in relation to biodiversity.

Several contributors also draw attention to the shifting political landscape shaped by the rise of right-wing populist parties and the increasingly brazen political influence of ultra-wealthy individuals who have little interest in regulations that might curtail their sources of income – even when those very sources are directly implicated in environmental degradation and social injustice. Makere Stewart-Harawira (on biodiversity), Katerina Teaiwa (on phosphorus), and Erle Ellis (on novel entities) all explore these entanglements in different ways. Notably, Allan (2018), in his already mentioned work on *scientific cosmologies*, discusses anti-scientific sentiment and the ascent of right-wing politics as key reasons why cosmologies such as the planetary boundaries framework may be losing political traction – even as the socio-environmental dynamics and normative challenges the framework seeks to describe persist.

29.4 Glimpses of an Ethics of Planetary Boundaries

This volume sets out with a clear intention: to reveal a wide and diverse a panorama of ethical reflection on the planetary boundaries. As editors, we have deliberately refrained from prescribing how the contributors ought to understand *ethics*. Instead, each author was free to highlight the aspects that they consider most significant. In this way, the book brings together a constellation of different approaches and ways of thinking about planetary boundaries. This summary can only cast a few glimmers of light on this wealth of perspectives – always with the guiding question in mind: What kind of ethics do we need to address the planetary boundaries?

Since the term *ethics* can mean different things in different contexts, it is helpful to clarify its philosophical meaning as a postlude to thinking about how different chapters in this volume proceed (Düwell 2013; Tugendhat 1993; Williams 1985). In philosophy, we usually speak of ethics when the reflection on morals is not merely intuitive or rooted in everyday judgement, but pursued systematically (Ricken 2012; Stoecker et al. 2011). Morals, in this sense, refer to the shared set of beliefs within a community about what is of value (values) and what ought to be done (norms). Normative ethics, in turn, does not stop at describing or reconstructing these beliefs; it probes deeper – asking for the reasons that sustain these beliefs and justify their validity.

What does it mean to *do ethics*? Onora O'Neill (2009: 220–221) distinguishes between two fundamental approaches: one treats normative claims as *objects of study*, the other as the *proper aim* of ethics. O'Neill clearly argues for the latter. Yet, for this volume both perspectives are relevant, for it matters first to identify the moral concerns in relation to which moral principles must be formulated and justified. For any future engagement with ethics, it is crucial to be aware that whichever of these two options one chooses, different language games and ethical skills come into play.

To regard normative claims as *objects of study* means to analyse and reconstruct them, particularly those implicit norms that shape specific fields of action. Such an approach is often driven by an emancipatory impulse: through a critical theoretical lens it seeks to uncover injustices embedded in these fields. This volume, for example, includes contributions that explore planetary boundaries through the lenses of political ecology, feminist theory, postcolonial critique, or critical Indigenous Studies – such as Nicole Wilson on water, Miriam Gay-Antaki on climate, or Makere Stewart-Harawira on biodiversity.

The second approach highlighted by O'Neill aims at grounding normative claims, clarifying the contexts in which they apply, and reflecting on political contexts or institutions designed to avert moral dilemmas where possible. Examples of this approach can be found in Christine Winter's call for a relational ethic, Jeremy Schmidt's discussion of ethical experimentation, or in Konrad Ott's argument for a deep anthropocentrism. Ott's contribution provides an instructive example of how competing ethical perspectives can be carefully weighed. He examines their respective strengths and weaknesses and adopts a well-reasoned position considering a concrete case, that is, ocean acidification. As discussed earlier, the planetary boundaries framework has provoked justified critique because of its anthropocentric focus.

These aspects of the book relate closely to an intense debate within the fields of environmental and nature ethics. As Angelika Krebs (1999: 137) compellingly argued, this debate is often framed as a stark opposition: on the one hand, a narrowly instrumental form of anthropocentrism, which reduces nature to a resource for limitless human satisfaction; on the other, a form of absolute physiocentrism, which posits nature's intrinsic value as beyond question. Krebs points to the possibility of more nuanced positions that acknowledge nature's manifold value for flourishing human lives, while also recognizing that humans, as moral agents, attribute value to nature as part of their moral cultures. For instance, Konrad Ott brings a mediating position into focus. Several other contributions in the volume highlight similar approaches – for instance, Uta Eser on biodiversity, Christine Winter on ozone, and Alice Benessia on land use change.

The discussion of environmental and nature ethics reflects a core characteristic of an ethics of planetary boundaries as such: the nuanced weighing of ethical positions in response to specific moral problems, aiming to provide a well-reasoned basis for action. An example for this is the contribution by Andreas Bachmann and Ariane Willemsen on novel entities. They critically examine ethical principles found in planetary boundaries literature – such as the harm principle, the precautionary principle, and risk-based approaches. Drawing on their experience from ethical advisory work within political organizations, Bachmann and Willemsen demonstrate the potential for nuanced ethical reasoning and affirm that ethics can and need to provide guidance even in the face of uncertainty.

Perspectives that take their cue from social ethics turn our attention to the political and legal structures that frame human action. For instance, the contributions by Neelke Doorn on water and Christine Winter on ozone explicitly draw upon the tradition of political philosophy. Social institutions shape the conditions under which action unfolds, and as such hold the potential either to enable responsible conduct or, at the very least, to dismantle the conditions fostering unjust practices, such as hermeneutical or contributory injustices (Dotson 2012; Pohlhaus 2019). In this volume, two contributions examine the ethical dimensions of international legal instruments designed to address the challenges posed by ozone depletion, by Louise du Toit, and air pollution, Leslie-Anne Duvic-Paoli. Another social-ethical focus is offered by Joana Pereira and Carlota Houart. They critically analyse global and national bioeconomy policies that impede a fairer use of land in the Amazon region and propose *forestzenship*. Similarly, other authors argue for recognizing non-human nature as an integral and systematic part of human social organization – for instance, Rachel Haug Fossbakk on oceans, Christine Winter on ozone, Makere Stewart-Harawira on biodiversity, or Alice Benessia on land-use change. Such considerations might also be captured by what the political scientist John Dryzek called *socio-ecosystemic reflexivity*, that is,

> the incorporation into human institutions of better ways to listen to ecological systems that have no voice; and an ability to re-think what core values, such as justice, mean in the context of an active and unstable Earth system. (Dryzek 2016: 9)

From all the contributions in this volume, one insight becomes unmistakably clear: any ethical engagement with planetary boundaries must be, at the very least, interdisciplinary – if not indeed transdisciplinary. The challenges are so complex that they can only be meaningfully addressed by working collaboratively on questions and methods across a variety of epistemic agents. Such collaboration arises necessarily from the very problem structure of these challenges, which Silvio Funtowicz and Jerry Ravetz (1993) described as *post-normal* – that is, when 'facts

are uncertain, values in dispute, stakes high, and decisions urgent'. In this volume, Konrad Ott describes ocean acidification as a post-normal problem. Such inter- and transdisciplinary collaboration leads to forms of knowledge that are epistemically better. At the same time, it is a matter of epistemic justice, for all epistemic agents have a right to participate in those processes of knowledge-making that concern their lives and their worlds (Meisch 2024a, b; Meisch et al. 2022). This requires two potentially opposing skills: the ability to understand the knowledge claims of other epistemic agents, as well as to unlearn, relearn, and critically question established forms of knowledge and habitual ways of thinking. For ethics, this also means embracing an old, yet ever-pertinent, insight: no one is free of prejudice – but we can train ourselves to recognize and critically examine our biases (Mieth 1993). In this spirit, for instance, Miriam Gay-Antaki, Tanguy Sandré, and Jeanne Gherardi (all on climate) call for an especially self-critical and reflective stance.

The texts by Katerina Teaiwa on phosphorus and Alice Benessia on land-use change stand out for their particular moral stance, as they express both grief and pain at the devastation that has already occurred, as well as hope that healing the natural world may still be possible through active, creative work and knowledge-sharing. Their moral approach resonates with the call to 'stay with the trouble' (Haraway 2016) or with the conscious willingness to confront the precarious (Tsing 2015). They write from the vantage point of embodied experience with a damaged natural environment and the new, meaningful relationships that can emerge from it. In doing so, they also draw our attention to the fact that the destruction of the natural environment is not something external to humans but is experienced in and through human bodies – it passes through them, quite literally, as Adam Dickinson (2018) poignantly reveals elsewhere in his poetry collection *Anatomic*. There, he gives voice to his horror at the traces of what he calls 'petro-culture' has left within his own bodily tissues, mapping where and how novel entities have inscribed themselves into the material fabric of his being. In such a context, Katerina Teaiwa and Alice Benessia invite us to critically contemplate the conditions under which human flourishing remains conceivable in contexts of socio-environmental degradations.

29.5 Knowing What the Question Is

The planetary boundaries framework has become a key reference point for reflecting on, and guiding, state action in response to the central challenges of the human environment. Originating in the Earth system sciences and enriched by insights from other fields, it has also been adopted in legal studies (e.g. French & Kotzé 2021; Kim & Kotzé 2021) and political science (Biermann & Kim 2020). Its

influence now extends across a wide spectrum of political institutions at local, regional, national, and global levels – most notably the UN – as well as business organizations such as the World Business Council for Sustainable Development and NGOs including WWF (Rockström et al. 2024a). A major achievement of the framework lies in broadening the narrow focus on climate change to encompass other critical challenges of global environmental policy, while fostering collaboration across diverse disciplines. Yet, precisely because it has assumed such prominence, it cannot be taken as an unquestioned perimeter of justice but requires constructive and critical ethical reflection – an inherently inter- and transdisciplinary task to which this volume seeks to contribute.

Let's return to Douglas Adams's *The Hitchhiker's Guide to the Galaxy*. The argument of this conclusion has been that there are ethical consequences if the planetary boundaries framework does not actually know its own question. How, then, do the authors in this volume position themselves in relation to this issue?

Obviously, no straightforward yes-or-no answer was to be expected. The contributors' responses vary considerably, and it was never the aim of this volume to arrive at a definitive or univocal judgement. What unites all contributions, however, is the shared recognition that the epistemic and ethical dimensions of the planetary boundaries framework are deeply intertwined and cannot meaningfully be considered in isolation. This, in turn, raises the question of what forms of ethical reasoning are needed to address the framework's normative and evaluative challenges.

The glimpses into the wealth of perspectives gathered in this conclusion allow us to trace certain threads, outlining forms of ethical reflection from which an ethics of planetary boundaries might be woven. Readers will notice that the contributions, although largely written independently, enter a kind of conversation: complementing one another, at times challenging each other, and together opening space for dialogue. It is striking how ethical arguments developed with respect to one planetary boundary can yield valuable insights when transferred to others. In this sense, the volume is best understood as a convergence of ongoing conversations – conversations in which different ethical approaches critically and constructively test their foundations, examine their core concepts, and jointly explore possibilities for moral action in pursuit of futures that matter, and fair ways of achieving them.

References

Adam, B. (2024). Tempering the Not-Yet: Towards a Social Theory for the Anthropocene. *European Journal of Social Theory*. 27(2), 191–208. DOI: 10.1177/13684310231221251.

Adam, B. (2023). Futures Imperfect: A Reflection on Challenges. *Sociology*. 57(2), 279–287. DOI: 10.1177/00380385221113478.

Adam, B. (2005). *Timescapes of Modernity: The Environment and Invisible Hazards*. London: Routledge.
Adams, D. (1992). *The Hitchhiker's Guide to the Galaxy. A Trilogy in Four Parts*. London: Heinemann.
Allan, B. (2018). *Scientific Cosmology and International Orders*. Cambridge: Cambridge University Press.
Biermann, F., & Kim, R. E. (2020). The Boundaries of the Planetary Boundary Framework: A Critical Appraisal of Approaches to Define a 'Safe Operating Space' for Humanity. *Annual Review of Environment and Resources*. 45(1), 497–521. DOI: 10.1146/annurev-environ-012320-080337.
Brand, U., Muraca, B., Pineault, É., et al. (2021). From Planetary to Societal Boundaries: An Argument for Collectively Defined Self-Limitation. *Sustainability: Science, Practice and Policy*. 17(1), 264–291. DOI: 10.1080/15487733.2021.1940754.
Bremer, S., & Meisch, S. P. (2017). Co-production in Climate Change Research: Reviewing Different Perspectives. *WIREs Climate Change*. 8, e482. DOI: 10.1002/wcc.482.
Dickinson, A. (2018). *Anatomic*. Toronto: Coach House Books.
Dotson, K. (2012). A Cautionary Tale: On Limiting Epistemic Oppression. *Frontiers: A Journal of Women Studies*, 33(1), 24. DOI: 10.5250/fronjwomestud.33.1.0024.
Dryzek, J. (2016). Institutions for the Anthropocene: Governance in a Changing Earth System. *British Journal of Political Science*. 46(4), 937–956. DOI: 10.1017/S0007123414000453.
Düwell, M. (2013). *Bioethics: Methods, Theories, Domains*. London: Routledge.
French, D., & Kotzé, L. J. (eds.). (2021). *Research Handbook on Law, Governance and Planetary Boundaries*. Cheltenham: Edward Elgar.
Funtowicz, S., & Ravetz, J. (1993). Science for the Post-Normal Age. *Futures*. 25(7), 739–755. DOI: 10.1016/0016-3287(93)90022-L.
Hanusch, F., & Meisch, S. P. (2022). The Temporal Cleavage: The Case of Populist Retrotopia vs. Climate Emergency. *Environmental Politics*. 31(5), 883–903. DOI: 10.1080/09644016.2022.2044691.
Haraway, D. (2016). *Staying with the Trouble: Making Kin in the Chthulucene*. Durham, NC: Duke University Press.
Kim, R. E., & Kotzé, L. J. (2021). Planetary Boundaries at the Intersection of Earth System Law, Science and Governance: A State-of-the-art Review. *Review of European, Comparative & International Environmental Law*. 30(1), 3–15. DOI: 10.1111/reel.12383.
Krebs, A. (1999). *Ethics of Nature: A Map*. With a Foreword by Bernard Williams. Berlin: De Gruyter.
Lövbrand, E., Beck, S., Chilvers, J., et al. (2015). Who Speaks for the Future of Earth? How Critical Social Science Can Extend the Conversation on the Anthropocene. *Global Environmental Change*. 32, 211–218. DOI: 10.1016/j.gloenvcha.2015.03.012.
Lugones, M. (1987). Playfulness, 'World'-Travelling, and Loving Perception. *Hypatia*. 2(2), 3–19. DOI: 10.1111/j.1527-2001.1987.tb01062.x.
Meisch, S. P., & Schmidt, J. J. n.d. Planetare Grenzen als wissenschaftliche Kosmologie. Wissenschaft, Ethik und ihre Bedeutung für die Konstitution internationaler Ordnungen. *Zeitschrift für Internationale Beziehungen* (accepted).
Meisch, S. P. (2024a). Epistemic Justice: An Ethical Basis for Transdisciplinary and Transformative Sustainability Research. *GAIA – Ecological Perspectives for Science and Society*. 33(4), 381–388. DOI: 10.14512/gaia.33.4.9.

Meisch, S. P. (2024b). Extended Peer Communities: Creating Good and Fair Knowledges. *Futures*. 163, 103455. DOI: 10.1016/j.futures.2024.103455.
Meisch, S. P., Bremer, S., Young, M. T., et al. (2022). Extended Peer Communities: Appraising the Contributions of Tacit Knowledges in Climate Change Decision-Making. *Futures*. 135, 102868 DOI: 10.1016/j.futures.2021.102868.
Mieth, D. (1993). Norm und Erfahrung. *Zeitschrift für Evangelische Ethik*. 37(1), 33–45.
O'Neill, O. (2009). Applied Ethics: Naturalism, Normativity and Public Policy. *Journal of Applied Philosophy*. 26(3), 219–230. DOI: 10.1111/j.1468-5930.2009.00446.x.
Ortega, M. (2006). Being Lovingly, Knowingly Ignorant: White Feminism and Women of Color. *Hypatia*. 21(3), 56–74. www.jstor.org/stable/3810951.
Pawlak, A., & Joll, N. (2012). The Funniest of All Improbable Worlds – Hitchhikers as Philosophical Satire. In N. Joll, ed., *Philosophy and The Hitchhiker's Guide to the Galaxy*. Basingstoke: Palgrave Macmillan, 236–268.
Pohlhaus, G. (2019). Varieties of Epistemic Injustice. In I. J. Kidd, J. Medina, & G. Pohlhaus, eds., *The Routledge Handbook of Epistemic Injustice*. London: Routledge, 13–26.
Potthast, T. (2015). Ethics and Sustainability Science beyond Hume, Moore and Weber: Taking Epistemic-Moral Hybrids Seriously. In S. P. Meisch, J. Lundershausen, L. Bossert, et al., eds., *Ethics of Science in the Research for Sustainable Development*. Baden-Baden: Nomos, 129–152.
Rayner, S., & Cantor, R. (1987). How Fair Is Safe Enough? The Cultural Approach to Societal Technology Choice. *Risk Analysis*. 7(1), 3–9. DOI: 10.1111/j.1539-6924.1987.tb00963.x.
Ricken, F. (2012). *Allgemeine Ethik*. Stuttgart: Kohlhammer.
Rockström, J., Gupta, J., Qin, D., et al. (2023): Safe and Just Earth System Boundaries. *Nature* 619: 7968, 102–111. DOI: https://doi.org/10.1038/s41586-023-06083-8.
Rockström, J., Donges, J., Fetzer, I., et al. (2024a). Planetary Boundaries Guide Humanity's Future on Earth. *Nature Reviews Earth & Environment*. 5(11): 773–788. DOI: 10.1038/s43017–024-00597-z.
Rockström, J., Kotzé, L., Milutinović, S., et al. (2024b). The Planetary Commons: A New Paradigm for Safeguarding Earth-Regulating Systems in the Anthropocene. *Proceedings of the National Academy of Sciences*. 121(5), e2301531121. DOI: 10.1073/pnas.2301531121.
Rudner, R. (1953). The Scientist Qua Scientist Makes Value Judgments. *Philosophy of Science*. 20(1), 1–6.
Sörlin, S. (2025). Planetary Environing: The Return of Boundaries as a Category in Global Environmental Governance. *The Anthropocene Review*. DOI: 10.1177/20530196241308865.
Star, S. L. (2010). This Is Not a Boundary Object: Reflections on the Origin of a Concept. *Science, Technology, & Human Values*. 35(5), 601–617. DOI: 10.1177/0162243910377624.
Stirling, A. (2010). Keep It Complex. *Nature*. 468, 1029–1031. DOI: 10.1038/4681029a.
Stoecker, R., Neuhäuser, C., & Raters, M.-L. (2011). Einleitung. In C. Neuhäuser, M.-L. Raters, & R. Stoecker, eds., *Handbuch Angewandte Ethik*. Stuttgart: J. B. Metzler, 1–11.
Tugendhat, E. (1993). *Vorlesungen über Ethik*. Frankfurt: Suhrkamp.
Tsing, A. (2015). *The Mushroom at the End of the World: On the Possibility of Life in Capitalist Ruins*. Princeton, NJ: Princeton University Press.
Williams, B. (1985). *Ethics and the Limits of Philosophy*. Cambridge, MA: Harvard University Press.

Index

Printed in the United Kingdom by TJ Clays Ltd.